Teaching Technology
To
Children

TEACHING TECHNOLOGY TO CHILDREN

Beverly K. Minton
Elementary Teacher, Milan Area Schools

Gene D. Minton
Professional Development Division
Society of Manufacturing Engineers
Dearborn, Michigan

DAVIS PUBLICATIONS, INC.
Worcester, Massachusetts

Printed in the United States of America
Library of Congress Catalog Card Number:
ISBN: 0-87192-189-8

10 9 8 7 6 5 4 3 2 1

Contents

Preface

Technology has always been the core around which our daily living evolves. However, like any tool or technique, it can be misused. The production of implements of war, the pollution of our environment, and the depletion of natural resources emphasize the extremely important choices in how we use our technology.

How do we ensure that correct choices are made? The answer is in educating our children at an early age about technology—what it is, how it affects us, and how it can be properly applied to solve problems. If this is accomplished, our young people will be better prepared to assume the responsibilities of our highly complex, technical society.

The study of technology should be an important aspect of learning at this level. It does not require the introduction of new subjects into the curriculum. On the contrary, it is best taught through an integration of existing subjects, such as social studies, math, science, language arts, and the creative arts. Since technology is an integral part of our lives, weaving the thread of technology through the program of study most closely approximates reality.

This book covers the historical development of technology, the analysis of technology systems, a rationale for the study of technology, basic concepts concerning how students learn, designing a program of study, and preparing for an activity-based classroom. Teaching becomes exciting for both the students and the teacher when the content is directly related to our daily living and the methodology is activity based. This will assist the teacher in developing technology as a foundation for a viable instructional program. As such, the book becomes both a textbook and a valuable resource for every elementary classroom.

Since the beginning of human existence, the application of technology has enabled humans to gain an understanding of and control over their lives and their environments.

PART I

Technology: The Study of Human Accomplishments

CHAPTER 1

Introduction to Technology

In this chapter you will learn:

1. **How technology is used to satisfy our basic needs.**
2. **The difference between tools and machines.**
3. **The difference between simple machines and complex machines.**
4. **How technical knowledge changed the social structure of early humans.**

THE EVOLUTION OF TECHNICAL KNOWLEDGE

The present is one of the most challenging and exciting periods in human history. Knowledge and its applications have reached the highest levels yet. The results of this knowledge can not only be seen around the world but they also penetrate the outer reaches of space.

To fully appreciate the extent of today's knowledge, we must examine its evolution since the beginning of recorded history. Two aspects may be used to analyze how technical knowledge has evolved: (1) the development of technical knowledge and (2) its applications.

The Development of Technical Knowledge

From the earliest times humans have sought to satisfy three basic needs: food, shelter, and clothing. Early humans were hunters and gatherers. They caught animals and foraged for wild plants and when a group exhausted the local supply of food, they moved to another location. Unlike these early humans, we today control the growth of both plants and animals to ensure constancy in our food supply and variety in our food stuffs.

To shelter themselves, early humans took advantage of natural formations, but they also made crude, usually temporary, homes from materials found nearby. Because they were made of local materials, the structures varied greatly, depending upon the environment in which they were built. Today, most shelters are permanent and may be made from a variety of materials that have been transported around the world.

Early humans had the same basic needs we have today. The difference is in the level of technology that is applied to satisfy those needs.

The first clothing was probably skins of animals. Later, when a knowledge of weaving was developed, natural fibers were used. Like food, these materials were obtained through hunting and gathering. Many of our fabrics today (such as plastics) begin as natural materials but are processed into a different form by changes made in the chemical structure. As a result, the basic characteristics are changed, and they are called synthetic materials.

These needs are satisfied by applying energy. The conversion of natural resources into a usable state requires the application of energy in one form or another. One of the first significant events in the development of energy sources, and a turning point for civilization, was the discovery of fire. For the first time human beings could exercise control over their environment. Prior to this there were very few differences between humans and animals in their interactions with their environment. This new knowledge afforded immediate advantages and gave rise to new ideas and techniques which provided long-term advantages. Methods of satisfying basic needs began to change. The new source of controlled energy was utilized in many ways, creating changes in nearly all facets of life. The discovery and control of fire supplied the impetus for evolutionary change which has continued to the present time.

Controlled heat and light provided the first opportunity to create an "artificial" environment; that is, when the weather outside was cold and dark, a fire could provide warmth and illumination inside. We may assume that this, in turn, affected the evolution of various social structures. By extending the

Converting natural resources to human use requires energy. This was just as true for early humans as it is today.

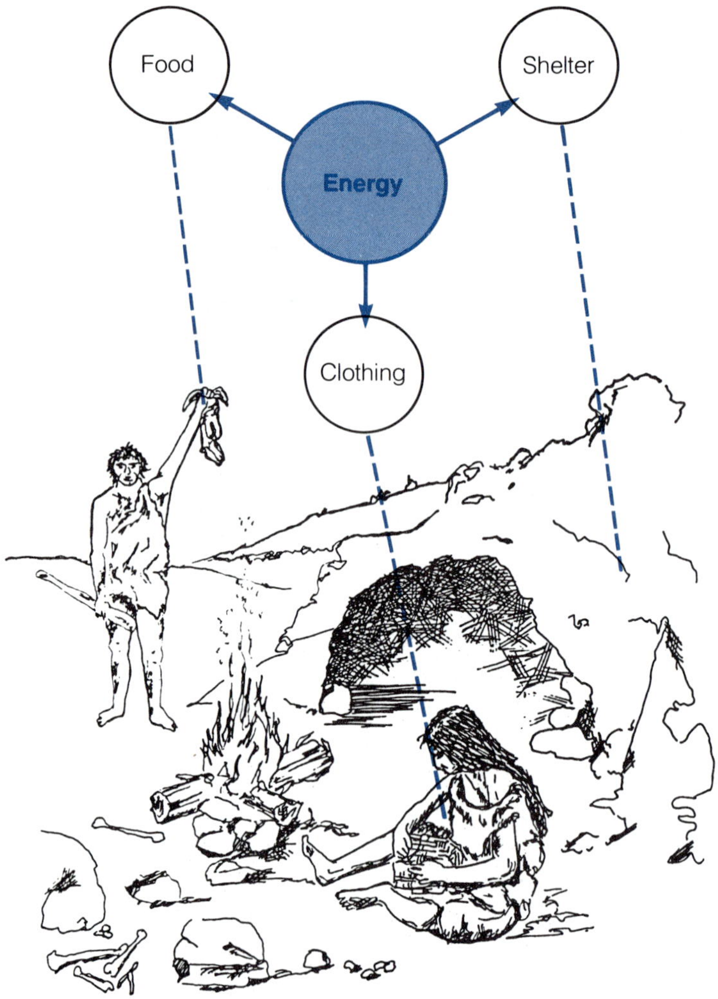

number of hours of light and heat in each day, it allowed for more contact between people and, perhaps, for more time to develop an awareness of other important aspects of life, such as religion and art.

The new energy source also provided protection, giving humans a decided advantage over those animals that presented a constant danger. Methods of food preparation and preservation also changed. Cooked food undoubtedly had an effect on health because now a wider range of food could be eaten. One of the most important advantages of controlled fire, however, was the ability to heat materials for making tools. This allowed the materials to be cut, shaped, and molded into

Fire enabled early humans to satisfy their basic needs more easily, thus providing more time for family interaction and religious and artistic pursuits.

forms that had previously been difficult or even impossible to achieve. A direct result of the new forming techniques was the development of new tools. Most of these tools made use of mechanical energy which allowed more work to be accomplished. From that point on, the application of mechanical energy became the single most important method of accomplishing work.

Humans have always depended upon the natural resources of the environment, which requires a knowledge of that environment. As they accumulated more knowledge concerning the resources and the development of energy it became easier to satisfy their basic needs. Early humans, for example, had to understand the habits of animals in order to hunt successfully. Today, we also rely on our knowledge of the environment. We must have a knowledge of weather patterns to grow crops in different parts of the world, for example. To make products and to build shelters, we must know about the characteristics of different materials.

The Application of Technical Knowledge

Humans have been defined by, among other things, their tool-using capacity. As more knowledge was acquired concerning the environment and its natural resources, new and better tools were developed. Also, as knowledge concerning the laws

Cutting and scraping tools are called* flake tools *because the structure of the stone allowed sharp edges to form as flat pieces were flaked off. The chopping tool is a sharpened piece of stone set in the socket of an animal horn and forced into a wood handle, while the smooth stone of the striking tool is tied to the stick with strips of animal hide.

Early Tools

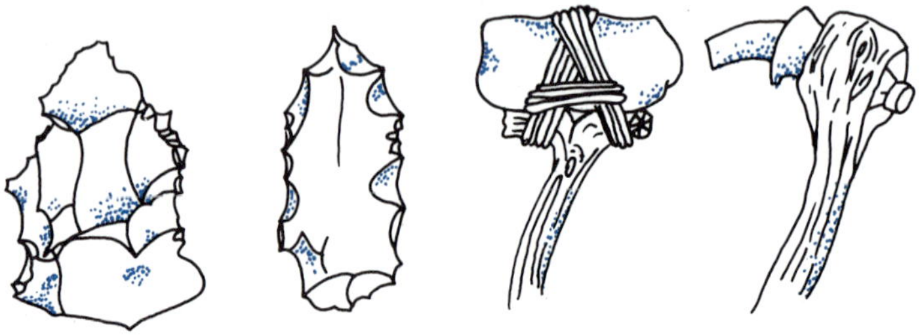

Cutting Scraping Striking Chopping

of physics was coupled with needs, more complex machines evolved.

The first tools were designed with one purpose in mind: to get work done. A hand-held stone might have been used to break other stones. A wooden stick might have been useful to kill animals for food. When these two tools were combined, the resulting complex tool became more effective for both uses. As time passed, tools changed not only because of the kind of work they performed, but because they made that work easier and faster.

In *Scientific Technology And Social Change,* Gene I. Rochlin states:

> The history of technics . . . encompasses far more than just the study of tools and their use. A tool is only an extension of an innate animal capability. Civilization and culture began with the discovery of machines, in the general sense, of devices able to perform functions that a single human body cannot perform. Fire, being generally available, was not so much invented or discovered as it was controlled. The earliest inventions, such as housing, clothing, domestication, or the wheel, attest to a unique capability for abstraction that goes far beyond the purely mechanical and that truly characterizes *homo sapiens* alone. Some machines, perhaps most, arise more from the propensity for luxury and (relatively speaking) than from necessity. (Rochlin, p. 8)

As more was learned about mechanical energy, it became easier to design tools which would accomplish more work while making human input less strenuous.

The evolution of technical knowledge can be easily understood when one analyzes the application of tools to accomplishing work. Up to this point in the text, the term tool has been used to describe all implements of work. Another term which people many times use interchangeably with *tool* is *machine*. Lewis Mumford, in *Technics And Civilization,* states the difference between the two is in the amount of control required by the user.

> The essential distinction between a machine and a tool lies in the degree of independence in the operation from the skill and motive power of the operator: the tool lends itself to manipulation, the machine to automatic action. The degree of complexity is unimportant: for, using the tool, the human hand and eye perform complicated actions which are the equivalent, in function, of a well developed machine; while on the other hand, there are highly effective machines, like the drop hammer, which do very simple tasks, with the aid of a relatively simple mechanism. The difference between tools and machines lies primarily in the degree of automatism they have reached. (Mumford, p. 100)

The mechanical advantage of this crane allows a load to be lifted at any location along the boom because of the heavy counterbalancing weight at the far right end.

In this text we wish to differentiate not so much between *tool* and *machine* as between *simple machine,* which includes tools, and *complex machine.* When work is accomplished without the aid of a tool or machine, muscle energy is utilized. When an extension of the body is used (such as a hammer held in the hand) a mechanical advantage is obtained, and more work can be done with the same output of energy. Any object used to gain mechanical advantage may be called a *machine.* The physical ability of humans to accomplish work without mechanical help is severely limited. Therefore, we would define a machine as a device which accomplishes work by extending the limited physical and/or mental abilities of the user. There are two basic types of machines: simple machines and complex machines. Simple machines are the lever, wheel and axle, pulley, inclined plane, wedge, and screw. Each is used for a specific task. In addition, they are considered in the category of a simple machine because they usually are a single-purpose device and are usually hand-held. Complex machines are made by combining two or more simple machines so that they function together. Many simple machines, for example, are used on a bicycle, a complex machine. The working together of combined simple machines improves the ability to do many types of work and increases the amount of work accomplished.

As the complexity of machines advanced, so did the technical knowledge and accomplishments. New applications gave rise to new needs, which in turn required additional new and improved machines.

THE DISTRIBUTION OF TECHNICAL KNOWLEDGE

The early growth of technical knowledge was very slow. Techniques of accomplishing tasks were usually passed to each successive generation within the family or community. There were little outside stimuli to help generate new ideas. The primary method of communication was word-of-mouth; there was little flow of information in the form of mass communication. Two factors which have figured in the distribution of technical information are: (1) population mobility and (2) changes in community social structures.

As trade increased, some people became craftsmen, because each individual could not produce the quality and quantity of products needed.

Population Mobility

Early humans were a relatively mobile population. When their primary concerns of food and safety from animals and other humans became a problem, they simply moved to a new location. Transportation was by foot, even after the domestication of animals.

Trade was also an important factor in the distribution of knowledge through transportation. Transportation and trade went hand-in-hand. The goods traded reflected the materials and techniques of each community or culture, and demand for particular objects stimulated the development of the craftsman (or specialist), who made a living by providing a specific product. Technical knowledge distributed among craftsmen became specialized knowledge.

Changes in Social Structures

Social change was very slow during the several thousand years of early human evolution. As a result, it is impossible to determine how much effect one change had on another. It appears, however, that the domestication of animals and the cultivation of crops were closely related to the growth of communities. As people started living closer to each other, there was more cross-fertilization of ideas. This led to improved techniques of accomplishing work.

As living styles changed, so did the everyday needs. The trend toward an agricultural livelihood required new tools and growing methods and more permanent living structures. As a result, a more closely knit civilization developed which made use of collective technical information.

Each significant step in the development of technology had an effect on the evolution of technical knowledge; each one led to another. The development of the wheel and the domestication of animals, for example, made travel and trade a part of everyday life. This led to the development of transportation routes which allowed easy movement from one community to another. Early rafts and dugouts evolved into larger ships, which soon had sails. These developments provided the opportunity to travel and trade with people living on other continents.

As the ability to travel from one environment to another evolved, humans around the world were brought closer together. Consequently, the development and application of technical knowledge began to grow at a faster rate.

SUMMARY QUESTIONS

1. Name two ways our technology today is like the technology of early humans. Name two ways in which it is different.
2. Why is energy important to technology?
3. Why are natural resources important to technology?
4. Explain the difference between a tool and a machine.
5. What is the difference between a simple machine and a complex machine
6. Name three factors that contributed to the spread of technical knowledge in its early development.

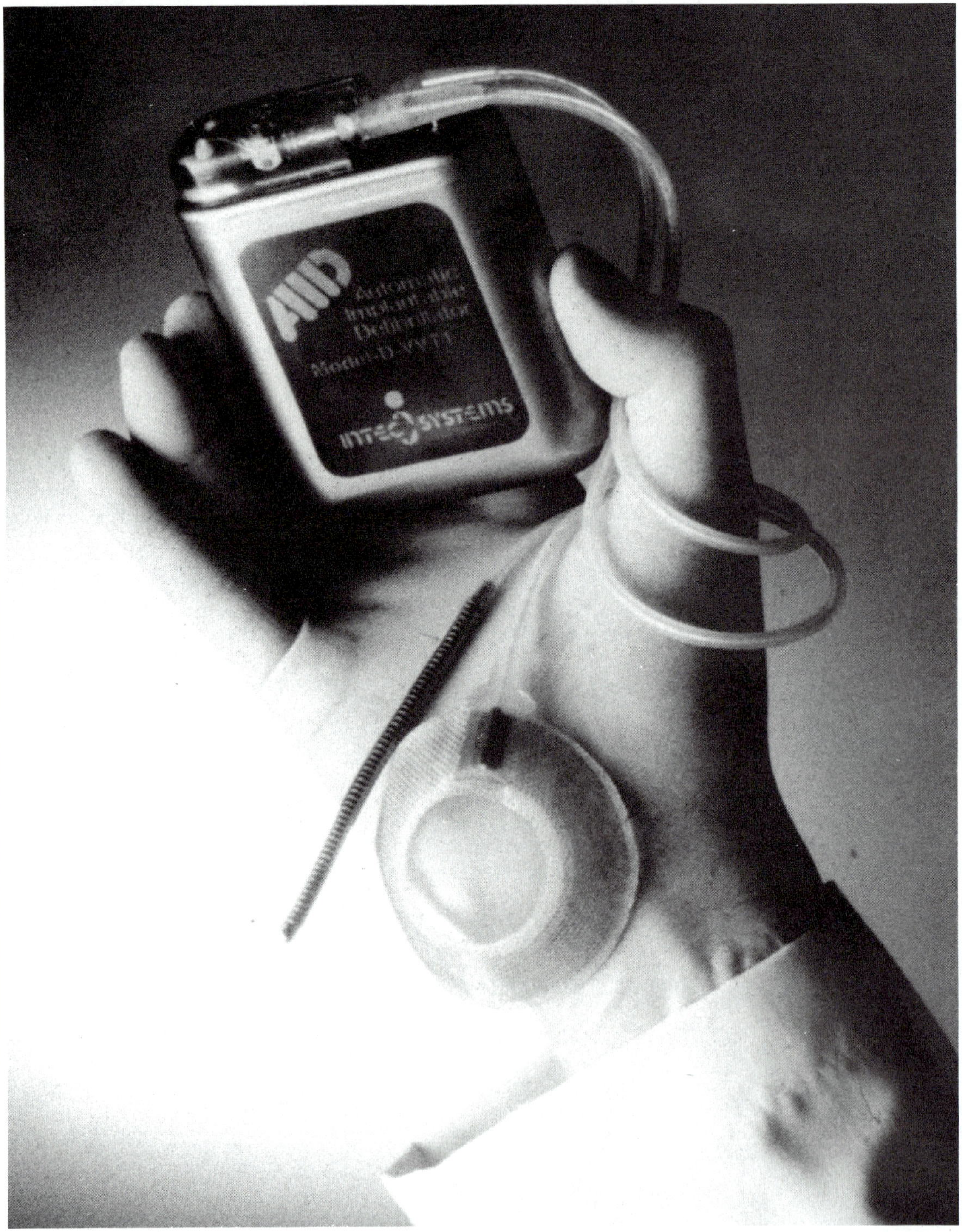
AID
Automatic
Implantable
Defibrillator
INTEC SYSTEMS

CHAPTER 2

The Impact of Technical Knowledge

In this chapter you will learn:

1. The definition of technology.
2. How technology evolves.
3. Why technology is constantly evolving.
4. How technology provides positive benefits to the quality of life.
5. How technology produces negative aspects that affect the quality of life.

TECHNOLOGICAL CHANGE

Technology is a multidimensional term. Although it can be defined in a brief sentence, it needs an accompanying explanation to fully convey the interrelatedness of the various aspects. The definition requires an examination of three elements: technique, purpose, and effect.

Technique. The word *technology* is derived from the Greek term TECHNOLOGIS' which means art or craft. It also carries the implication that there is a system in the activities involved. The term *technology,* then, connotes the application of tools and materials in a specific, systematic manner. As a result, the term *technology* may be just as appropriately applied to early human efforts to build a raft as to today's efforts to build a nuclear-powered submarine. The primary difference is in the level of applied knowledge. Early technology was based on knowledge learned by trial and error. This, however, does not imply that the knowledge was faulty or inferior. It simply means that change in techniques came very slowly because there was little prior knowledge on which to draw.

Although early technology started from trial-and-error efforts, the accumulated technical knowledge over thousands of years evolved into systematic classifications. Science is the organization of knowledge into classifications, and technology has become synonymous with science. Advancement in one means advancement in the other. There are two basic types of science which reflect separate purposes: pure and applied. Pure science, sometimes called theoretical science, is concerned only with developing and testing new theories. Its primary purpose is to increase knowledge. Many ideas are tested without having a specific problem to solve. The new knowledge is then put to use through applied science. This type of science solves problems in employing the knowledge as well as problems encountered after the knowledge is in use. Applied science brings together the systematic knowledge of many scientific fields, such as the physical sciences, biological sciences, behavioral sciences, and mathematics.

Purpose. Nearly all activities of early humans were directed toward satisfying the basic needs of food, shelter, and clothing, in short, toward survival. Tools made survival possible. Humans have been characterized by many anthropologists and historians as "tool users." This is not what sets them apart from animals, however, because it is a well known fact that some animals are tool users. For example, chimpanzees use sticks to retrieve objects that are out of reach.

What sets humans apart from animals is the way they use their knowledge. Humans are able to analyze situations and apply a variety of techniques, thereby selecting the best possible solution. The purpose is not only to "accomplish" but also to "improve." Without this motivation to improve, technological advancement would be very slow.

As time went on, the tendency to improve techniques and their applications grew. The accumulation of knowledge made available more possibilities for change. The purpose of satisfying basic needs remained the same, but the needs and applications became better defined. For example, today's technology not only provides food, it provides a variety for a balanced diet, methods for analyzing and measuring nutrient content, and many techniques for preparing and preserving food.

Because technology is an integral part of society, there are additional purposes other than satisfying the basic needs of food, shelter, and clothing. Technology provides jobs, for example, thus allowing people to make a living without having to grow their own food, make their own shoes, or build their own homes. Technology enhances methods of production, allowing many products to be made in a very short time. The purpose of using technology in this manner is to provide the products for all who want them at a price they can afford.

Mass production provides jobs for millions of people and provides products that would not be possible or would be prohibitively expensive if produced by individuals working alone.

Effect. Advancing technology causes constant change that affects all facets of life, both directly and indirectly. Direct effects are usually felt immediately. For example, the development of a medicine, a new mode of travel, or a new work-saving device is usually available as soon as it is judged safe for general use. Indirect effects, on the other hand, may take as long as several hundred years to evaluate. Examples of indirect effects might be changes in occupational requirements, changes in social organization, and long-term effects of technology on the environment.

In the past, more knowledge meant more emphasis on improvement. Gradually life was made easier for people. Time became available for leisure, for pursuing the arts, and for speculating on human existence and purpose.

Many nontechnical fields of knowledge, such as fine arts, religion, and philosophy, arose and served to guide decisions concerning technology and its application. As these new interests developed, the quality of life began to improve at a faster pace, and the quest for new knowledge and techniques became a part of human evolution.

These three elements; technique, purpose, and effect, are very much interrelated and are very much a part of human existence. If they are combined into a single statement, the result is a complete definition of technology. *Technology is the systematic application of skills and knowledge in using tools, materials, and techniques for the purpose of satisfying the needs of individuals and society.*

The Snowballing Effect

In looking at the development of technology since the beginning of recorded history, we are immediately struck by one important fact: the tremendous acceleration of knowledge in only the last few years. The impact of technological change was very small over the first several thousand years, and its effect on society was minimal until the time of the Machine Age. The machine age represents the movement toward advanced use of tools and machines. This time period is said to mark the beginning of the Industrial Revolution, when the society began to change from agriculture to industrial and move from home manufacturing to factory production. The level of technology reached the point where machines were not only able to produce more goods, they were able to produce other new and improved machines. At this time in the course of human existence the snowballing effect becomes very apparent. Each new development made several more developments possible. The primary reason for the beginning of the Industrial Revolution was the development of steam power. This one technological breakthrough generated change in nearly every technical field.

Sequential development. Science, as noted earlier, is an integral part of technological development. The systematic classification of knowledge grows as each new bit of information is

Steam power rapidly spread to nearly every type of production. One steam engine could take the place of many horses, as this 1890 farm tractor illustrates.

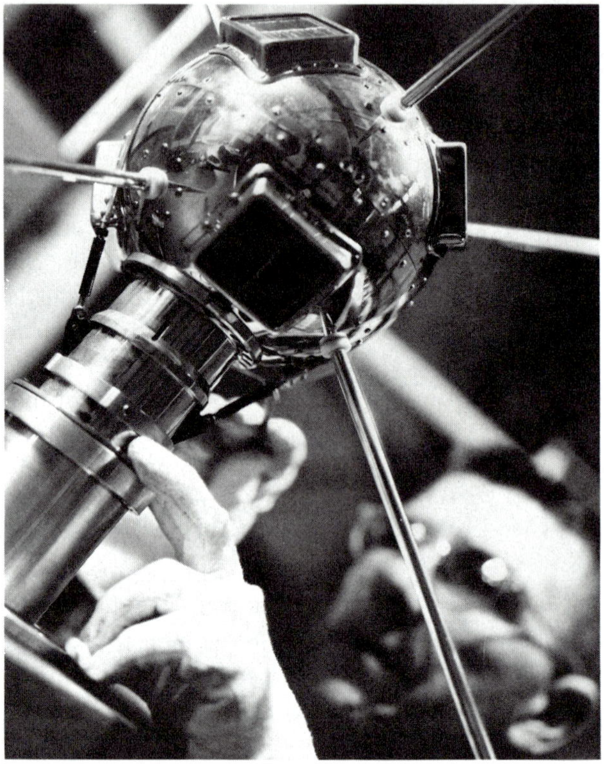

The first satellites to orbit the earth were not much larger than a grapefruit. They represented previously accumulated knowledge and also led to future knowledge.

added. This growth is a result of building on what was previously known. There are very few completely new discoveries and inventions. Most are based on knowledge and techniques already developed. For example, in 1879, Thomas Edison was successful in developing the electric light bulb. It was the first time in history that humans were able to produce continuous light. However, there was considerable knowledge acquired concerning electricity between the first recorded mention of it several hundred years before the time of Christ and Thomas Edison's discovery. An analogy might be drawn between the sequential process of technical development and the assembling of a jigsaw puzzle. Each small piece represents a bit of information. As new observations and discoveries are made, they are linked with other bits of information. Very little progress can be seen until there are enough puzzle pieces linked together to form a picture. The same can be said of technical knowledge. Very little can be accomplished until there is

A drawing such as this was found in the sketches of Leonardo da Vinci, dating back to about 1500. His theoretical wooden, two-wheeled vehicle had in-line wheels, a flexible chain, gears with teeth, and pedals for transmitting energy. It became the first bicycle in about 1817 (more than 300 years later). By the time these theories were incorporated, the bicycle was made of metal.

enough information available to test the theories and apply them to a practical purpose. Each new discovery comes about as the result of combining several other discoveries. Thus, as sequential development of knowledge takes place, the speed of development accelerates.

Simultaneous development. Another interesting aspect of technological progress lends credence to the observation just made concerning sequential development. This is the fact that a number of technological developments came at approximately the same time at different locations in the world. Although the same basic scientific principles were used, the development of the applications was not necessarily a joint effort. A dramatic example of simultaneous development is the space race. Both the United States and the Soviet Union launched the Space Age at nearly the same time: in 1957 Russia fired the first satellite into orbit, and the United States accomplished the feat in 1958, just four months later. The world prestige that could be gained from space exploration was tremendous. As a result, there was very little cooperation between the two countries in the early space efforts. It is interesting to note, however, that both countries accumulated the necessary knowledge to accomplish the space probes at approximately the same time. Even though the satellites and launch vehicles looked different, and their manned vehicle reentry systems were different, both countries' space attempts were successful. The speed at which the technical knowledge was accumulated over the years appears to have been approximately the same.

Cross-field Application. Many of the discoveries made by early pioneers in science, such as Aristotle, Archimedes, and Leonardo da Vinci, were primarily theories. Over the years many of these theories have been put to use in fields totally unrelated to those envisioned by the pioneers. For example, the study of gases made by Robert Boyle, an Irish chemist, around 1660, lead to the discovery of theories concerning oxygen (chemistry), which were used in the study of living organisms (biology). All of these theories were brought together, with thousands more from nearly all scientific fields, in the effort to carry humans to the moon. As a result of cross-field

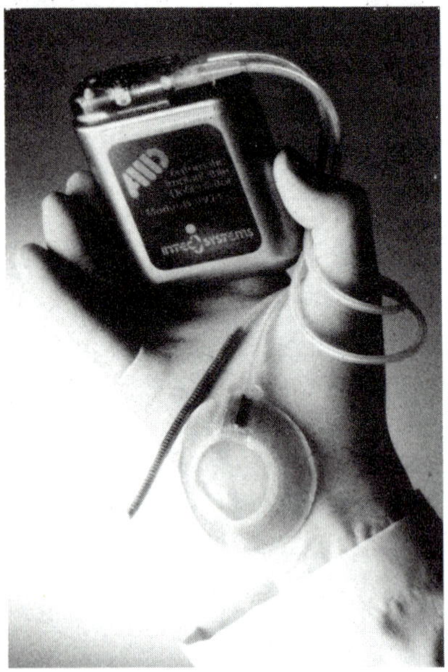

This microcomputer has been engineered to be implanted into heart patients to monitor their heart beat. If the beating stops or becomes irregular, it applies a corrective electrical charge until the normal rhythm is regained.

applications, new developments in one field made progress possible in other fields. This compounds the speed of technological advancement.

The combining of disciplines is a relatively new development that has come about as a result of cross-field applications. The combining of biology and engineering seems rather remote at first glance. Yet, these fields have combined their technical knowledge to provide significant applications in such areas as medicine, transportation, and communication. In fact, a new field, called bionics, has been created by combining biology, engineering, and electronics. This field holds promise of providing artificial joints, limbs, and other body organs to be used as replacements. As the technology of medicine becomes more sophisticated it will be a relatively common occurrence for an engineer to be present in the operating room to assist the doctor. Obviously, this requires both the doctor and the engineer to have a knowledge of the other's field.

Spin-off. A new product or theory that is a by-product of an original application is called a spin-off. The combining of such fields as biology, communication, power, and astronomy in the space program provided research which normally would not have taken place. As a result of combining many hundreds of specialized fields, significant discoveries came about almost faster than they could be implemented. Many of them we now use every day without realizing they are the result of space technology research. Examples of direct or indirect returns from our space exploration include the following:

- robots that can see, hear, and talk
- pocket telephones
- energy-efficient electric cars
- palm-size television receivers
- cable television that produces prints
- television with a two-way communication system
- various artificial body organs
- electrical devices that allow handicapped people to hear, see, and speak
- new medicines and cures for a variety of diseases
- reusable space vehicles
- appliances and tools, such as vacuum cleaners and lawn mowers, with a memory

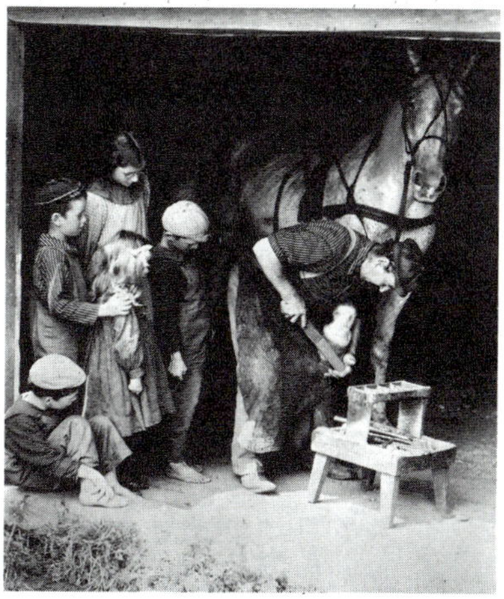

In the early twentieth century, blacksmiths enjoyed a thriving business in cities and towns across the country. By mid-century, changing technology had eliminated much of the need for their services.

- business, such as shopping and banking, conducted from a home computer
- unconventional foods from the sea
- new building materials that provide different shapes and designs for homes and other structures

There is a good possibility that all of these items will be a part of our daily lives by the year 2000, we already have the available technology.

Spin-off is a relatively recent phenomenon when compared with the thousands of years of technological progress. The earlier applications of technology required the combining of fewer fields and less research. Also, they consisted of a more simple application than the space program (such as building a machine, constructing a road, or developing a form of transportation). New spin-offs are being developed every day as a result of the research in the space program and other technical pursuits.

The Causes of Change

Why is technology always in a state of evolution? The answer to this question can be found by examining two conditions, one involving the psychological makeup of humans and the other their physical needs.

Because humans have the capacity to think and analyze, they are able to predict the results of an action. Theories which were thought out "on paper" by Albert Einstein had a tremendous effect on the entire world, for example. His theory of relativity, expressed in 1905, lead to the development of atomic energy. His theory was based on mathematical equations which produced the predicted results.

In 1942, with the advent of the Atomic Age, his theory concerning atomic energy was proved to be correct. It is said when he was five years old he was very much impressed with the needle of a compass because it always pointed north. In later years he said that even at this age, he felt there was something unknown which was behind all things which caused them to function. Humans are naturally curious about their environment. Without this curiosity and the ability to think in abstract terms, new ideas would come about only by accident.

The second condition that causes change is the demand for new and better methods of satisfying needs. As society

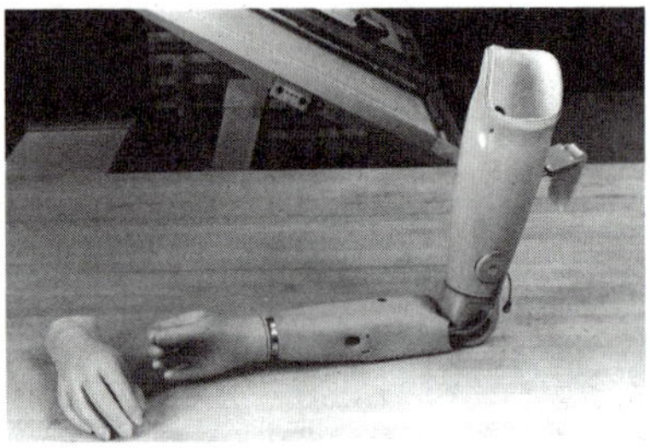

The artificial arm looks real (notice the fingernails on the plastic glove) and acts as a real arm. The electronic circuitry causes movement through impulses transmitted from the brain to remaining arm muscles. Both the elbow and wrist joints are moveable.

evolves, needs change. The increase in world population has called attention to new needs, some of which have become critical, in the last few years. The needs for increased food production, for improved methods of producing goods, and for developing new energy sources are examples. Other examples would be the need for a better method of disposing of solid waste, the need for improved transportation systems in and around urban areas, and the need for more inexpensive methods of constructing homes. These current needs will eventually be taken care of just as past needs have been down through the course of history. The differences between the past and the present are the increased speed at which the needs are currently evolving and the impact they have on a wider segment of the world population. When the ability to speculate and apply analysis to conditions and actions is coupled with the desire to improve the methods of accomplishing tasks, change is constantly occurring.

THE WORLD TODAY

Opinions about technology and its impact upon society reveal both positive and negative attitudes. Jacques Ellul, in *The Technological Order* writes:

> The technical world is the world of material things; it is put together out of material things and with respect to them. When Technique displays any interest in man, it

The windmill represents the passing of an old technology. The demand for more water is now satisfied by the construction of new canals and waterways.

> does so by converting him into a material object. The supreme and final authority in the technological society is fact, at once ground and evidence. And when we think on man as he exists in this society it can only be as a being immersed in a universe of objects, machines, and innumerable material things. Technique indeed guarantees him such material happiness as material objects can. But, the technical society is not, and cannot be, a genuinely humanist society since it puts in first place not man but material things. (Ellul, p. 90)

Ellul feels there are contradictions between civilization and "Technique," or technology. Rather than Technique being a part of a self-controlled human life, it "monopolizes all human forces," causing people to be controlled by technology. It is his contention that technology enslaves humans by making them conform, separating them from the real purpose of work, and causing social problems through alienation of people. He feels that humanism and technology are not compatible.

But on the other hand, R. Buckminster Fuller, in *Utopia Or Oblivion,* presents an opposing view as he discusses his work.

> My task as inventor is to employ the earth's resources and energy income in such a way as to support all humanity while also enabling all people to enjoy the whole earth, all its historical artifacts and its beautiful places without one man interfering with the other, and without any man enjoying life around earth at the cost of another. Always the cost must be prepaid by design-science competence in modifying the environment. (Fuller, p. 104)

Fuller's attitude is that technology serves the needs of human progress by improving life. This is accomplished by controlling and modifying the environment in which we live.

These two diverse points of view serve to illustrate the dichotomy between the benefits of technology and the detrimental effects which may develop. Technology has advanced the quality of life far beyond human expectation of only a few years ago: at the same time it has presented problems that may take centuries to eradicate.

If one looks at the level of technology as applied throughout the world, it is readily apparent that a wide range

Traffic jams are certainly not a modern phenomenon. The short-term benefit of being able to drive to town in a car soon became a long-term benefit, as mass transportation strategies were devised to ease congestion.

exists. The standards of living and the life styles of people in countries with advanced technology differ greatly from those who live in the so-called underdeveloped countries. There is one common element, however, and that is that advancing technology will affect people, both in the short and long term, regardless of their culture's level of technology or rate of change.

Because technology is an extension of human thinking, abilities, and actions, it is a dynamic, evolving element in the life of everyone. The changes that occur may not benefit all people equally and may actually hurt some. For example, the transition from animal horsepower to mechanical horsepower helped improve transportation systems and increase agricultural production, but it left buggy makers and blacksmiths with a decreased demand for their trade, and small acreage farmers were unable to compete with larger, mechanized farms. While the benefits of mechanization may be classified as long term, the detrimental effects of job loss may be considered short term. It is possible, however, to have short-term benefits and long-term disadvantages. The use of pesticides to protect our food crops is necessary if we are to produce enough food to feed the population of the world, but this short-term benefit may also be accompanied by long-term detrimental effects on birds, small animals and other forms of wildlife. In the

Natural rubber comes from liquid latex drained from rubber trees. These sheets of latex are being processed at a plantation in Malaya.

following sections, we shall look more closely at both the benefits and detrimental effects of technology.

The Benefits of Technology

The benefits of technology are most easily appreciated by comparing the standard of living today with that of the past. In 1900 the average life expectancy of people living in the United States was 47.3 years. By 1950 it was 68.2, and by 1978 it had risen to 73.3. By the year 2000, it is expected to approach 80. Advances in medical science have allowed a longer productive life. Science and technology have worked hand-in-hand to provide medical equipment and medicines which help to heal the body, and, in many instances, even replace damaged or outworn body parts.

Advances in agriculture have increased food production by providing larger equipment, new varieties of crops which resist disease, and new methods of processing the raw materials. Although there is still a critical worldwide food shortage, the current world population would never survive if we had to rely on food production levels of a hundred years ago.

No longer do people have to spend all their waking hours at work to provide for their basic needs. Because of technology there is time to enjoy many leisure activities, engage in

The effect of changing technology can be seen in many of our urban areas. In many cases, society has not been able to provide an adequate environment for people because long-term effects were not considered.

community and social work, or pursue an avocation with the same fervor as a vocational pursuit. In years past, traveling a few short miles (km) by horseback was a long trip. We now are able to fly nearly anyplace on earth at any time.

People of the world have also been brought closer together by many types of communication systems. Help comes immediately from all over the world when a natural disaster strikes a country. In the past, there might have been no help at all because neither communication nor transportation systems were immediately available.

For the most part, technological change is a response to an immediate need, and, thus, the solution tends to be implemented with a short-term benefit in mind. Many times, however, short-term benefits stretch into long-term benefits because they cause further changes. This chain reaction contributes to the "snowballing effect" mentioned earlier in this chapter. The development of synthetic rubber provides an example of a short-term benefit which became a long-term benefit. Before World War II nearly all rubber was made from the sap of rubber trees found in Southeastern Asia. During the war, Japan captured these areas, stopping supplies of rubber to the United States and its Allies. Although synthetic rubber had been in the experimental stages for several years, the production cost was too high and much of the rubber was unsatisfactory. The immediate need provided impetus for extensive research and development. In a very short time a major synthetic rubber industry was formed to provide the much needed rubber for home and abroad. After the war, the industry continued to grow, producing new and better rubber products that are now used all over the world.

Long-term benefits usually have an impact on a larger number of people because the benefits become a part of the cultural structure. For example, the car has become a part of nearly everyone's life. It is not only a means of transportation, it also provides the freedom to choose the speed of travel. Television is a medium of communication, however, it also represents entertainment, an educational tool, and (to some) a friend and companion. When technology becomes a part of the culture, as exemplified by the car and television, the benefits are extended to nearly everyone, both directly and indirectly.

The Detrimental Effects of Technology

The choices made in applying technology have not always proved to be the most beneficial. There are many instances where evolving technology has caused detrimental effects.

The impact of technology has affected social organization, values, and life styles. As the movement from an agrarian culture to a manufacturing one took place, the cities grew very rapidly. Today the cities face the problems of population density and urban decay. High unemployment, high living costs, and overcrowding of people in substandard housing are some urban problems. Many economically depressed areas have experienced a general breakdown of traditional social and moral standards. Yet, technology replaces much of our hard, dangerous labor and provides for more of our basic needs. The effect is a changing lifestyle for all people, but whether that change is positive or not in the life of an individual is to some extent determined by individual choices.

Probably one of the most far-reaching effects of technological change has been population mobility. In the past, people tended to grow up and remain in the same community most of their lives; the strong community and family ties were not severed. These strong ties provided guidance in social and moral values and contributed to a very stable community and family structure. When transportation technology developed to a point where nearly anyone could move anyplace at any time, this structure started to break down. Changes occurred as peo-

This once-productive grape vineyard resembles a graveyard. Thousands of formerly fertile acres have been abandoned because of the buildup of salt in the soil due to irrigation.

ple moved into and out of communities. The traditional structure could no longer provide the informal guidance and discipline it once had. As either a direct or indirect result of technological change, such elements as crime rates and drug abuse increased. People began exhibiting a lack of concern for others and even a rejection of society in general. Many psychologists feel these trends are the result of a feeling of alienation because the close ties are no longer there.

Since technology usually evolves as a result of need, there are far fewer short-term detrimental effects than long-term detrimental effects. With the exception of technology applied to war use, most short-term effects on the general population are felt in local areas, affecting small groups of people or individuals. For example, when a community relies heavily on one industry, such as mining, the financial status of the local industry affects the entire local area. If advancing technology causes the mining industry to close, the community suffers because of job losses and resulting financial hardships. Sometimes these communities completely disappear.

Adverse effects may result not only from the death of an industry but also from an influx of new industry into an established setting. Many farming communities have had to give up their farms as industries and urban areas continued to expand.

Many modern products are designed to resist rust, moisture, heat, light, and chemicals—natures's way of breaking down materials. Solid waste dump sites may very well remain for hundreds of years slowly polluting both soil and water.

These communities experience instability because of the new population mix and the conflicting styles of rural and urban living. Such instability may take the form of inefficient community planning and problems in local government and financial decision making.

The inner city is also affected by growth. When decaying urban areas are leveled and new buildings are built, the people who live there are usually forced to move. Many times these people are the ones who cannot stay and they cannot relocate because both choices are beyond their financial means.

As with long-term benefits, long-term detrimental effects usually have an impact on a larger number of people. We now live in what may be called a global community. That is, the population of the entire world is under the effects of the same technology. For example, we have world trade involving nearly all goods, instantaneous communication around the globe, and travel to nearly any location on earth. Because of this network of technology, it is possible for adverse effects of technology to be felt in all parts of the world.

Probably the two most pressing long-term problems are the applications and disposal of toxic substances and radioactive material and the depletion of natural resources. The first concerns us because we live in a "chemical" culture. Much of the clothes we wear, the materials we build with, even the foods we eat are the result of chemical processing and manufacturing. Some chemicals are harmless, some are highly toxic, and many which we use every day have not yet been identified as either harmless or dangerous. Because many dangerous chemicals take years to break down to a harmless state, the chances of ingesting them through plants, meat, and other foods is great if proper disposal procedures are not used. Some of these chemicals in the environment may cause health problems not only today but also for many years to come. As nuclear power continues to be developed there naturally is more radioactive material being produced, raising the possibility of higher levels of radiation throughout the world. As with toxic substances, long-term radiation pollution could last several hundred years.

The concern over the damage or depletion of natural resources has grown as a result of poor decisions made in applying technology and of the increased awareness of the limits of natural resources needed for energy. Damage to natural resources results when we dispose of waste in water and underground landfills, when we disperse pollutants into the air, when we allow surface erosion to take place, and when we plan improperly for construction of buildings, transportation systems, and housing communities. The necessity for increased amounts of energy to supply our advanced technological needs has also started to deplete worldwide natural resources. These natural resources were developed over a period of several million years. Because of the time required for development, they are considered irreplaceable.

In the past, decisions concerning the use of technology affected relatively few people, because there was a choice in being a part of it. All people now, however, are affected by technology even when they have no choice as to their own involvement. Decisions made today about the development and use of technology in one country will very likely have an effect on other countries of the world. Although there are many short-term and long-term benefits, there are also short-term and long-term detrimental effects. It is most important that decisions be analyzed in relation to global implications, and be balanced with both short- and long-term effects in mind.

SUMMARY QUESTIONS

1. Explain the relationship between science and technology.
2. Describe the "snowballing effect."
3. Name and briefly describe the four ways technical knowledge comes about.
4. Name one product of technology that has produced a short-term benefit and one which has produced a long-term benefit. Briefly explain.
5. If technology is controlled by humans why do negative factors occur that lower the quality of life?
6. What would happen if we cut back on technological development and returned to the past? Name two positive and two negative results.

CHAPTER 3

Charting Human Progress

In this chapter you will learn:

1. **How technological innovation has affected the way humans lived in various time periods.**
2. **How energy development has played a major role in human society.**
3. **How the Industrial Revolution changed the social structure.**
4. **What effect the two world wars had on the development of technology.**
5. **How our lives today are greatly enriched, in comparison with times past, as a result of evolving technology.**

TECHNOLOGY AND THE PAST

Since the dawn of civilization, humans have been using technology. The crude tools of early humans were made of stone and bone; their primitive spears and their methods of using them were as much a part of their everyday lives as our more advanced technology is in our own lives today. A comparison of the two time periods may also be made in regard to how changing technology has an effect on the socioeconomic structure. Although the degree of change and the length of time required for the change to occur might be quite different, the impact of technology was, and still is, very dramatic.

In order to quickly analyze some of these developments, the time line from the beginning of recorded history to today has been broken into four segments: early civilizations; pre-Industrial Revolution; Industrial Revolution; and 1900 to the present.

Early Civilizations

Early humans left a fairly detailed record of their history even though they could not read or write. Many different sciences such as archeology, geology, chemistry, physics, mathematics, and the social sciences, are being combined today to recover and analyze this record which consists of bones, tools, and other early artifacts. As a result, time periods have been identified based on the use of materials to apply technology. The early ones, called the Paleolithic, Mesolithic, and Neolithic periods were characterized by the use of stone, while the following periods were characterized by metals. The discovery and use of different metals may be identified in the names, such as the Bronze Age and the Iron Age.

Discoveries within the last decade indicate there were several different types of hominids, or humanlike creatures, which date back as far as five and a half million years. This period to 8000 B.C. is called the Paleolithic period, or Old Stone Age, because the tools which remain from that time consist of crude stone implements used in hunting, farming, preparing food, and making clothing. Prior to about one million B.C. pebble tools were used. These were stones which were chipped to a sharp edge on one side and probably used as cutters and scrapers. After one million B.C. an increased variety of tools appeared, such as hand axes, stone knives, and stone-tipped spears. Toward the latter part of the Paleolithic period, the bow and arrow came into being. This was probably the first application of a mechanical device with stored energy.

Hunting large animals required several, but not all, members of a group. Division of labor resulted, with the criteria for hunters being skills and strength.

These early peoples were hunters and gatherers who moved frequently, depending upon where food was most plentiful. Early humans did not band together in large groups because there was not enough wild food in one area to serve very many people. The dispersion of the population limited communication which in turn hindered the development of technology. However, in areas where there was large game, such as the prehistoric mammoth (which was much larger than the modern-day elephant), people did live together in groups. Hunting was carried out on a cooperative basis, which probably led to the division of labor, because not all men, women, and children were needed as hunters.

Probably the most important event in the development of technology, even to this day, was the control of fire around five hundred thousand B.C. This single discovery brought many changes. Stones could be more easily shaped by heating and cooling, wooden spears could be dried and hardened, wild animals could be kept away from living areas, and herds of animals could be driven into water or bogs where they could be more easily killed.

Control of fire and the change in eating habits likely had a direct physical effect on the people. Cooked food was more digestible and more varied. This provided a more balanced diet and allowed the young and the very old to survive more easily. The control of fire may even have had an indirect effect on physical appearance. Although the bodies of early and modern humans would be very much alike (allowing for their shorter stature and larger bones) the head structure would be quite different. Their jaw was low-slung and very heavy, with large teeth and much muscle for biting and chewing raw meat and plants. The top of the head was very small and sloping because the brain cavity was very small. Over a period of many thousands of years the brain increased in size, causing the forehead to straighten. Because food could be softened by cooking, the heavy jaw became unnecessary, and it as well as the teeth decreased in size. In this way technology played an early role in the evolution of human physical appearance, and it continues to do so today.

The Neolithic period, or New Stone Age, lasted from about 7000 B.C. to 3000 B.C. During this time humans started raising animals and crops. New tools appeared, such as the sickle and the hoe, and pottery began to be made. People lived more closely together because raising food provided security. It was no longer necessary to rely entirely on chance to obtain food, as it was with hunting. Settlements were larger, and homes were bigger and more permanent. The close proximity of people required laws and rules of government. As a result, governing bodies were probably formed to protect the rights of individuals and the community. In those areas where social organizations did not develop, the technology did not evolve nearly as fast.

Toward the end of this period, shortly before 3000 B.C., technological development began to accelerate—very slowly, but the new developments indicated the application of knowledge to the solution of problems. For example, the discovery of pottery meant that water could be boiled. Previously, animal skins, which were used to hold water, would catch fire if placed over an open flame, so hot stones were dropped in the water to heat it. After the development of pottery, water could be boiled directly over a fire, and raw materials could be soaked in boiling water for processing. In areas where the climate was cooler, weaving was developed because plant and animal fibers were more available as a result of farming. Early technicians developed the wooden plow to turn the soil and used oxen to provide the necessary power. This allowed more land to be farmed and helped increase the quality of crops.

Irrigation also came into practice in areas where water was scarce.

Also during this time, probably the second most important development in the evolution of technology took place: the invention of the wheel. With the advent of the wheel, transportation became easier as carts and wagons were used to carry heavy loads. The wheel was adapted in many other ways to make work easier. The result of these and many other developments was a greater division of labor. Some people became craftsmen in specific trades to provide the needed products and services. By this time not all people were either hunters or farmers.

Two points should be noted at this time. The first point is that the Mesolithic period, or Middle Stone Age, came between the Paleolithic and the Neolithic periods. This was the transition period between hunting and farming when people were involved in both. For this reason it is commonly thought that there were many types of hominids, and it can be assumed that some developed much faster than others. The ones who did not develop improved tools of technology probably did not survive. Second, the coming and going of the glacial ages allowed movement from one part of the world to another. When much of the earth's surface became frozen, the ocean waters receded, connecting some continents by dry land. When the ice melted, the water level rose and many people were isolated in a new environment. Because of differences in need and in natural resources, some types of technology developed in one part of the world but not in others. The people of northern Russia and Alaska, for example, had very different needs from

Early fastening methods consisted of either tying with flexible strips of material or using wood pegs driven into a hole, much as we use nails today.

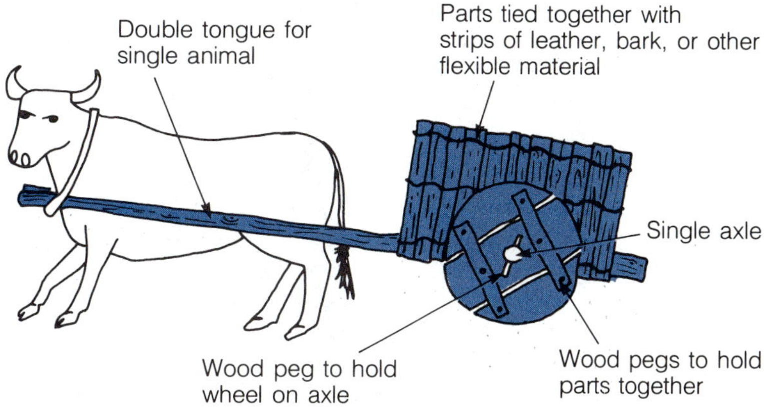

people in South Africa, and different technologies developed in these regions in response to particular needs. The Bronze Age lasted from about 3000 B.C. to 1200 B.C. Civilization had become complex. Some one-time farming communities had developed into cities, and farming ceased to be the main source of income. Craftsmen superseded farmers as they produced goods and provided services for others, including the farmers on the edges of the urban areas.

During the Bronze Age, the basic material of technology changed from stone to metal. Although many soft metals were used, bronze (made by combining copper and tin) was used extensively for tools and utensils. The other metals such as gold, silver, and pure copper were used for art objects, jewelry and decoration. Bronze has a number of characteristics which made it useful at that time. It has a low melting point, it can be made relatively sharp, and it will not rust. This new material gave impetus to a rate of change that was more dramatic than at any time in the past. This speeding up of the evolutionary process in technology and its effect on culture is shown by the fact that this era passed quickly (in less than 2000 years). Some areas entered the Bronze Age earlier than others, some passed through this stage quickly, and some skipped it entirely, jumping from the Stone Age to the Iron Age.

The technology of fire, the wheel, and bronze, culminated in the beginning of what is called modern civilization. The standard of living was the highest, up to that point, of any time in history. For the first time people had the freedom to

Smelting (the melting of ore to extract metal) was accomplished by building fires in crudely shaped stone furnaces.

develop such interests as art and sports. Transportation, although crude by our standards, provided food for the urban areas, and larger, more permanent construction began to take place.

This new level of technology, however, also brought with it some of the social problems still facing us today. Two problems stand out as significant: the establishment of dynasties (a line of rulers coming from one family) and the establishment of a class system. As a result of the new, large governments, many people lost control of their own destinies. The quest for more land and more wealth inevitably led to war between dynasties and empires. Bronze provided weapons of war which could not be stopped by those who had not yet reached the same level of technical development.

Even in primitive societies there emerged those who were leaders and those who were followers. With the addition of technology, those with the mental capacity and ability to lead took advantage of the new opportunities and tended to become more successful. As a result, the distance between the rich and the poor and the strong and the weak became greater. Either a formal or informal class system developed, even to a point where there may have been aristocracy and slavery within the same system.

The Bronze Age brought many changes. Weapons and tools were better constructed, armor was developed for protection against metal weapons, and cooking and eating utensils were developed which were functional and easily decorated. Shields such as seen here were often used for communication by reflecting the sun.

One of the most notable achievements of this period was the development of writing. Although humans had been drawing pictures on cave walls for many thousands of years, there was no system of symbols that could convey a continuous flow of thought. Writing systems usually move through an evolutionary process that starts with pictures and develops into symbolic characters. Several systems were practiced, some of which have been deciphered today. Many have been preserved, because before paper (which deteriorated rapidly) writing was done on clay and bronze tablets. These have lasted through the centuries and give us a record of human events. Although not everyone could read, writing provided for a certain standardization of laws, responsibilities, and even weights and measures. It was also during this time that language and literature began to flourish, placing education in a position of elevation and respect.

The area around the Nile and Tigres and Euphrates Rivers, including Mesopotamia and Egypt, is considered the birthplace of civilization. The level of early technology is evidenced by pyramids and statues, tablets of writing, and various other artifacts found in tombs.

Near the end of the Bronze Age iron was discovered, launching what is called the Iron Age at about 1100 B.C. Actually this age has lasted into today, because iron is still one of our primary manufacturing materials. Even though iron ore was abundant in the regions of early settlement, the processing was not developed earlier because the technology was not available. The melting and working of iron requires higher heat than bronze. With more knowledge about how fire burns, for example, the role of oxygen in combustion, the process evolved.

Iron quickly replaced bronze because it was harder, held a cutting edge better, and was less expensive because of the availability of the iron ore. The new iron tools allowed other tools to develop at a faster pace because of their superior quality.

From about 1100 B.C. to the time of Christ, there was tremendous intellectual growth. The center of intellectual life was Greece. Within a very short period, philosophy, literature, history, art, poetry, drama, sculpture, architecture, and science were flourishing. No other period before or after has been as productive in so many fields. The significant point here is the development of science. Until this time technology was based on trial and error, with progress resulting from readoption of older methods. With the beginning of science and the gradual accumulation of knowledge concerning nature and the universe, technology found a new base in intellectual application. This gave rise to new inventions that had started as theories. Sometimes, however, the technology was not available to test the theory; some theories were not even tested or developed until hundreds of years later when the level of technology could support it. Even today, theory often outdistances the level of application until the theory is tested.

After Greece, Rome came into prominence, making great advances in both architecture and engineering. The Romans were more concerned with applied science than with theoretical science. Some of their bridges, roads, and aqueducts were so well constructed that many are still used today. Before and shortly after the time of Christ, Rome ruled about one quarter of Europe, the northern coastal area of Africa, and much of the Middle East. There has been much speculation about the cause

Although this list is by no means complete, it represents a summary of scientific progress. Without these significant developments, the Industrial Revolution would never have occurred. The technology that followed involved the application of these scientific theories.

Selected Scientific Contributions to the Industrial Revolution

400 B.C.	Hippocrates believed that diseases are caused by natural, not supernatural, phenomena.
300 B.C.	Aristotle's studies in classification provided a basic foundation for science. Euclid formulated mathematical theories.
200 B.C.	Archimedes performed experiments that led to the formulation of many physical principles.
A.D. 100	Galen studied and developed theories of anatomy and physiology.
1500	Leonardo da Vinci proposed many theories dealing with the physical world.
1543	Copernicus espoused the theory that the earth and other planets revolve around the sun.
1600s	Galileo applied mathematics to the study of scientific experiments. Boyle developed the application of scientific methodology to chemistry.
1687	Newton published his theories on the laws of mechanics.
1730	Linnaeus developed the scientific method of classifying plants and animals.
1774	Priestley discovered oxygen.

or causes of the fall of the Roman Empire in A.D. 476. Certainly, preoccupation with military endeavors would account for the lack of scientific advancement being made by the time the empire finally collapsed. Because the Romans had no interest in science, new technology floundered. The same was true with most other parts of the civilized world.

The Pre-Industrial Revolution Period

Two segments constitute the period we refer to as pre-Industrial Revolution. The first is the Middle Ages, which lasted from the fall of the Roman Empire to about A.D. 1500. Little progress was made in science and technology. Through trade, many cultures exchanged ideas and inventions, but there was little emphasis on learning the sciences. The most notable contribution of Europe to technology during this time was the development of Gothic architecture.

Differences in technology and in the quality of life of various parts of the globe were probably more pronounced at that time than at any other. Some parts of the world advanced, some regressed as a result of wars, and some remained completely uncivilized. China was probably the most technologically advanced country in the world, having developed such technologies as gun powder, the magnetic compass, and printing with movable type.

The second segment of the pre-Industrial Revolution period was the Renaissance, or the Age of Reason, covering from approximately 1500 to about 1750. While the Middle Ages were characterized by little technological progress, during the Renaissance, as befitting the age's name, technology was "reborn." A new cultural awakening discarded ancient ideas and customs, revived an interest in learning, and ignited an intellectual revolution. The Renaissance was the beginning of modern times and it brought with it a renewed interest in science and technology.

Not since the Greeks had science and technology enjoyed such a close relationship. The Renaissance, with its emphasis on learning, pioneered our modern scientific method. This involves the observance of rules and the organization of steps to reach scientific conclusions. It was probably the most important contribution of the period because it led to a systematic organization of knowledge.

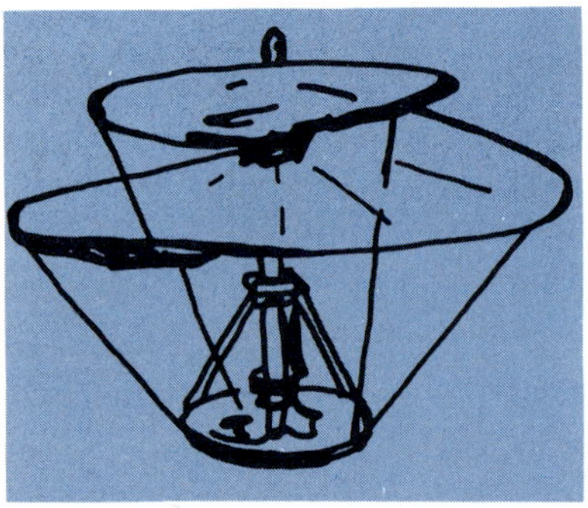

This illustration depicts Leonardo da Vinci's idea of a flying vehicle which was sketched around the year 1500. The theory was that as it rotated it would be drawn into the air. Because the development of a new technology requires the simultaneous development of many related technologies, the idea was beyond the scope of possibility. It was not until 1940 (nearly 450 years later) that Igor Sikorsky built and flew the first successful helicopter.

In addition, several scientific instruments were developed, such as the microscope, slide rule, sextant, air pump, adding machine, and timing mechanisms. These and other instruments like them led in turn to other scientific discoveries that were applied to technology. During this period of time new developments in ship construction made world trade possible, food production increased, and manufacturing methods changed drastically with the advent of new materials and processes.

One development that was to prove especially significant for the future course of Western civilization was the harnessing of steam power. Although the theoretical use of steam had been known for several centuries, Thomas Newcomen's steam engine in 1712 was the first practical and efficient application. Its use in the British coal fields to pump water from the deep mines proved its value and potential. Although wind and water power were still used, they were limited because of location and weather conditions. This new source of power could be controlled, could be used anywhere as long as there was fuel, and could operate continuously. It was this development that began the movement toward the Industrial Revolution.

One of the major industries of this time was mining. New technology was applied to older technology in processing various materials, such as iron, copper, zinc, tin, and lead. The fuel needed for processing these metals was primarily coal. As the demand for products became greater, new processes were developed that required more raw materials.

Weaving also became a major industry. Although weaving had been done mechanically in the past, the demand for more cloth required a more efficient machine. They were still, however, hand-operated.

The major technological developments took place in those kinds of work that involved acquisition of raw materials and the manufacturing of products. There was little innovation in construction, transportation, or communication, primarily because of the lack of efficient energy sources. In construction, the same natural materials (stone, wood, and so forth) were still cut and assembled by hand. In transportation, animals were the main source of power on land and wind the main power source on water. Communication was either by writing or talking, because many of the scientific theories of physics had not yet been formulated for creating sound and visual images.

Although the manufacturing of products increased, both

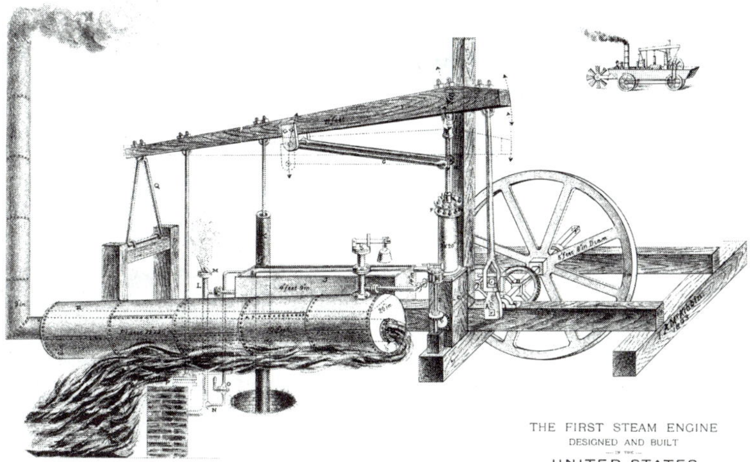

This steam engine was the first (1801) to be designed and built in the United States. It looked much like Thomas Newcomen's engine, nearly twenty-five feet (750 cm) in length.

in terms of new products and quantity, the highly structured manufacturing organization we have today had not yet appeared. Home industries, sometimes referred to as "cottage industries," were the norm. Much of the manufacturing took place within the craftsmen's homes, where there was little division of labor, except between adults and children. Many times children learned the trade as they grew up. As a result, many children followed in their parents' footsteps, becoming craftsmen in a particular trade.

During the latter part of the 18th century, formal schools became an established part of society both in Europe and in the American colonies. Emphasis was on reading, writing, and arithmetic. Most students attended school less than eight years, after which they began to learn the practical skills needed in later life. The girls usually worked at home while the boys learned a trade under their father or with a master craftsman. Manufacturing had not yet reached a level permitting young people to freely move about, changing jobs when they pleased.

One result of the "cottage industries" was a tightly structured society. The home and community served as a complete unit of moral and educational development for those

Cottage industries lasted into the beginning of the Industrial Revolution. The people who produced the goods did not market them, neither were they employed by the distributor; they just supplied the goods for a specific price.

within it. Values were consistent within a community and even within the broader context of society.

Industrial Revolution

The Industrial Revolution is usually thought of as extending from approximately 1750 to 1900, but there is no definite starting or stopping point. It was a period of accelerated growth in technology that had never been experienced before, and it had worldwide impact, touching even the most primitive cultures.

Scientific and technological growth requires a political and social climate that encourages innovation. England had all the prerequisites in the mid-eighteenth century, and it was in this country that the Industrial Revolution began. Conditions were even more favorable in America as the colonies broke away from England and needed to compensate for loss of the source of goods and services. Also, Americans started moving westward and had to depend upon their own ingenuity for survival. President Washington, in his first message to Congress in 1790, said that new and useful inventions should be encouraged and that laws should be passed to aid their development. In 1790, the first patent laws were established which guaranteed inventors' rights and thus encouraged innovation.

Four major aspects characterize the Industrial Revolution. First, the development of steam power revolutionized society, affecting all phases of life. Before steam, there was only hand power, natural power (wind or water for example), and animal power. Steam provided the first continuous source of power that could be completely controlled. It was soon applied in production and transportation. Steam-powered boats and trains provided more raw materials for production, and factories became larger because steam provided more energy.

Second, the movement from a craft-based technology (primarily hand work) to a science-based technology (mechanical and chemical) provided more products in a shorter time. Machines became larger and more sophisticated to accomplish work that could never be done by hand. Science-based technology led to rapid changes. While craft technology changed slowly because any new change required the development of new physical skills, scientific knowledge changed rapidly by building on previous knowledge and application. Not only governmental action but also the founding of several en-

The simplest application of steam was to provide pressure, as this fire wagon, which used steam for water pressure shows. Steam is not used, however, to power an engine because it would take much too long to get a steam engine fired up.

gineering and science schools demonstrated society's interest in science and technology. In addition there were many professional societies, expositions, and fairs that displayed the latest innovations in technology and served to stimulate innovation in many different fields of technology.

Third, the movement from primarily an agrarian society to a manufacturing society caused urban areas to grow. Improvements in design and construction of buildings, bridges, roads, and water transportation systems were in answer to the new needs. So, too, were new laws and regulations written to deal with the stress on society as the population in the urban

This city street of the late 1800s shows a science-based technology, with the electric power lines and the electric railroad.

areas grew at a rapid rate and the role of the rural areas changed.

Fourth, the system of mass production changed the entire manufacturing process. It has been said that Eli Whitney and Samuel Colt developed the system of interchangeable parts, which was America's unique contribution to world technology. In 1798, Whitney devised a system by which many rifle parts were made so precisely that they were interchangeable. Prior to this, each piece was made for a specific gun. With the interchangeable system, several people could assemble many guns from a common stockpile of parts. In 1836, Colt perfected this system in producing the famous Colt Revolver (called the six-shooter), which became a trademark of the West. This concept was soon adapted to the manufacture of many other products, thus enabling more products to be produced at a faster rate. This method of manufacturing, which gave rise to the factory system, brought about several changes. Women began taking their place in the work force on a formal wage basis, jobs were available for both skilled and unskilled workers, and people began making more money, which helped raise the

Most factories operated on water or central steam power. An overhead belt system powers these machines rather than individual motors.

standard of living. Because families no longer had to be self-sufficient, there was more time for leisure activities. Many modern conveniences appeared in the urban areas, such as gas lighting, indoor plumbing, and central heating. Canning processes preserved food that could be kept indefinitely, and ready-to-wear clothing could be purchased whenever it was needed.

Between 1776 and 1900, more changes occurred in America than had ever taken place in any country in history. Beginning as a rather disorganized group of colonies during the American Revolution, America became a world leader by the turn of the twentieth century and proved that technological innovation is the key to providing a better life for more people.

1900 to the Present

The beginning of the twentieth century marked the start of an era in which more changes have taken place than at any other time in history. The chart that follows illustrates the speed at which technology is moving. In little more than eighty years we have made more discoveries, have developed more technology, and have passed through more descriptive "ages" than any people before us. Although it is impossible to arrive at a specific number of years to describe the rate at which new developments are happening, general estimates suggest that we are doubling our technical information about every ten years. The cumulative result of this growth rate over a period

Beginning Dates of the Ages in History

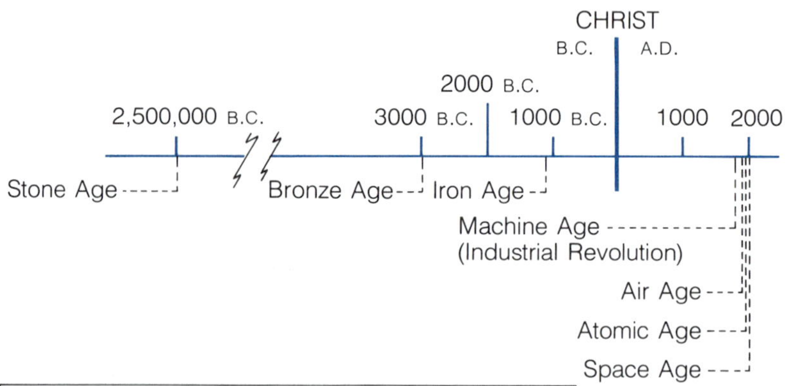

This illustration provides a graphic view of the increasing speed of human accomplishments as technology continues to evolve.

*SCALE: 20 mm = 1,000 years
NOTE: Without the break in the timeline, the distance between the beginning of the Stone Age and the beginning of the Bronze Age at this scale would be nearly 170 feet (49,940 mm).

of several decades can hardly be imagined. In a period of one lifetime, approximately seventy-five years, humans have gone from not being able to fly, to landing on the moon traveling close to 20,000 miles per hour (32,186 km/h). Assuming a continued rate of change, within the next seventy-five years the advancement should be more dramatic than the past seventy-five years.

From 1900 to the present, the "ages"—the Air Age, Atomic Age, Jet Age, Space Age, and so on—are so close together that one is superimposed on another. Individual ones have not gradually phased out because they are integral to all the others. New developments in one have an immediate, direct relationship to the others, causing change to come more rapidly. For example, the Air Age produced aircraft with wings, the Jet Age produced jets and rockets. All three technologies have been used in developing the Space Shuttle vehicle. Computer systems developed for space flight are now being used to make commercial aircraft safer. Atomic energy not only produces enough electricity to create new and better transportation vehicles, it also powers some of the vehicles directly. Submarines, for example, have been powered by atomic energy for several years, and it holds great promise as the power source for exploration into deep space. The unique element in these ages since 1900 is that each one in itself has caused a small revolution in how people live.

As we have said, technology develops to satisfy a need. It is not surprising, then, that periods of stress are accompanied by an upsurge in technological innovation. Such was the case during World War I and World War II. The Civil War period produced many innovations but primarily in the area of military hardware, such as submarines, machine guns, and land mines. Although industries were geared up for war production, there was not the spin-off and application of innovations to peacetime uses after the Civil War as there was after World Wars I and II. The latter two wars were scientific wars. World War I was called a chemist's war and World War II a physicist's war. Both wars, however, involved many of the sciences.

The American technology that developed during World War I involved much chemistry, because previously most of the chemical industries were established in foreign countries. When the war cut off the supply, American chemists began a large-scale research program, which led to new developments

World War II saw rapid advances in manufacturing. Many advances involved more efficient production, because the emphasis was on producing a high number of units in the shortest possible time.

in such fields as metal alloys, food preservation, nutrition, disease prevention, packaging, and paper products. In addition, optical glass, dyes, medicines, and many new methods of processing raw materials came about in a very short time. Equally important developments came about in transportation and communication, including aircraft design, radio communication, and photography.

By the time World War II broke out in 1941, the innovations of the first war had been implemented for civilian use. The radio had become a part of everyone's life, along with the automobile, daily newspapers, and modern conveniences in the home. When World War II began, the experience learned in mobilizing for World War I was useful. It was evident from the beginning that new technologies would have to be developed, because many sources of raw materials from foreign countries were not available.

Nearly every major production industry began research programs to provide new products and methods of production that would satisfy the war needs as well as domestic needs. From this came such developments as many types of plastics, synthetic rubber, improved radio communication, new metal alloys, and a tremendous growth in medical technology. Five areas in particular have benefitted from technology developed during World War II and have helped shape our society.

The first is air travel. Prior to World War II mass transportation by air was limited. Trains were more popular. The aircraft were propeller driven and very noisy. Although jet engines had been in use, they were not perfected. By the end of the war, jet planes had been established as the plane of the future. Now, millions of people fly by jet every year, some flying on a daily basis. Not only are people able to fly anywhere in the world, they are able to buy fresh food which is flown in daily from all parts of the world. Air travel has brought the world closer together. This one factor, more than any other, has probably contributed the most to what we now call our "global community."

The second is synthetic materials. If one tries to count the number of synthetic materials surrounding us at any particular moment; it is amazing what numbers are identified. A few examples would include: fabrics such as nylon, rayon, and polyesters; plastic laminates on table tops, floor tiles, wall paint, wall coverings, ballpoint pens, and book jackets; related synthetic materials called latex (rubber) might be found in shoes, chair seats, and erasers. We now live in a world of synthetic materials, including our clothes, buildings, and in some cases, our food. This has changed our lifestyle tremendously, because we are able to have more products at less cost. Many products we now use would not be available if only natural materials were utilized, because they are not adaptable to the many necessary design requirements. In addition, there would not be nearly enough natural materials available to provide even the basic needs of people.

The third is communications. Developments in electronics have provided us with so many products it is difficult to choose those which are most important. Probably television is the one which most people come in contact with. It is a mass communication system and a major factor in our sense of a "global community." Visual communication has done much to help people understand other cultures of the world. At home, television serves as a source of both entertainment and education. The computer probably has had the largest impact on our lives because of its use in research, production, transportation,

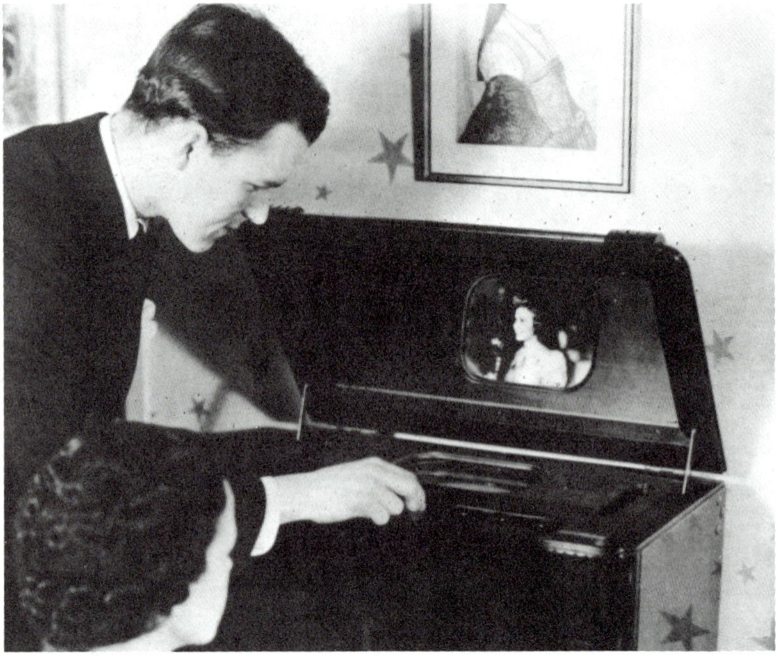

Early television sets could produce only a small image, but had to have a large cabinet because large electronic tubes were used. The tubes have since been replaced with very small microprocessor chips.

and communication. Even though we do not come in contact with it directly as much as with television, it is very much part of our lives. Its application in medical research and technology today cannot be adequately measured, either in terms of benefit or cost, because of the number of lives it has saved.

The fourth is energy. World War II brought about developments in atomic energy. Peaceful uses of atomic energy have been primarily directed toward providing the large amount of electricity needed for our level of technology. However, its full potential has not been realized, and research is continuing in developing a wide variety of applications.

The fifth is social change. Technology provided new jobs with the opportunity of advancement for those who desired to improve their position. Women proved their capabilities for work in technical fields during World War II by performing jobs ranging from running heavy equipment to serving as aircraft test pilots. This placed them in a permanent position in the labor force. Today there are more women in the labor force than at any other time in history. A system of public education provides training in technical fields, which helps people prepare for professions in technology. This provides the opportunity for training, or retraining, in preparation for changing jobs at any point in one's life.

During World War II women trained for jobs traditionally held by men.

Social and cultural changes since 1900 parallel the rapid pace of technology. Cities grew as industrial centers were located near them and people moved from the farms to where the jobs were. When the cities became crowded, the movement of people to the edges of cities created suburbs, sprawling out further and further from the city core. In addition, people became more mobile, moving from one end of the country to another. As a result, educational systems became more standardized and living styles more alike. There developed a norm that exists from north to south and east to west in American life.

TECHNOLOGY TODAY

The majority of people living in the world today have the highest standard of living experienced by humans in recorded history. Technology has served to relieve humans of the critical, continuous burden of providing clothing, food, and shelter, although humans still have to work to provide these needs. Many kinds of work are available, providing security, safe working conditions, and the opportunity for advancement for those who seek it.

Homes today are far beyond the imagination of people who lived only a few short years ago. There are many people living today who can remember when such items as electric lights, indoor bathrooms, air-conditioning, insulated windows, refrigerators, and wall-to-wall carpeting were nonexistent. Construction techniques have allowed homes to be built according to the person's wants and needs, with a variety of choices including style, size, and location.

Medical care has reached a high level as a result of the combined efforts of science and technology. People today are the healthiest of any time in history and they are living longer on the average than any humans ever lived. Preventive medicine has completely eliminated some diseases, specialized surgery is able to replace some body parts, and plastic surgery is able to restore body features to their original visual state. Dentistry is able to save natural teeth for many years and construct exact duplicates if the natural teeth cannot be saved.

Also contributing to good health is the availability of more nutritious food. Nearly any kind of food is available the year around, providing a well-balanced diet. Large equipment for planting, harvesting, transporting, and preparing food to be sold allows huge quantities to be sent all over the world.

One of the elements related to the standard of living which many people feel is important is the amount of leisure time. In the past, nearly all available time was spent in working. Now, as a result of applying technology, jobs are not only physically easier but the number of actual working hours is much less.

The last indicator of the level of standard of living is probably the most important. This indicator is the freedom of choice. As technology advances, more opportunities are available for more people. Wages are higher, allowing people to choose where they want to live and what life-style they would like. They are able to travel where they please and return when they please. There are also many educational and recreational opportunities as well as avocational pursuits. There is a large choice of products from which to choose, whether one is buying bread or a new car. There is even an opportunity to purchase items through credit when money is not immediately available.

SUMMARY QUESTIONS

1. This chapter identified four time periods through which humans have passed. List two important aspects of society in each period which illustrate the results of technology at that time.

2. Why was the development of the steam engine so important in the evolution of technology?
3. Name three ways the Industrial Revolution changed our society.
4. Why has World War II been called a "war of physics"?
5. In what way does technology contribute to human freedom?
6. Name a product which we consider important to our daily lives, and identify how people of the past satisfied the same need before the product was available.

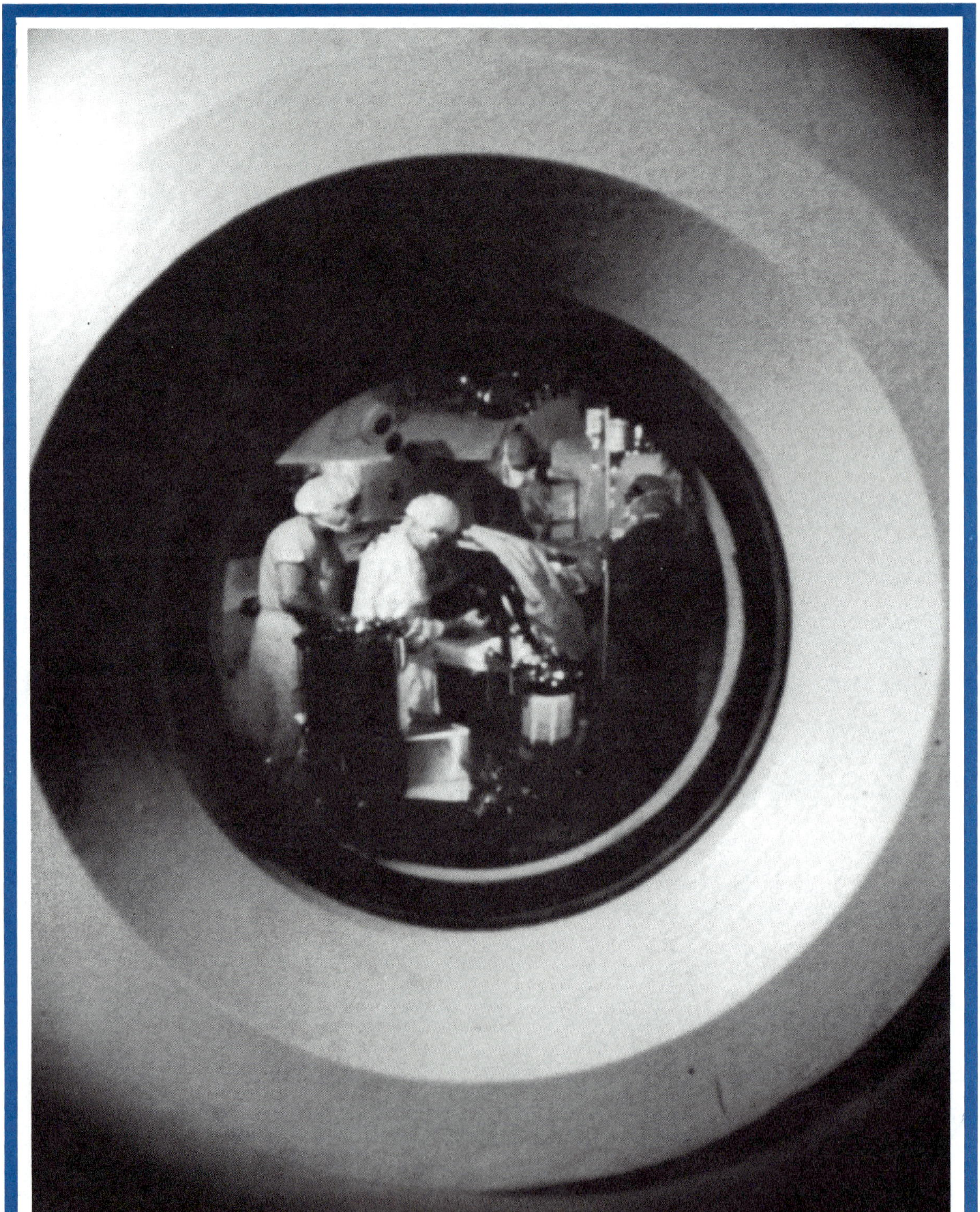

CHAPTER 4

The World in Which We Live

In this chapter you will learn:

1. Why most countries of the world today have a higher standard of living than they have ever had before.
2. How technology has improved working conditions in the last 200 years.
3. Where the application of technology has had an effect on the natural resources of the world.
4. How technology affects both the psychological and sociological aspects of life.
5. How people can prepare for a future that involves rapid change.

THE BENEFITS OF TECHNOLOGY

The advancement of technology since human ingenuity first fashioned artifacts has served to improve the quality of life. Much technology that had destructive purposes during the past wars has been redirected toward peacetime uses, thus saving lives and providing for the needs of everyday living. In some cases technological innovations have had a direct beneficial effect, such as the discovery and production of smallpox vaccines. In other cases the benefits have been indirect, such as the development of the diesel engine which is used in most of the large trucks on our roads. Because we see so many trucks each day, we tend to take them for granted, but they represent the fastest, most flexible, and most efficient freight transportation system in the world.

In most countries, each day marks new advances in technology, ensuring maintenance and improvement in living standards. Although extremely broad in scope, living standards may be analyzed by examining three basic elements that are necessary for survival.

Advances in Medicine

The combination of science and technology has provided for longer, happier, and more productive lives by applying technical knowledge to health needs. Advances in the medical field have paralleled the growth of technology since the Industrial Revolution. Each discovery about how the body functions ulti-

There are more than one million tractor-trailers in the United States today. The diesel engine generates more power, burns less expensive fuel, and has fewer engine parts which need replacement compared with gasoline engines.

mately led to a need for new medical tools. These developments gave rise to new knowledge, thus creating further needs for improved applications of technology. Medical advances might be broken into phases.

Diagnosis. In the past when people were ill, the doctor had to rely heavily on the patients' description of how they felt. Today, there are many methods of diagnosis, ranging from the simple stethoscope to advanced technology using X-rays and computerized equipment that easily provides many bits of information at one time. Thanks to these advances, it is now possible to identify medical problems even before the patient becomes ill.

Treatment. Probably one of the most significant areas of advancements in treatment has been surgery. An operating room is literally a technological maze. In addition to technology that controls the room's lighting, temperature, and humidity, there are all kinds of equipment for monitoring the patient's condition and for supporting life. Another significant area of development is the production and administration of medicines. Technology provides thousands of medicines from general, nonprescription aspirin to very specific medicines made for an individual patient. One of the newest methods of administering medicine is the injection gun which uses no needle.

Current health care facilities combine and apply a very high level of science and technology.

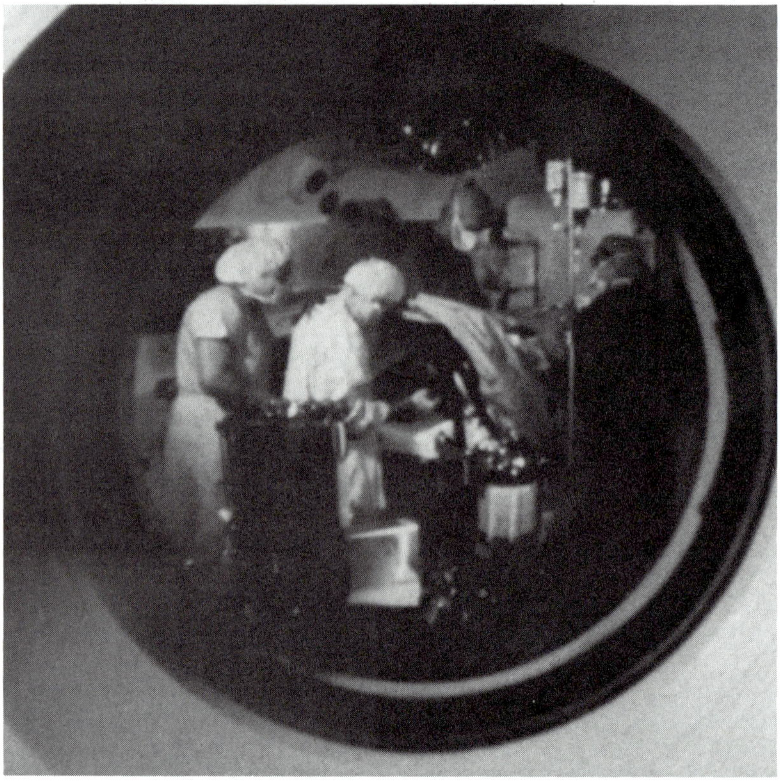

Replacement of Body Parts. New developments in electronics have opened up a new field of medical treatment involving the replacement of damaged or inoperative body parts. Today, there are thousands of people with heart disease who live perfectly normal lives because they have an artificial valve or a pacemaker. Hearing aids, contact lenses, artificial teeth, and a number of devices to help people talk are common examples of this kind of applied technology. Artificial limbs are another example. Some now can connect body tissues, such as muscle and nerves, to the hardware, making the part function much like a living limb.

Advances in Food Production

The production of food is probably one of the most important industries of the world. Food is, of course, necessary for life, but the fact that it is also considered a product means that it provides jobs for many people. In the United States alone, farming and related fields employ more than three and a half

Citrus fruits can only be grown in warm climates. These grapefruit are being processed into canned goods for sale throughout the world.

million people, and more than a million and a half are employed in processing. And these numbers do not include transportation and marketing, which is also a major part of the food production system.

Because of the many different growing climates throughout the world, certain foods can only be produced in areas where the climate is conducive to their growth. In the past, people in some parts of the world did not get a balanced diet because only local food was available. Now, thanks to increased food production and world transportation systems, they are able to eat a more balanced diet. This has served to improve general health and to extend life expectancy.

Advances in Work Opportunities

Most people have to work to meet their responsibility of providing food, shelter, clothing and other essentials for themselves and their families. Technology has changed the nature of that work and has brought with it greater freedom of choice and diminishing working hours.

In the past, people did not always have the opportunity to choose their life's work. Skills were generally learned with the family, and the family craft or service was carried on as the children grew older. Also, the opportunity of moving to an-

In the past these heavy products might have been moved by hand. Today, this worker has a much easier and safer job: the fork lift carries the products, and the steel cab, hard hat, and eye goggles provide protection.

other location to seek a job was limited, because long-distance travel was difficult. Today, people may select their work, based on many factors. This freedom of choice also provides the opportunity to change to another line of work at any time the person desires, providing the job requirements are satisfied.

The number of hours worked each week has diminished considerably in the past two hundred years. In the 1700s, when this country was primarily involved in agriculture and crafts, the average workweek was about seventy-two hours. By 1900, it had fallen to about sixty-five to fifty hours per week. Shortly thereafter it dropped to forty hours per week, and this has remained a standard until today. Some organizations, however, have now adopted the thirty-five hour workweek. Today's workers are able to provide for their needs while they pursue interests in leisure activities, community service, and additional education.

THE DICHOTOMY IN TECHNOLOGICAL ADVANCEMENT

At first glance it would appear that with the level of technology now at our disposal, we should be able to eliminate the world's problems and create a Utopian civilization. There are three reasons why this situation has not happened and probably will not in the foreseeable future.

First, technology, being an extension of human knowledge, skills, and desires, cannot be separated from the other elements of society. The application of technology is dependent upon economics, religion, politics, and social values. The diversity of cultures throughout the world precludes a common view of the needs and goals of all peoples.

Second, cultural needs are constantly changing, requiring new technology and applications. Population figures and the shift in age groups, for example, are indicators of future needs. The population growth in our country has slowed considerably in the last ten years, and one hopes we are on the road to zero population growth. At the same time, however, the need for housing is growing at three times the rate of population growth because of the large number of adults in the twenty-to-forty age group. At the same time we are experiencing a housing shortage, many schools are standing empty. Only a few years ago, these schools were overcrowded as these adults were obtaining their education. The impact of this age group will be felt even more when they reach retirement age and are no longer in the labor force, but still requiring housing, goods, and services. As each generation comes along, their attitudes and needs are perceived differently, causing the social structure to change and, in turn, affecting how technology is utilized.

Finally, new technologies are developed to satisfy immediate needs, but sometimes potentially dangerous side effects can occur that were not apparent when the technology was initiated. These problems are not known until other new technologies used to detect the problems are developed. For example, the insecticide DDT became famous during World War II when it was used to stop an epidemic of typhus fever in Italy. Until 1972, it was used in nearly all parts of the world to destroy insects that caused disease and damaged food crops. It was then determined to be a potential long-term danger to both humans and wild life. Thus, new technologies serve two purposes; they satisfy current needs and identify problems with old technologies. Because of these three reasons, and others, problems may be found throughout the world.

World Poverty

Between 1950 and 1975, the world population increased by 59 percent, and the prediction for growth during the last quarter of this century indicates another 58 percent increase. The growth is taking place primarily in underdeveloped countries which are not able to produce the goods and services to take care of their people. As a result, a large part of the world population lacks education, medical care, and housing. If the trend continues, it is most certain that millions will face starvation in the next few years, because there are not enough available resources to feed the world population at the current level of technology.

Depletion of Natural Resources

Because of the rapidly growing world population, many natural resources are being used faster than they can be renewed. In the underdeveloped countries, less crop-producing land is available because of space needed for more people. In addition, much of the productive soil is wearing out because of

If the present predictions of world population growth hold true, rapidly advancing technology will be essential for human survival.

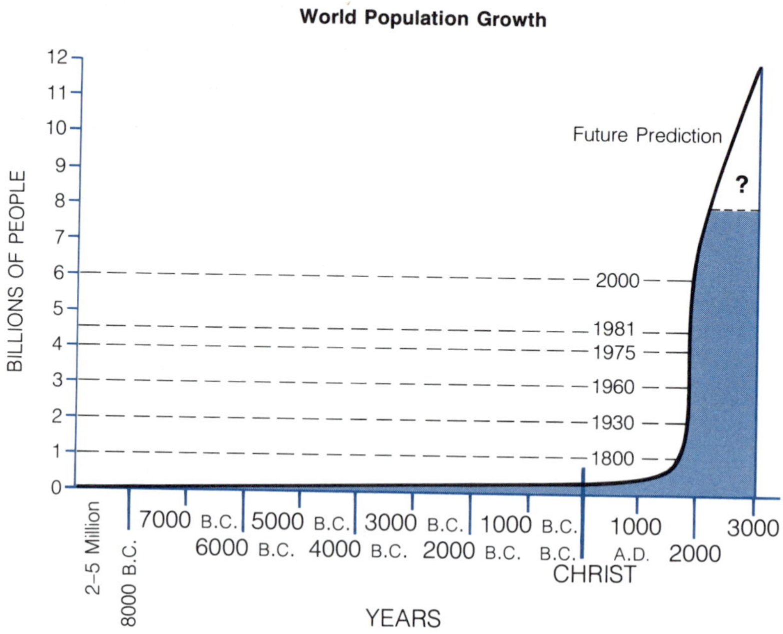

Productive soil must maintain a delicate balance of nutrients and water. Because of the removal of large quantities of underground water in some areas, cracks more than a foot (30cm) wide and several feet (cm) deep can appear.

overuse. Overgrazing, overcutting of trees, and overplowing of large areas have contributed to the encroachment of nearly all of the world's deserts into previously productive land. At a time when more arable land is needed, it is actually shrinking due to the process of desertification.

Another problem has developed in areas where, for centuries, irrigation has turned previously dry areas into productive land. The water which lies near the surface evaporates, leaving a concentration of minerals and salt in the soil. Over many years the level of salt begins to retard and even prohibit plant growth. This has become a serious problem in many countries which depend upon irrigation for food.

Developed countries also face problems. Most cities were first built near water, rivers, lakes, or oceans, because of transportation needs. But the presence of water made most of the surrounding land very fertile and farms sprang up around the cities. Today, the expansion of cities and suburbs, commonly called urban sprawl, is taking up some of our most productive land. As the population grows, there are other needs in addition to living space. New and bigger transportation systems are needed, more factories are needed to produce goods, and more space is needed for energy production. According to a U.S. Department of Agriculture land use survey, between 1967 and 1975 this country converted close to 6,201,000 acres (2,510,000 ha) to urban and other build-up uses. In eight years we "urbanized" a combined area nearly the size of the state of Rhode Island, and this trend will continue into the 1990's.

What makes the natural resource supply so critical is its direct relationship to the development of energy. Nearly all of our major energy sources, such as wood, coal, oil, and gas, are nonrenewable or, at best long-term renewable. At a time when an increased population needs more energy, the sources are rapidly being depleted.

Environmental Pollution

For some people environmental pollution has become synonymous with technology, but this is not always the case. Environmental pollution can be caused by nature itself; but it most certainly can also be caused by people who misuse the products of technology. Much of the world pollution is due to the processing and manufacturing of materials. Because products are made from raw materials which are natural resources and because they are changed by applying energy which is derived from natural resources, it is inevitable that changes in the earth's surface will occur. In addition, the mixing and separating of raw materials during processing causes new materials to be developed, some of which are usable and others not. For those that are not, waste disposal becomes a problem.

Many times industrial waste is in the form of chemicals which are suspended in heat exhaust. Additional research and money are needed to develop the proper technologies to take care of this problem.

Air pollution has become a worldwide problem as a result of burning fossil fuels (coal and other petroleum products). The sulfur and nitrogen oxides which are released in burning combine with other chemicals in the air to produce acids. These then fall back to earth with rain and snow, affecting all living things from fish to forests. Centuries-old buildings are deteriorating and crops in some parts of the world are decreasing because of the acid. Yet the demand for more energy grows, and so does the pollution.

The need for more area given over to transportation, manufacturing, and living has resulted in a loss of natural vegetation, which in turn has caused an erosion of soil. Water, one of our most precious resources, is also being polluted by chemicals through improper disposal and through natural runoff when it rains. At the same time, the water table (the level of underground water) is dropping as more people draw on it. It is conceivable that in the future there could be a worldwide shortage of fresh water.

Worker Dissatisfaction

At a time when work is easier and safer than any time in history, there is growing job dissatisfaction. Fewer workers feel their job is challenging and interesting, and many feel their skills will not be useful five years from now. Others feel they are locked into their job with little chance of finding a comparable position elsewhere. According to the U.S. Department of Labor, in 1969, 40 percent of the workers thought they could easily find a comparable job, while in 1977, only 20 percent shared this view. This trend has continued into the last half of the 1980's. There are a number of reasons for this, including the need for more education and retraining. As the level of technology rises, there is a need for upgrading within nearly all fields. Those without the ability or inclination to take advantage of retraining usually must remain in the less satisfying positions. The attitude of the worker toward life values also plays an important part. As one worker put it, "I hate my job, but I'm not going to change because I like the money." Unfortunately, this feeling is all too common. As a result of this attitude, many times the quality of work is low, and the production of goods and services usually suffers. Worker attitude and absenteeism is one of the major problems confronting society today.

Inability to Cope with Change

A condition described as "future shock" by Alven Toffler, in his book of the same name, may afflict people who experience too much change in their lives in too short a time. Symptoms might include an increase in work-related illnesses and absenteeism, various attempts at alternate lifestyles (such as rejecting traditional social roles), and a feeling of hopelessness concerning the future. Indeed a major cause of death in young people today is suicide. Two elements may be identified as major causes of change-related problems: the pace of life and uncertainty about where one stands in society.

A hectic, pressured pace of life contributes to impersonal feelings. Mobility takes people away from family and community. People try to fit as many activities as possible into each day. We seem to have developed the philosophy that success is based on the amount of money one has to spend. As a

Coping with rapid change is more of a problem in urban areas because all aspects of life—living, working , recreation facilities and transportation systems—are all intertwined and compacted. Change in any one causes change in the others.

result, interpersonal relations and a genuine concern for others is sometimes forgotten.

Technology brings classes of people closer together. As a result of both job and educational opportunities it is possible for nearly anyone to move quickly from one socioeconomic level to another. However, not all institutions of society (religious, political, educational, and so forth) are able to accept this movement, because they have not changed at the same rate. As a result, it is possible for a person to be accepted as an equal in one institution of society and at the same time be rejected in another. This inconsistency of change can cause difficulty in adjustment because it is a dynamic process.

TECHNOLOGY AND THE FUTURE

What will life be like in the next few years? One thing is certain: it will be vastly different from today. Advancing technology will have an even greater impact on our lives than in the past. The time period referred to in science fiction stories of the past, which included cars with self-guidance systems, robots, space stations, and laser beams, is now here. We are on the verge of a new era in which today's generations will be a part. It should prove to be a time of great change, a time of great potential for humankind, but it will also present many problems and require many important decisions.

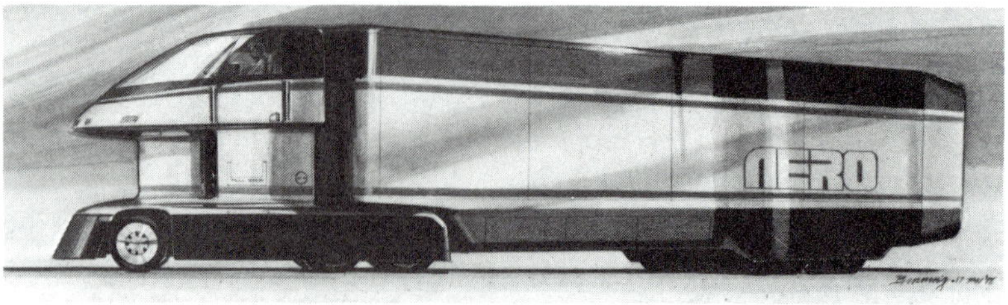

Goods will still have to be transported in the future. This design for a freight hauler would be much more efficient, because it is made of lighter-weight materials and has a more efficient energy system.

Two occurrences in the last decade will have a tremendous impact on our lives in the future. One has appeared as a problem and the other as a solution.

The Energy Shortage: A Problem

The purpose of technology is to accomplish work, and work cannot be done without energy. A high level of technology requires a tremendous amount of energy. The United States helps to supply the world with goods and services, because our level of technology allows us to produce more than most other countries. To give an example of the energy required to accomplish our level of production, it is estimated that our country has 6 percent of the world's population, while we consume nearly 35 percent of the world's resources. An ideal situation would be to have all countries develop the capacity to produce their own goods and services to take care of their own needs. In most cases this is not possible for two reasons: some countries are not blessed with natural resources needed in energy development for today's technology, and even with the resources, the cost of energy production is more than most underdeveloped countries can afford.

When energy needs are higher than production, the cost of energy goes up. When the cost goes up, it takes more energy to produce more goods to help pay for the increased cost. Therein lies our present-day problem. Our current technology of using fossil fuels served us well in the past. As the need for more energy grew, we produced more by using more of the natural resources. Now that the resources are dwindling, wells have to be drilled deeper, mines have to be dug deeper, and new processing methods have to be developed to tap less desirable resources. Naturally, all of this raises the cost of energy production. Some estimates of necessary energy costs by the year 2000 range from a doubling of the current price to one

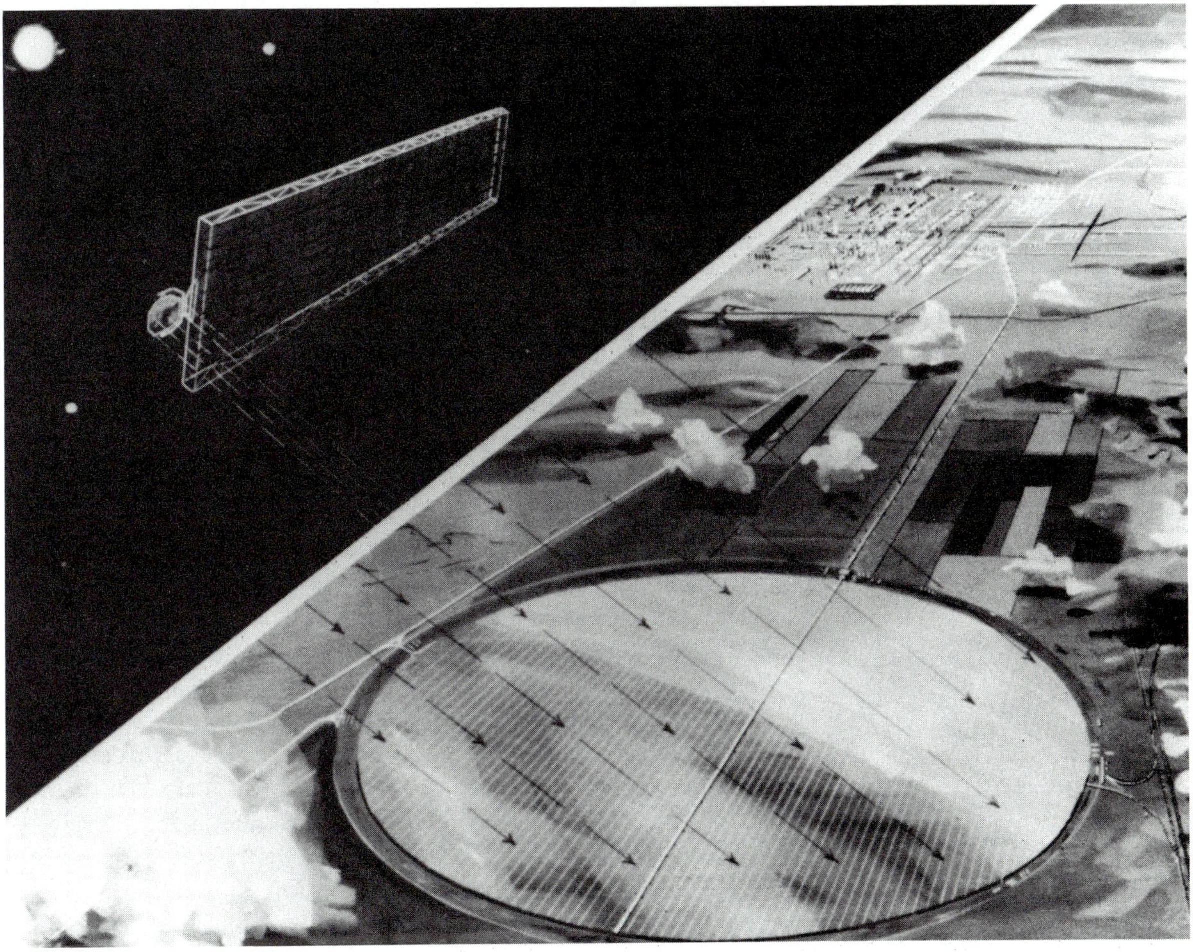

Future energy needs may be partially met by the Satellite Power System. NASA and the U.S. Department of Energy are jointly investigating the system which uses a satellite solar collector several miles (km) in length and an antenna on earth covering several hundred acres (ha). Light is converted into electrical energy by solar cells and beamed down to the antenna as microwaves. It is then converted back into electrical energy.

hundred times the current price. Either estimate will require a change in lifestyle for all of us.

Among some of the changes we are likely to see are cars with alternate energy sources (such as electricity), a decline in the number of cars because of more efficient mass transportation systems, and negative incentives for driving cars (such as taxes and/or tolls to travel where mass transportation is available). More rigid restrictions will be enacted on land use, including urban expansion and new construction. As a result of increased building costs and restrictions, there will be fewer energy-using highrises, with more residential housing taking the form of clusters where walls are shared to conserve energy. Architectural styles will change drastically to take advantage

Pedestrian movers such as this may become a common sight in downtown areas where cars may be completely prohibited. Pedestrians enter and exit while the vehicle moves at walking speed.

of energy-saving designs. In general, most decisions concerning what to buy, where to go, and how to go, will be made based on cost, determined by the amount of energy required.

The Computer Revolution: A Solution

Many have called the development of the computer a "revolution," because it promises to have an impact upon society which equals or exceeds the impact felt by the Industrial Revolution. The age of computerization is less than forty years old, and it already has changed our lives in medical technology, communication, transportation, and the production of goods and services.

The key element in a computer is a component called a microprocessor. It is a chip of material made of crystals which contains the circuitry needed for conducting logical and mathematical functions. It may be as small as one-fourth of your fingernail yet as powerful as the huge computers of yesterday which took up several air-conditioned rooms. An additional amazing fact is that in an age of rising prices, computers and other similar equipment are rapidly decreasing in price. Today, hand-held computers costing only a few dollars are more powerful than the most powerful computer available a few years ago, which cost several million dollars. The reason for this drop in price involves new discoveries in the field of electronics that allow low manufacturing costs. Microprocessors involve less labor than many other products to manufacture. They are made of relatively inexpensive materials, they take little energy to produce, and they are small, requiring less space for manufacturing and transportation.

It appears the pattern of decreasing costs and increased potential will continue. It is highly probable that in the future nearly all products which perform an active function will contain microprocessors. We already have many, including cars with electronic ignition, microwave ovens that can be programmed for time and temperature, and word processors that will automatically type a column of text. Family records, food inventory, and other similar records will probably be kept in the home computer. It is also possible the home computer will provide a direct link with health-care facilities where body functions like blood pressure and temperature could be monitored daily. There will be virtually no segment of our culture untouched by advanced electronics.

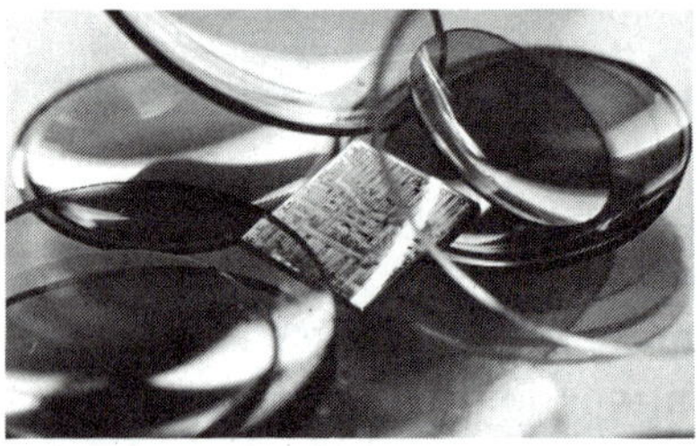

This small silicon chip is dwarfed by the tiny contact lenses. What makes this chip so amazing is that it is a microcomputer or "computer-on-a-chip," it carries the computational power of an electronic computer.

This dashboard has several readouts for a computer monitoring system. In addition to information about the engine, the driver is informed of such things as inside air temperature, the amount of fuel being used, the distance possible on the amount of fuel remaining, and distance remaining until the destination is reached.

Why has this phenomenon been called "a solution"? There are three reasons the development of microprocessors is important.

The first involves the relatively small amount of energy consumed in such a system. Little heat is generated, making the amount of electricity required very small. Little maintenance and few replacement parts are required. They require no additional installation systems for transmission of data (such as telephone lines).

Second, microprocessors make systems more efficient. Microprocessors can be used to automatically monitor and control occurrences such as home heating and cooling temperatures, car speed, air craft takeoff and landings, and traffic signals. When these and similar situations are controlled according to need, it cuts down on energy waste and provides an optimum safe, healthy environment.

Third, microprocessors provide increased knowledge-processing capabilities. Answers to questions may be obtained within minutes, while in the past it might have taken many people several months to complete the same number of calculations. It also is possible to predict future occurrences and conditions through simulation. This helps in avoiding decisions that could be potentially dangerous or costly.

The newest emerging field in computer science is artificial intelligence. In this application, information not only is produced, it represents the best problem solutions which are a result of analyzing several possible choices. The computer then becomes a decision-making tool, applying, in simplistic terms, basic functions associated with human intelligence. Computer systems represent a very important applied research tool as well as a theoretical research tool. The increased shortage of fossil fuels and the high cost of energy production will most certainly be eased somewhat by the application of microprocessors in research and development of new energy sources.

There have been many predictions about the future, ranging from a worldwide Utopian civilization to a picture of doom and destruction. The reality will lie somewhere in between, depending on choices made in the use of technology.

Since science and technology takes the form of self-sustaining growth (the snowballing effect), it reaches into all areas of life, whether it is intended or not. As this growth takes place it becomes more difficult to make decisions concerning the use of technology. As previously mentioned, the reason for this is the fact that beneficial effects in one situation may present detrimental effects in others.

There are two developments evolving at the present time which may help to make some of the decisions involving the application of new technology.

First, the combination of traditional fields has led to the cross application of knowledge. This leads to more communication and, thus, to a better understanding of the impact of the new technology under consideration. An encouraging trend is the increased involvement of the social sciences with engineering and technology. Decisions in such areas as urban planning, land use, transportation systems, and environmental preservation should certainly involve both.

Second, as technology evolves, specialized fields also evolve. As knowledge is developed, it leads to questions which require research in a specific topic. As more knowledge is obtained, what was once a small topic becomes a broad field of application. An example of this expansion of knowledge might be the development of plastics. The first plastics were the result of chemistry experiments. Today, we have many different plastics industries, and it is possible to prepare specifically for work in the field of polymer (plastic) chemistry. These new fields are developing at a faster rate than in the past because of the "knowledge explosion." In addition to increasing the flow of knowledge through these new fields, problems in the application of old technologies should be identified and solved much faster.

Determining New Directions

The future holds great promise if we make wise decisions in the use of technology. The shortage of energy, the depletion of natural resources, and the growth of world population require new thinking to determine the role of technology at home and around the world. In the past there was little integrative planning. If a need developed, the technology was devised to solve that particular problem. But a failure to analyze the need and plan for its solution has led to problems. For example, our culture has developed a "throwaway" lifestyle where many products have been designed to be discarded rather than repaired. At the same time, we have not developed disposal systems to take care of the accumulated waste. Burying the tremendous amounts of solid waste caused by our "throwaway" culture can pollute underground water veins, while burning it pollutes the air. In addition to pollution problems, many disposal methods require the use of our much-needed energy. In addition to a "throwaway" life-style, we have developed the attitude that "bigger" and "faster" is better because we believed we had the technology, unlimited natural resources, and unlimited energy. The limits of this belief became obvious during the last decade, and our society is reorienting its priorities and recognizing the necessity of long-range planning. Technological assessment and appropriate technology are two means to this end.

Technological Assessment

The process of technological assessment involves determining our present situation, including both needs and resources, and projecting future needs based on the findings. It assesses all effects, adverse as well as beneficial. This allows time for the development of guidelines as we move toward the future. The government has formed an Office of Technological Assessment, which reports directly to Congress on matters concerning present status and future directions. A number of other organizations, including many private industries, are also becoming involved in assessment, with an emphasis on technological forecasting.

Appropriate Technology

The determining factor in choosing the appropriate level of technology should not be what is possible but what is required.

Appropriate technology is the process of more closely analyzing needs and determining the most efficient technology for each situation. It challenges the concept that "bigger" and "faster" is better. It defines technology in three levels: low technology is the type used by our primitive ancestors where all work was muscle power; intermediate technology is defined

Characteristics of Technology Levels

Low	Intermediate	High
Applies the basic principles of science	Applies advanced principles of science	Tests and develops new scientific theories, finds new applications for known principles and applies them to highly specialized situations
Requires simple tools and facilities	Requires machines and complex support facilities (buildings, equipment, and so forth)	Requires sophisticated machines and equipment, support facilities, and many different technology systems which apply high technolofy in their own activities
Uses little processing of raw materials, other than heating cutting and shaping	Uses advanced processing methods which can change the general chemical composition and physical characteristics of materials	Uses highly specialized processing techniques which result in materials that have very specific characteristics
Requires little energy	Requires medium to high energy at a constant rate	Requires high energy initially, but has the potential of requiring less total energy than intermediate technology
Uses mostly muscle power, both human and animal, with high physical energy input and low power output	Uses tools and machines, with medium physical energy input and medium to high power output	Uses automated machines, with low physical energy input and high to ultrahigh power output
Provides little control over the environment	Provides limited control over the environment	Provides greater control over the environment, with the potential of being able to alter it, as well as creating new artificial environments
Can provide benefits to only those few people who come in direct contact with it	Can provide benefits to many people, even though they may not come in direct contact with it	Can provide benefits to people all over the world, representing all countries and all cultures
Provides the opportunity to solve immediate problems of a society	Provides the opportunity to solve current and short-range problems of a society	Provides the opportunity to solve current, short-range, and long-range problems on a global scale

as involving tools, machines, and techniques which still require human control and manipulation; and high technology involves such aspects as advanced chemistry and physics, push-button automation, robots, and self-guidance systems. Appropriate technology does not advocate one level of technology over the other. It simply states that one may be appropriate in a given situation, while another may be appropriate in a different situation.

The concept of appropriate technology is an important one for both developed and underdeveloped countries to keep in mind. It makes little sense for an underdeveloped country to develop an automated manufacturing system (high technology) that uses automated robots when the population has no jobs and no means of purchasing the products being manufactured. Conversely, it makes little sense for a developed country to dispose of waste by dumping it in rivers and burying it in soil (low technology) when a higher level of technology would allow appropriate processes to transform it into useful or nonpolluting material.

WORLD PRIORITIES

We no longer live in a world where countries are isolated and each is able to function as an entity unto itself. Improved communications and the importing and exporting of raw materials and manufactured goods has brought the people of the world closer together, helping to raise the quality of life in nearly every country. It is important that this continue, including the sharing of technical developments.

Even though the quality of life is much higher in most countries than in the past, the number of serious world problems to be solved is growing. Several have been previously identified in this chapter; the following compilation helps summarize the seriousness of our world situation:

- exploding population
- food shortage
- energy shortage
- depletion of natural resources
- rain and snow with high levels of acid

- depletion of fertile soil because of urban sprawl
- deterioration of urban areas
- pollution of water with toxic chemicals
- inadequate solid waste disposal methods
- soil erosion
- salinity in the soil (concentrations of salt as a result of irrigation)
- infertility of previously fertile soil
- desertification (previously productive soil becomes desert).

Advancing technology is the only hope of survival for millions of people on this planet. Some argue that technology has caused many of the problems and that traditional methods would not have caused the situation we face today. It must be remembered that the so-called traditional methods also represent technology. Many of our present problems have occurred because we have continued to use the older technology far longer than it should have been used. It is simply a matter of making appropriate choices in developing and choosing the technology.

Research

Most developed countries are hard at work, within the limits of their resources, at improving their level of technology. In our country we have some corporations that do only applied research, while others produce only ideas and theoretical solutions. Both provide a valuable service in identifying the best possible direction to take in solving previously unencountered problems. As the world population continues to grow there is

Improved equipment can help increase food production and provide for more efficient use of natural resources. This machine is excavating a new concrete irrigation canal which will be much more efficient than the previous clay-bottom ditches.

less and less land available for new development. There are no "western frontiers" or exotic lands in which to settle to make a new life as there have been in the past. With land and resources becoming more scarce each year, it is imperative that different, undeveloped frontiers be explored. There are three basic frontiers of research underway which may hold the solution of some of the world problems. Space exploration is one. We are reaping many benefits today from spin-offs of the research conducted in the space program. We are now undertaking space exploration not only to reach other planets but to solve some of our earth's problems. A coupling of earthbound technology with the possibilities of space technology never before available makes it difficult to comprehend what new discoveries will take place in the coming years.

A second frontier of research is in energy development. There is a direct relationship between natural resources and the amount of energy required to develop them. As the supply of natural resources becomes limited, the amount of required energy rises because it necessitates more equipment, longer hours, and the handling of more raw material. The production of minerals in the United States has increased 50 percent in the last fifty years, while the amount of energy required to reach this level has skyrocketed to 600 percent. There is little choice but to develop alternative sources of energy. The energy of the future will need to incorporate three characteristics: it should be easily converted to power, it should be renewable, and it should be nonpollutive.

A number of sources are currently being considered. Among those types of energy which may satisfy these three characteristics are solar, wind, and nuclear. There are also a number of other sources undergoing research, but these three appear to hold the most promise. Their major advantage lies in the fact that they are renewable, that is, they do not require the rapid consumption of materials which could eventually be no longer available.

The depths of the sea is the third frontier. Human existence has always been dependent upon large bodies of water. They affect weather conditions, provide transportation, and supply food. Nevertheless, marine research has not been conducted on a large scale. The shortage of traditional natural resources has generated a growing interest in the unexplored potential of the sea. Water covers approximately 70 percent of

Geothermal resources are another potential energy source. This geothermal well has been drilled deep enough into the earth to bring steam and hot water to the surface under pressure much like a controlled geyser. This energy source can generate both water and electric energy.

the total surface area of the earth, so it stands to reason there are many resources which we have not yet considered. Marine research may be broken down into three general categories: food production, mineral extraction, and energy development.

Preparation for Decision Making

Preparation for the future requires developing insights which allow correct decisions to be made. These decisions will be required of all people, whether on an individual level, such as occupational changes, or on a larger basis, such as world leadership.

We are rapidly approaching a fork in the road of history. The road we choose to take in the next few years might well determine the future of all humanity. There have been many forks in the road during the course of past history: some have led to brighter futures and some have led to small disasters. Survival of the disastrous choices, however, was usually assured because time provided an opportunity to make additional alternate choices which lessened the impact of the incorrect decision. For example, in the last two hundred years we built our country from forests which covered the land. For many decades there was a need for a tremendous amount of wood. There was much waste and no replanting. Because it takes many years for some species to grow to maturity we have very few hardwood forests left. However, the development of new materials such as various kinds of metals and plastics lessened the need for so much wood. This fact, coupled with the application of conservation methods, has ensured the continued existence of the once-endangered natural resource. Time provided new options and an opportunity to correct past mistakes. Now, however, there are two reasons why mistakes will be more severely felt than in the past: the speed of change and world involvement. New technologies are applied so quickly that they sometimes overlap. There is little time in which to test decisions and make needed corrections. The disposal of toxic chemicals, solid waste, and nuclear radiation present problems which overlap. These problems require immediate solutions, and it is imperative that correct decisions be made. The residual effects of these three immediate problems, when disposed of incorrectly, could last for many centuries to come.

The effects of advancing technology are now felt on a global scale (world involvement). Because of instantaneous communication, speedy intercontinental travel, and interdependence of countries for goods and services, we are truly a global community. This brings with it complex responsibilities and conflicting demands. For example, as developed countries require additional energy, the burning of such fuels as coal, oil, and gas releases more carbon dioxide into the atmosphere. The combination of carbon dioxide, water vapor, and a number of pollutants with the ozone layer tends to act much like a piece of glass which lets light pass through. This causes a condition called "the greenhouse effect." It lets light from the sun pass through, but captures the heat. As a result, the earth is becoming warmer. At the same time, some pollutants cause a breakdown of the ozone layer, allowing harmful sunrays to enter the

A view from an airplane reveals a heavy layer of natural and human-induced pollution over a large city, which is not always noticeable from the ground.

earth's atmosphere. Decisions which affect the delicate balance of our protective shield surrounding the earth is important to all people throughout the world.

How can people be prepared for the future so that correct decisions may be made? There are three imperatives for preparation. Each is necessary for individual adaptability.

Awareness of Culture. Culture may be defined as the complete environment in which one lives. This not only includes the natural and human-made environment, but the social structures and value systems. The interaction of all these elements make up our culture. Thus, to understand one's culture is to understand the various elements and their relationship to each other.

Accelerated technology produces forces that speed up social and cultural change. It is important that people understand these forces so that they are able to control the elements which should remain unaltered and adapt to those changes which are beneficial. When individuals possess this awareness, they are better able to control factors that make a more satisfying life possible. When groups of individuals work collectively toward this end, the entire culture is able to adapt to rapid changes.

Acceptance of Responsibility. Technology extends human physical and mental abilities. Because it is an extension, it is subject to the same controls necessary for any kind of simple human activity. Problems arise when technology is allowed to develop spontaneously and without proper control. When this happens, decisions are made which are not in the best interest of society. It is important that appropriate values be used when decisions are made.

For example, many people feel all nuclear energy development should be stopped immediately because of the potential danger of radiation. Others argue that the complete stoppage of nuclear development might very well prove to be a greater potential danger as the world population grows and energy supplies diminish. The acceptance of responsibility in this case would involve placing proper controls on the development of nuclear power so the technology does not exceed our ability to use it safely. The decision to exclude research and development of a potential technology application could be just as serious as including an inappropriate one. It would be just as wrong to advocate only the development of nuclear energy, excluding other possible, less potentially dangerous energy sources.

In some cases, conditions with serious consequences have evolved gradually and are accepted as normal. Highway fatalities, for example, fall into this category. Every year nearly 50,000 lives are lost on highways in the United States. Our society has decided that this is the price that we have to pay if we are to continue to have the freedom that the automobile provides. We still have not accepted the responsibility, even though we have had this technology for well over eighty years.

Technology causes value systems to change. Many people have expressed the opinion that technology is dehumanizing. Some have even gone so far as to "dropout" of society by rejecting modern technology so their daily work would be more "human." If one analyzes this basic value in its broader context, it is found that technology is just the opposite of dehumanizing. A number of reasons can be identified: (1) it relieves hard physical labor; (2) it makes dangerous work much safer; (3) it produces new, more exciting jobs, and (4) it allows more time for recreation and creative activities. *The difference between dehumanizing and humanizing technology is determined by the acceptance of responsibility in directing its use.*

An Eye to the Future. Many people find that the past is much more comfortable than the future. Marshall McLuhan, in his book *The Medium Is the Massage*, suggests that we march into the future while we are looking into a rear-view mirror. That

Compared with these early automobile assembly-line workers, today's workers have better working conditions, shorter hours with higher pay, fringe benefits such as health and accident insurance, and representation in decision making.

is, we are always wishing things would be like they used to be, or at least stay as they now are. This does not help us adapt to the changes the future brings. It is necessary to change our concepts of several basic elements.

Our concept of time produces a lag between reality and our acceptance of reality. For example, we now have the technology to begin placing space colonies in orbit around the earth, but to many people this is a "Buck Rogers" dream of the far-distant future. The world of transcontinental personal communication, home computers, travel faster than the speed of sound, and three-dimensional pictures made by a laser beam all seem to be a part of the future, when in reality they are elements of our present. A greater awareness of our culture and a realization that rapid technological change increases the speed at which the future becomes the present will help in cutting down on the reality time lag.

Our concept of our lives in the future may need to be altered to keep pace with changes in our society. An informal study conducted on a number of high school students revealed a shocking lack of awareness when asked about what their lives would be like in the future. They were asked to predict what the world would be like in the years to come and then predict what their personal life would be like in the future. They saw such things as rapid strides in medicine allowing people to live longer, the use of antigravity cars, space travel, robots, advanced computers, and many changes in our social structure. Yet they did not incorporate those changes in their own life. The story was strikingly different. Alvin Toffler summarized the dichotomy this way:

> I must emphasize that the teenagers making these forecasts were incontestably bright, lively, and probably more sophisticated than their counterparts in smaller cities. Yet, no matter how turbulent a world they pictured, no matter how many new technologies might appear or what political revolutions might take place, the way of life foreseen for themselves as individuals seldom differed from the way of life possible in the present and actually lived by many today. It is as though they believed that everything happening outside one's life simply by-passes the individual. The respondents, in short, made no provision for change in themselves, no provision for adaptation to a world exploding with change. (Toffler, p.11)

This lack of a feeling of personal involvement makes adaption to change much more difficult.

An attitude which is based on "letting tomorrow take care of itself" usually leads to difficulty in adjustment. Tomorrow comes rapidly, and it certainly requires changes. Preparation for life's work, identifying ones likes and dislikes, and preparing major alterations in our daily lives cannot be done the same day they become necessary. In our age of technological change, planning becomes more important than ever. In the past, preparation for many jobs could be obtained while working in that position. If one desired a change, it would be easily made, and, if not, the old job would be there for as long as the person desired. In the future, most jobs will require at least certain basic skills and most likely more advanced skills before starting in the position. It is also a near certainty that the job will either change drastically, or even be eliminated, before the person retires. Planning for change allows the changes to be taken in stride. As someone once wisely said, "The future belongs to those who plan for it."

SUMMARY QUESTIONS

1. Explain how technology can have both positive and negative effects at the same time.
2. Identify five major problems facing the world today. Explain why they are considered major problems.
3. What is a microprocessor? Why is it considered to be such an important element in our lives today?
4. Why is technological assessment and the application of appropriate technology important in determining future directions of technology?
5. Identify two reasons why it is important to be aware of one's culture.
6. Explain the term *reality time lag*.

CHAPTER 5

Technology and Elementary Education

In this chapter you will learn:

1. The purposes of education.
2. Why the past, present, and future are important elements in classroom study.
3. How our society shapes elementary students.
4. How physical and mental maturity affect learning.
5. Why developmental tasks are important to physical and mental growth.

RELEVANCY IN TODAY'S EDUCATION

Effective learning helps students interact well with their environments. It stands to reason, then, as environments change, learning has to change also. Relevant education refers to the development of knowledge and experience which reflect the culture of today and tomorrow. Some types of knowledge, such as the basic laws of nature, do not change at all. Some change rather slowly, such as the value systems of a society. Still others change very rapidly, such as emerging technologies which affect the lives of everyone.

Reflection of the Technological Society

It has been said many times that education reflects society. That is, the learning that takes place in school reflects what the students need in order to take their place in society as contributing members. There has always been a time lag in education between subject matter and the realities when students complete their education.

A few decades ago, the estimated lag of ten-to-fifteen years was not considered critical. Changes were not radical, and the speed at which they arrived was slow enough to allow people to become acclimated to them, even if they were not prepared for them. The time lag has rapidly expanded, probably doubling. In order to provide an educational program that reflects today's needs, schools have to constantly incorporate change into the curriculum. An extremely critical element in making these changes is teacher awareness of our technological society.

Most classrooms of the 1930s reflected a relatively inflexible society: the desks are permanently attached to the floor; there is no space for activity-based learning; and the illustrations on the chalkboard indicate the traditional study of policemen and firemen.

Preparation for Change

Although in the last several years various innovative teaching methods and programs have been implemented, most still retained the same basic subject content and objectives. Now, however, education finds itself on the verge of tremendous change. Because of the changes that have taken place in the last decade in economics, government, industry, and society, education must move to provide programs that reflect these changes and their implications for the future. Even though the changes are taking place on the national level, they are having a direct effect on each individual. Life-styles change in response to these social forces.

To prepare students for changes to come, education must provide experiences that are relevant not only for today but for the future. This means education has to reflect the current technological culture, not that of twenty years ago. Nell Eurich, writing on the topic of education and the future, states:

> The future arrives with increasing velocity, and we have come to expect startling announcements of new developments daily. Obviously the future will not offer us an environment basically similar to the present, with only slight modifications. It will be radically different from anything we have known. The role of educators as custodians of the past must, therefore, be cast in a different

The adaptive element of education is probably the most difficult to teach because it has to incorporate the future which has not yet been seen or experienced.

light. Their past-orientation for future purposes requires fundamental adjustment.

It is entirely possible that students in the schools of today may, in their lifetime, have the opportunity to live in a space colony completely separated from their earthly environment. Education that accurately reflects the current technological culture will automatically reflect the future toward which it is moving.

Purpose of Education

Education can be said to have three general purposes: transmission, development, and adaption. Transmission serves to perpetuate social and cultural values, regardless of the level of current technology. Primitive cultures, with a relatively low level of technology, educate their young children to carry on the accepted traditions just as advanced cultures do. This benefits society and the individual. Transmission occurs in an organized system that develops guidelines and levels of expectation for the children.

The development of young people is multidimensional. Although physical, mental, social, and value development may be discussed separately, they are, in reality, very much interrelated. Simultaneous development in each area is very important. Failure in the development of even one may prevent or distort an individual's contribution to society. Education attempts to develop the student in all aspects of growth by providing a wide variety of activities and experiences.

Adaption is extremely important in a culture that has developed a high level of technology because of the snowballing effect of change. The adaptive element in education helps the individual cope with a life of constant change. Of the three purposes of education, adaption differs the most between primitive cultures and cultures with a high level of technology. In primitive cultures there was little need to concentrate on the adaptive because there was little change. What adaptive skills there were developed secondary to other skills. In today's education a specific effort has to be made in the classroom to encourage the flexibility of adaption.

These three purposes—transmission, development, and adaption—guide experiences in the classroom that help the students expand their awareness both inwardly and outwardly. Expansion outwardly involves learning about the environment of which they are a part. Expansion inwardly involves learning about themselves, physically, mentally, and emotionally. The process of becoming aware of one's environment is known as *culturalization;* developing an understanding of oneself is known as *liberation.*

Culturalization. Culturalization is accomplished by looking outward and becoming aware of those things that are part of

The familiar classroom projector touches the lives of most elementary students daily. Very few, however, have seen the inside of one or know how it functions. Culturalization involves being aware of those things around us.

our daily lives. It does not necessarily mean becoming a "cultured person." A narrow definition of a cultured person is one who is knowledgeable in the "classical arts." However, the person may be uninformed about the culture in which he or she lives. To undergo a process of culturalization is to acquire a knowledge of one's culture that allows one to adapt to changes as they occur. There are many people today who feel rejected, demoralized, and even hostile toward society. When changes in society directly affect them, they feel frustrated because they do not understand the rapidly changing culture in which they live. The five following examples present typical comments disaffected people might make about technology. A short response to each illustrates, in part, basic concepts about technology and its functions in our culture.

(1) "*Technology creates too many risks; we should go back to using some of the older methods which were much safer.*"

Technology evolves as a result of need. It is true that change involves some element of risk, but as problems arise, technology can be used to solve them. If we could return to the technology of the past, we would find the risks replaced with dire consequences, such as starvation, uncontrolled diseases, and complete isolation of parts of the world.

(2) "*Our products cost us too much today; they should be the same price they were in the past.*"

The cost of a product involves much more than the cost of materials and labor to produce it. Part of the cost finances the research and development necessary for the creation of new products and the improvement of current ones. Without research and development we would not have such products as television sets, synthetic fibers for clothes, and a wide variety of medicines. A large part of the cost also goes for transportation. The raw materials are transported to the processing plant and then to the manufacturing plant before they become products. After manufacturing they are transported to distribution centers and then to stores where they are to be purchased. In the past, we had very few choices of products, and, in many cases, it took several weeks before they even reached the store. Today, we are able to choose what best serves our individual needs, and it usually may be purchased on the same day it is needed.

(3) "*Technology causes too much pollution.*"

Technology per se does not cause pollution; rather, most pollution can be attributed to poor decisions made in applying technology. The process of producing goods involves extracting raw materials from the earth and changing the structure and characteristics. Waste from this process may result. Technology can be used to prevent these circumstances from becoming a problem if the decision is made to do so.

(4) "*Technology should be used to solve our many social problems.*"

Technology has limitations. It is based on the application of the laws of nature which cannot be altered. It is much easier for technology to solve technical problems than to solve social problems. The social problems which develop are the result of human reaction to situations and not a result of the use of technology. It is highly probable that many of the complex social problems we are concerned with today could not be solved with the application of technology alone.

(5) "*Automation takes jobs away from people.*"

Automation is usually applied where the work is either dangerous or repetitious. Automation creates new jobs, which tend to be more challenging and rewarding. Changing jobs, however, often requires additional training and skills. People who understand this are able to adapt to the changes, continue to be productive, and enjoy the increased responsibility.

It is easy to see why people who lack a basic understanding of the culture in which they live feel defeated when the system goes against them. Because our culture is based on evolving technology, it is imperative that students be made aware of the forces that help shape their everyday lives. The study of technology is, in essence, the study of culture.

Liberation. While the process of culturalization involves understanding one's surroundings, the process of liberation involves looking inward and developing an understanding of one's self. This also is a part of a total study of technology. The

Liberation involves identifying one's own talent, abilities, and interests. If this is accomplished as students pass through school, they will be more flexible, more confident, and more ready to make wise career choices.

liberation process helps individuals identify where in the culture they are the most comfortable, where they may contribute, and where they are able to realize their fullest potential. This process consists of three general categories of development.

The first is developing knowledge of one's self. This involves becoming aware of mental and physical potential, determining likes and dislikes, and evaluating social skills. The key word here is *potential,* because as growth takes place, skills and attitudes change. An awareness of one's potential is extremely important in determining life goals. Second, one must develop knowledge of one's responsibilities. As we move from childhood to adulthood, we gradually assume more responsibility. Small children usually start learning this by picking up their toys and helping with small jobs around the home, eventually acquiring more duties. Many people feel that technology has caused industry and government to become so big that individuals can no longer affect the social system. As a result, they feel no responsibility to anyone but themselves; even the community is excluded, because many times it is represented by a governing body. If more people understood the interrelationship of technology, industry, and the social structure, they would assume a more active role in improving the system by exhibiting a stronger sense of responsibility.

Third, one develops concern for the quality of life. Although the quality of life today has room for improvement, few people would prefer a return to the past. We have the highest standard of living of any time in history, but continuing this trend depends upon how the evolving technology is utilized. Nearly all decisions made in applying technology are based on social considerations, because people are involved. Problems concerning the quality of life, such as pollution, industry-induced health problems, mental strain, and feelings of alienation, can usually be attributed to poor decisions on the part of either an individual or society in general. An individual concerned with the quality of his or her life will choose a life-style that is comfortable for the individual. Inherent in this choice is a knowledge and appreciation of technology that enables wise decisions to be made.

ELEMENTARY EDUCATION TODAY

The three basic purposes of education—transmission, development, and adaption—should most certainly be part of any elementary education program. A definite emphasis should be placed on orienting students to their culture. This orientation involves developing an awareness of technological change and a knowledge of how these changes affect our lives. Both the benefits and the problems of technology should be identified.

Studying the Past and the Future

The content and methods used in elementary classrooms today cannot be based entirely on our world as it now exists. A total picture of our culture and technology's effect on it means that both the past and the future have to be part of the learning experience. Some programs spend much time on the past, most programs focus on the present, only a few study the future. Yet, the speed of technology can sweep some aspects of our lives into the future. For example, the students in a typical elementary classroom might study about how the early settlers cooked their food in fireplaces. In their own home, however, they may help cook the family dinner in a microwave oven. Or, the class might study about how things work, not going beyond the simple machines of levers, wheels, and pulleys, while after school students may play with toys that contain complex machines and microprocessors. This is not to say that students should be studying computers. It does indicate, however, that

It is important to study the past. Rarely, however, is the past studied in terms of technology. Few students realize that in a period of about ten years the car completely changed our entire society.

the students should be aware of computers and recognize where they might be used.

Traditional subject matter usually involves studying periods of time. Some popular units of study are the Vikings, Pilgrims, Indians, pioneers, and Civil War. The technology of those times is only touched upon, if covered at all. In addition, most of the periods occurred before the Industrial Revolution of the late 1800s. Developments since then have had a tremendous effect on our lives, and understanding these developments is necessary for making decisions in the future. Units such as the development of the car, or of a specific material such as plastics, or the history of flight from Kitty Hawk to the moon would provide for the development of many concepts dealing with technology. Of course the concepts chosen to be developed would have to be appropriate for the specific grade level.

Even material that ostensibly deals with the present may be inadequate from a technological point of view. This cannot be blamed entirely on the teacher, because most teaching is done from textbooks, printed materials, and supplies which have been written at various time periods. An important concept for teachers to remember when teaching the present is the fact that what is current now is already changing. In other words, the future is really starting at the present time.

Although studying the future already involves speculating as to the new forms technology will take in our lives, "hardware and techniques" are not the overriding concerns. The concepts of time and change should receive the greater proportion of attention.

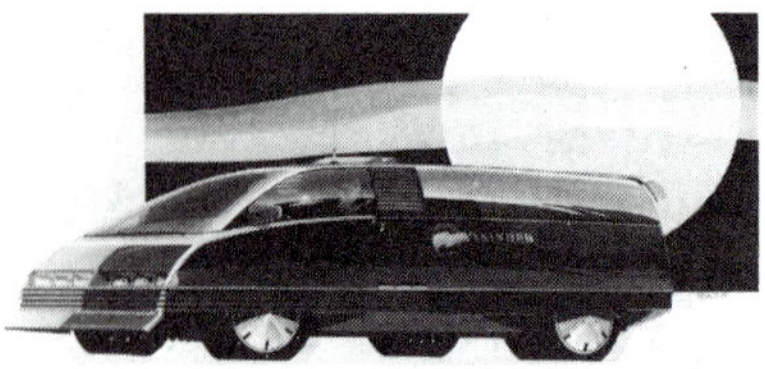

The hardware of techology will change drastically in the future. The important concepts to be learned do not deal with just hardware, however, but with how the changes will affect our lives.

Time. Each person has his or her own concept of time. Small children tend to view a certain period of time as much longer than adults would. To children a year may seem to be an extremely long time, while to an eighty-year-old, a year may seem very short. The difference is in the amount of experience one has with the passage of time. It is essential that students develop a time perspective as they prepare for the future. Lack of foresight makes people live only for the present. If they believe tomorrow will take care of itself, and it doesn't, they become frustrated and feel the "system" is working against them. Teachers can help children to realize that the future is not really far away by comparing the time periods being studied to time periods the students would comprehend, such as their own age, the age of their parents, or the length of the school year.

Change. A proper attitude toward the future involves the realization that it will be different from the present and the past. Students must recognize that planning for the future is necessary and that certain paths must be chosen from a variety of options. The elementary level is not too early in the students' education to start developing the realization that there will be many new jobs in the future. Most of these jobs will require some kind of knowledge and skills that will be gained in school or through special training. Lifelong learning may very well characterize the lives of today's children. Early awareness of these concepts and possibilities will give more meaning to the years of education that follow elementary school.

An Analysis of Today's Students

Although students of today share many similarities with students of the past, they are unique in some regards. The need for studying technology in the elementary grades can be illustrated by examining the students' level of awareness and understanding of their environment.

Environmental Influences. Today's elementary school students have more opportunities for receiving information than students of the past. Four sources of information in particular can be identified which provide for increased opportunity: television, the family, socioeconomic status, and everyday contact with technology.

Television is a part of the students' lives outside of school. The classroom should not be isolated from our use of technology; these students are studying current events by watching and discussing the daily newscast.

Various studies have estimated that an elementary student spends at least four hours a day watching television; approximately one-half of the total time spent on school classroom work. Television exposes children to a tremendous amount of knowledge concerning our society, as well as those of other parts of the world. Most students of today are able to identify, by sight, famous people and places throughout the world; they have seen them, where students of the past could only read about them. Today, through television, students can observe events as they happen, while students of the past could only read about occurrences long after they took place.

Today's family is much more involved in things around them. Sharing the experiences of various family member's provides much information. In many homes both parents are working in different jobs, there is more time for leisure activities, and the average educational level of the family members is higher than in the past. These factors provide a wide range of opportunities for learning.

Although our society still has problems of poverty and minority rights to deal with, generally speaking, the people of our country still have the highest standard of living of any culture in history. This higher living standard even allows opportunities for people in the lower socioeconomic level to obtain goods and services which were previously attainable by only the middle and upper levels. As a result of this upward shift, other opportunities are available, such as education, travel, and job opportunities. This has increased the opportunity for all elementary students to eventually take a more active role in society.

Students of today come in daily contact with technology. Such items as color television, electric toothbrushes, and pocket calculators are very much a part of their lives. Even going to the moon is accepted matter of factly, as are many other major technological developments of the last few years. This daily contact allows students an opportunity to apply the hardware of technology and exercise at least a small amount of judgment in its use.

High Level of Awareness. The high level of awareness of technology may be attributed to both environmental influence and human nature. Elementary-level students have always displayed a great deal of natural curiosity. Students have always had an interest in objects that perform a function, such as a mechanical or electrical device. Many children's toys move, light up, or make sounds. It is not at all uncommon for some of today's elementary-level students to recognize a difference in the quality of fidelity in stereo sound systems or to know the horsepower of a particular car. In addition, students enjoy hands-on experience because they are naturally active. They are happier and learn much more efficiently when they are actively involved. We learn about our environment through our senses of sight, hearing, smell, touch, and taste. The more senses that are involved in an activity the more complete the learning. When a student is interested in a particular area, he or she usually wants to become physically involved.

Low Level of Understanding. Ironically, at the same time that a student shows a high level of interest in technology, he or she can also have a low level of technical understanding. For example, when the electricity goes off, when the television set will not work, or when the car will not start, an individual who knows nothing about how electrical or mechanical devices function may feel frustrated. Three factors can contribute to the frustration we sometimes feel with our environment.

The first is a lack of understanding of the relationships

Students of today are living in an age of advanced technology which provides not only the necessities but play and recreation as well. As a result, students are highly aware of the hardware of technology.

between such disciplines as science, math, and social studies. One reason for this is the way we organize learning experiences in the classroom. Specific times are set aside for certain subjects, and other subjects are usually not discussed at that time. For example, most teachers at one time or another plan a unit on weather, which usually includes such things as electrical storms, cloud formations, and precipitation. Yet, many times at the conclusion of the unit the students are not able to answer such questions as: how does a barometer work, how does a weather satellite function; and how is lightning different from the electricity in our electrical outlets? When the students return home after school they might turn on the television set (electricity) and watch the weather forecast, which shows the viewers a satellite map of a storm center (weather-predicting instruments). Students might also spend many hours working on math problems and never realize math is used to design and build cars or to keep a spacecraft in proper orbit around the earth. The study of technology involves all fields of study as they relate to each other. This breaks down the traditional boundaries between fields and helps the students see critical relationships.

Second, failure to recognize basic principles of science in everyday contacts with technology can lead to frustration. Many times students fail to use the hardware of technology properly, because they do not understand the application of scientific functions. For example, most elementary students have toys and games that require batteries for operation. Few students, however, really know how a battery generates electricity. Most students ride bicycles, but few really know how the pedal brake makes them stop, or how a multispeed bicycle can change gears while it is moving. Almost everything we come in contact with today contains plastics, from building materials to our clothes. Yet, few elementary students know what raw materials are used in producing plastics. Science is not a discrete subject that can only be dealt with in relation to a textbook. Scentific theories can be related to technological applications which surround students every day. Too many times classroom learning develops the awareness, but not the understanding of specific scientific functions.

Bicycles are very much a part of the lives of most young people, although few know how the brakes function or how the gears operate to change speeds.

A third handicap to understanding technology is an inability to relate technology to one's life or to recognize only obvious connections. For example, if a group of students were asked to relate how technology might be bad, most probably would reply, "it causes pollution." If asked about how technology helps us, most would probably say "It provides products for us." These answers are indeed valid, but the subtle interrelationships of technology and social studies are seldom highlighted in the classroom. Most students, for example, would be unable to discuss why our major cities suffer from urban decay or in what ways space exploration has affected our lives in medicine, communications, and agriculture.

The combination of how things function around us and how they affect our lives determines the quality of our lives. Understanding both elements is essential in helping each individual make wise decisions in molding the kind of life he or she desires.

Physical and Mental Maturity

Two closely interrelated aspects of growth command attention: (1) physical growth, which deals with body structure, and (2) mental growth, which deals with thought organization and analysis capability. Most behavior theorists accept that the nurturing of a person depends upon the interaction of these two elements and the surrounding environment.

Efficient learning requires involvement. This kindergarten student is learning about sound and how it travels. She is listening to another student who is talking from another room through a water hose with plastic funnels taped to each end.

Physical growth reflects heredity and environment while mental growth depends on the amount and rate of physical development and on exposure to stimuli that require mental responses. Heredity provides the "blueprint" for growth; the environment provides nourishment for the realization of the potential.

Leaders in the field of education and learning theory, from as far back as Comenius (1592-1670), Locke (1632-1704), and Rousseau (1712-1778) to twentieth-century writers, such as Dewey, Piaget, and Bruner, agree that effective learning involves readiness, involvement, and guidance.

How well a student may learn a skill or concept depends upon the readiness of the individual to respond to the stimuli. Because learning is a wholistic process, it involves both physical and mental aspects. Learning capability varies according to the development level of the two aspects.

Physical readiness deals with accomplishing tasks that

may require a stronger or more appropriate stimulus. For example, a child learning to ride a bicycle has to experience mental learning, but this is incomplete until physical readiness allows the child to get on the bicycle and attempt to ride it. As students grow older, their physical maturity allows them to accomplish more highly complex tasks. This accomplishment contributes to the learning process. However, this is not to say that all learning depends upon physical readiness; one may learn from reading a book without direct physical involvement.

Experts disagree whether intelligence improves with maturity, or whether it is unchanging and knowledge increases with exposure to the environment. We do know, however, that there is a wide variation in the rate at which individual students develop and that the level of development determines the rate and amount of potential learning. Mental readiness is much more difficult to access than physical readiness in that it is not easily observed. It is also more complex, because it involves physical development, emotional development, and previous experience.

The ability to learn may be compared to the information processing ability of a computer. The early computers were only able to store small amounts of information and produce simple answers. As computer technology evolved, the information processing capability grew; today computers are able to process large amounts of information and solve complex problems. Just as succeeding generations of computers increase in processing capability, students move through developmental phases which allow higher levels of learning to occur.

As a result of what is known about learning readiness, it is felt that teaching very young children (even prekindergarten) about technology is entirely feasible, as long as it is presented on a level commensurate with their physical, mental, and emotional development.

Like most other living organisms, the involvement of the human organism with its environment depends upon and occurs through the sensory organs. Learning is a process whereby bits of information are gathered and processed by means of the senses (seeing, hearing, smelling, tasting, and touching). It is wholistic in that each of the senses contributes to and becomes a part of the total process, determining the level of learning. For example, if students were to study about the giant redwood trees of the western United States, they would learn more completely if pieces of redwood (obtained from a local lumber yard) were available to the students in the classroom. They could see the color, feel the texture, hear it by tapping on it, smell the aroma, and (maybe) even taste it. In this way all five senses would provide input, rather than only reading about it (seeing). In addition, if students could saw the wood, sand it, and pound nails into it, they would learn much about the characteristics of this wood, as well as the general characteristics of other woods. This in turn would reinforce the learning of facts related to the study of such topics as ecosystems, lumbering, and wood products.

A second benefit derived from active involvement with the environment is a sense of accomplishment. This could be as insignificant as observation or as dramatic as the acquisition of a basic skill. *Success makes succeeding successes more accessible.* This little tongue twister makes a very important point; when a student succeeds, he or she develops a desire for more success. The feeling of pride and self-assurance one has after accomplishing a task goes a long way toward causing that person to reach for higher levels of accomplishment. If students are allowed to become actively involved with tools, materials, and processes in the classroom, more complete learning will certainly take place.

The teacher as the source of guidance in the classroom,

Students need to have the opportunity to feel successful in many different situations. This can be accomplished by providing those experiences which allow them to become involved at a level commensurate with their abilities.

becomes the central core of the learning process. It is true that learning can be spontaneous and will occur in a natural way over a period of time. However, if learning is structured into a sequential set of experiences that are commensurate with the students' abilities and developmental levels, it is more complete and occurs at a faster rate. It is the teacher's responsibility to recognize these different abilities and developmental levels so that the most appropriate type and sequence of experiences may be called upon.

Developmental Tasks

As every individual passes through the various stages of development from the time of birth until old age, certain achievements are required. These are termed *developmental tasks.*

Biological developmental tasks involve developing skills, body control, and knowledge of one's abilities as growth takes place.

Skills and Behavior of Biological Developmental Tasks

Skills	Behavior
Small muscle control	Able to communicate by writing and drawing
Depth perception and hand-eye coordination	Can shape objects of clay or assemble a model car
Participation in games	Able to run, throw a ball, or jump a rope
Awareness of increasing physical capabilities	Able to safely ride a bicycle on the street (strength, perception, and so forth)
Ability to perform more than one physical activity at a time	Able to reach high above eye level and staple letters of a word on a bulletin board (keeping balance, holding letters in one hand, holding stapler in the other hand, and attaching the letters in a straight line)
Awareness of changing sleep requirements	Judges how much sleep is required as growth changes take place
Adjustment to longer work periods in and out of school	Accepts longer times between recesses and assignments which require several different work periods
Awareness of necessary personal hygiene habits	Wears clean clothes and washes before meals

They are accomplishments that are necessary in order to live a happy and satisfying life. Robert J. Havighurst, in *Developmental Tasks and Education,* defines these accomplishments:

> A developmental task is a task which arises at or about a certain period in the life of an individual, successful achievement of which leads to his happiness and to success with later tasks, while failure leads to unhappiness in the individual, disapproval by the society, and difficulty with later tasks. (Havighurst, p. 2)

These tasks, although overlapping, may be divided into four groups: biological, sociological, psychological, and cultural.

Biological tasks are the most universal and the ones over which we have the least control. They deal with physical development. For example, an infant goes through the stages of crawling, walking, and then running. These take place at about the same time for all infants, because a certain level of physical maturity is required. At the other end of the life span, an elderly person may have to learn how to walk more carefully to prevent falling. Of course, there are many such tasks between these two extremes, which come in sequential steps.

It is important for the elementary teacher to be aware of these tasks, because the students are accomplishing many during the course of their elementary-school experience. Large muscle skills are developed first, with small muscle and finer hand-eye coordination evolving through all of the elementary grades. This indicates the need for an activity-based curriculum where students are involved in various manipulative operations. The satisfaction of accomplishing these developmental tasks, as well as other types of learning, produces satisfaction in the total learning process.

Sociological tasks are imposed by society and usually take the form of expectations and guidelines for conduct. Unlike the biological tasks, these may vary considerably from society to society, as well as within societies.
middle-class, democratic society. These may vary considerably as one moves to either the upper or lower strata of society. However, the implications for the study of technology and for effective functioning in a technological society should be the same for all levels. Technology affects all people, and all people affect technology.

Sociological developmental tasks involve the interaction of an individual and those expectations of conduct established by society.

Skills and Behavior of Sociological Developmental Tasks

Skills	Behavior
Relationships with peers of both sexes	Able to cooperate with others, both individually and in groups
Relationships with adults other than parents	Able to communicate and identify with adults, such as teachers, scout leaders, and neighbors
Sex roles	Knows what is acceptable behavior for each sex under various circumstances
Respect for authority	Obeys laws and rules at home, at school, and in the community
Gradual freedom from parents and family	Displays expected behavior without supervision and gradually accepts more responsibility
Moral values	Has a social conscious (right and wrong)
Qualities of leadership and followership	Identifies personal qualities of leadership and followership needed for each situation and is able to seek the most comfortable level of participation in the group
Awareness of the part various institutions play in our society	Realizes the purpose of attending school and taking advantage of opportunities in other institutions within the community (community programs, religious organizations, for example)

Psychological tasks are somewhat less distinctive, because they are of a more personal nature. They include such elements as values, feelings, and aspirations. They are usually formulated in conjunction with both the biological and sociological tasks and depend, to a certain extent, on how well those tasks are accomplished.

The implications for the study of technology are probably as strong in the psychological developmental tasks as any of the other groups of tasks. The ability to cope with change, to contribute to society, and to lead a satisfying life depends upon the psychological state of each individual. An understanding of our technological society and how it functions is crucial to the accomplishment of psychological developmental tasks.

Psychological developmental tasks revolve around the feelings one has concerning self; a positive image of self produces positive attitudes in all areas of life.

Skills and Behavior of Psychological Developmental Tasks

Skills	Behavior
Ability to cope with everyday stress	Able to accept failure and criticism
Ability to work alone	Functions well alone and works independently without constant reinforcement
Attitudes that help deal with peer pressure	Makes choices based on own value system
Acceptance of skill and talent shortcomings	Realizes that one can never be the "best" in everything
Self-confidence	Not afraid to try new things and not afraid of making mistakes
Interest in the future	Experiences are evaluated in terms of possible future career choices
Empathy and respect for feelings of others	Shows concern for others by offering help and including them in the peer group
Initiative	Sees things which should be done and proceeds to do them without being told

Cultural tasks deal with those skills considered necessary in a particular culture. This is different from the sociological tasks in that cultural tasks do not necessarily deal with social organization and expected behavior. Rather, cultural tasks center on the skills which are necessary to function in a technology-oriented culture. These may even vary within a culture, but most are basic to everyday living. Many cultural tasks may be accomplished without additional direction, but specific help many times enhances skill development and eases movement on to other tasks. Studying technology in terms of how products are made, how they operate, and how they may be used, is a definite help in accomplishing technologically oriented cultural tasks.

Characteristics of Developmental Tasks. There are four characteristics of developmental tasks which are important for the teacher to remember. First, developmental tasks are sequential. This means that the accomplishment of each one has a bearing on those which follow. It is important that the teacher utilize learning experiences that are appropriate for the level of tasks being mastered.

Cultural developmental tasks involve skills and knowledge required for functioning in the culture in which one lives.

Skills and Behavior of Cultural Developmental Tasks

Skills	Behavior
Knowledge of various types of communication devices	Uses the telephone, addresses letters, and selects desired television channels
Understanding of standard symbols used in society	Able to read traffic signs, lights, and other safety controls
Knowledge of how everyday products should be used safely	Knows how to use simple household appliances, such as electric toasters, lamps, and radios
Awareness of the value of money	Knows the different denominations of coins and bills
Knowledge of how to purchase products	Counts change and keeps receipt
Awareness of how food should be prepared	Knows what foods have to be cooked and how they should be stored
Understanding of basic scientific principles	Able to make use of simple machines, such as hand tools (screwdriver, hammer, and so forth)
Skills in reading and writing	Able to read about and interpret current events when written at own grade level

Second, developmental tasks are accomplished individually. Although instruction might be as a group, each student is unique in the rate and level at which the task may be attained. It is important to remember this point when determining activities and the time required for their completion.

Third, there are some forces which not only determine the task but also set the time in which it should be learned. Many of these forces are cultural. The typical educational level for learning a basic computer language, for example, is now set in the secondary school. In the future, this skill might be taught during elementary school. Some tasks may also change. The shift in the last few years in sex roles most certainly has altered society's expectations of both boys and girls.

Finally, success or failure in accomplishing developmental tasks affects other developmental tasks. This is not only true within the same group of tasks (such as sociological), but it is true across groups. For example, failure to achieve the basic task of reading could affect perception of self, manual-skills development, and social interaction with peers.

Learning Environment and Developmental Tasks. How can the teacher help students accomplish these important tasks? Although many suggestions could probably be made, the following indicates those that are critical:

- Choose classroom content that is relevant to today and the future. Our technological culture is constantly changing, requiring changes in content.
- Choose activities that will help students accomplish their tasks. This many times requires individual or small-group activity, such as grouping students by reading ability for reading instruction.
- Make the activities challenging. They must be relevant to the tasks being undertaken and presented at a level of difficulty where the students are able to attain success.
- Be alert to changes. There most certainly will be a wide range of tasks underway within any given classroom. Changes in the kinds of tasks and the accomplishment of those tasks may be slow or very rapid. In addition, the teacher should be alert to inappropriate or premature tasks. If a student moves through a sequence of tasks without accomplishing some which are critical, he or she will be hampered in the attempt to develop new tasks and may become frustrated.

SUMMARY QUESTIONS

1. Briefly explain the three purposes of education.
2. Identify two limitations of technology.
3. Why is it important to study the past?
4. Why is it important to study the future?
5. Name three factors dealing with a low level of understanding science and technology which contribute to frustration in our society.
6. Name five developmental tasks required of elementary students.

The city of the future at night, with high-rise apartments, professional centers, parking, entertainment areas, and many transportation systems will depend on the continued growth of technological systems, which, in turn, depends very heavily on the available energy resources.

PART II

Technology: The Scope of Study

CHAPTER 6

The Content Structure of Technology

In this chapter you will learn:

1. The definition of a system.
2. How graphic models may be used to clarify the organization of a system.
3. How technology may be analyzed as a system.
4. How the technology system functions in our environment, becoming a part of our daily lives.

THE SYSTEMS APPROACH

When problems are complex it becomes difficult to see all the various elements and how they affect each other. Nearly all fields of study from sociology (organization of people) to engineering (organization of hardware processes) use a system approach in order to analyze problems.

Defining the Term System

A system may be thought of as an organization. A tool by itself is a simple machine. A complex machine, such as a bicycle, utilizes several simple machines. The lever, pulley, wheel, and inclined plane can all be found on a bicycle. They work together to control the bicycle in starting, stopping, and changing a speed. This organization of machine elements may be called a system.

The term *system* may also be applied to a sequence of events. For example, baking a cake not only requires ingredients, it requires an organized step-by-step procedure. The organization of timing and sequence of adding and mixing the ingredients may be called a system. If one step is left out, it generally has an effect on the end product. Thus, each element contributes to the whole organization.

A system, then, has three basic requirements: (1) specific organization; (2) multiple elements; and (3) assistance from every element. A *system* is defined as a compilation of elements that are organized into a structure so that each contributes to the larger organization of which it is a part.

Baking a cake is an application of a system.

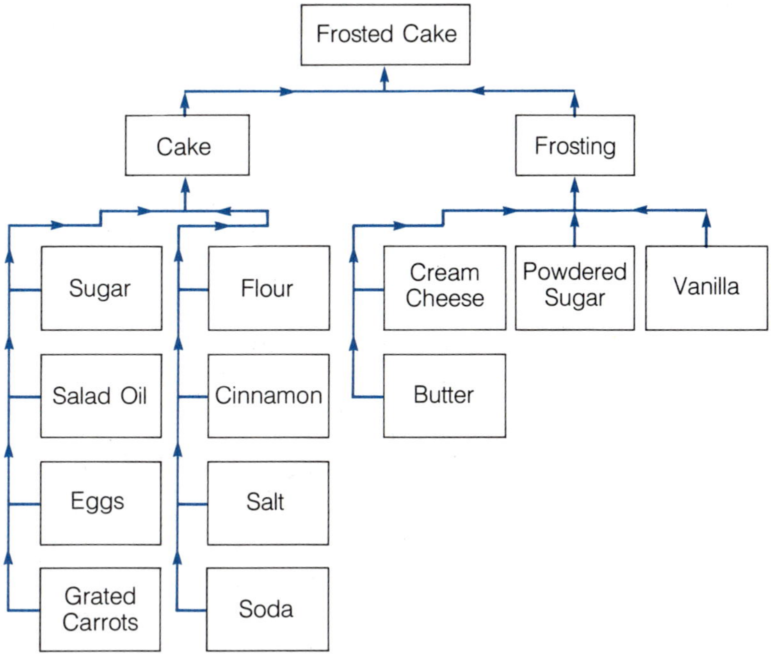

Using Models

The chart that follows shows a simple system of analyzing energy. The illustration may be called a model, because it represents a particular structure or organization of a system. There are many other ways of analyzing energy, depending upon the purpose of the system. As a result, there may be several models, each showing a particular function. For example, under the heading of *energy* may be the two elements of *stored*

This model illustrates one way of analyzing energy.

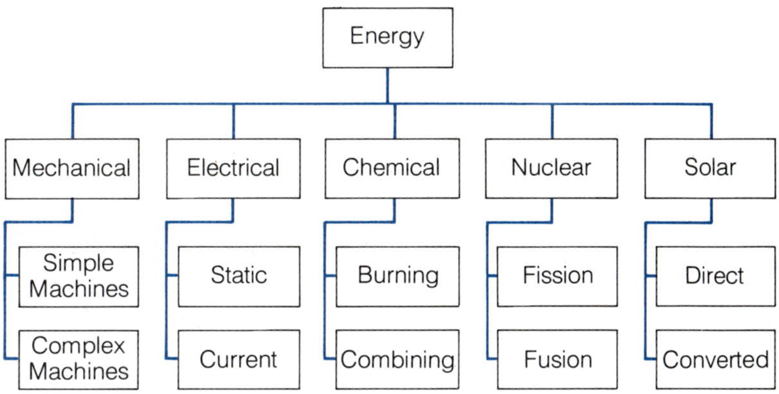

energy and *released energy*. Energy could also be considered in terms of *natural energy* (solar, wind, and so forth) and *induced energy* (mechanical, nuclear, and so forth).

Energy could be broken into *present energy* and *future energy*. The thing to remember is that a model of a system is designed to help solve a problem, and the structure is determined by the problem. There are four distinct advantages of using a model when analyzing a problem: it allows the identification of elements that are to be studied; it shows the relationships between elements; it presents a clear picture of the total organizational structure at one time; and it may serve as a developmental guide if changes are to be made to the system.

Subsystems

When a system is composed of several elements, it is highly probable that further elements exist within the original elements. These subelements form a group called a subsystem, because taken by itself, this smaller group has the same organizational structure as the complete system. Thus, each element, regardless of where it is in the structural hierarchy, is a part of a subsystem on which the total system is built.

The shaded blocks in this model represent a line of progression through the levels of the system. The same movement could be seen if the levels under each of the unshaded blocks were identified.

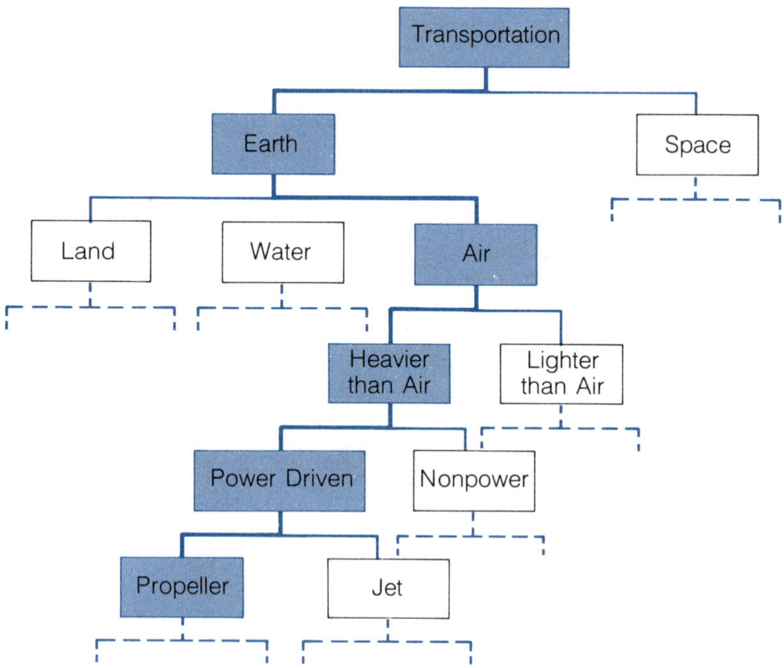

The illustration on page 118 shows one method of identifying the transportation system. Each element within a box is the major element of everything below. Thus, every element is part of that system. At the same time, that system functions as a subsystem of those elements above. Transportation could also be taken to a higher level where it becomes a subsystem of technology.

There are three characteristics of subsystems which are important in understanding their function.

- Each subsystem is designed to fulfill a basic purpose which is necessary to the total system. As the needs of the system change, so do the subsystems.
- Each subsystem is unique to all others. There is no duplication of the same characteristics or processes.
- Each subsystem is connected to other subsystems in some form of interaction. It either contributes directly to the functioning of other systems, or it supports them indirectly by helping support the total system.

THE TECHNOLOGY SYSTEMS

Technology is a broad and often vague term. It means different things to different people, given their background and experience with technology. An assembly-line factory worker, a college student, a housewife, and an aerospace engineer would likely have very different views of technology and how it affects their lives, because each has had different contacts with it. However, in many ways, the effect of technology on each is the same because they are a part of the same technological society.

The systems approach is an aid in defining technology. In this way, by using a systems approach, we can identify and place elements in a structure and recognize the interrelationships. To see the structure of the systems, we use various models.

The technology system involves much more than just tools and hardware. It involves techniques, materials, and people, all of which are organized for the purpose of satisfying human needs. Because technology is a human endeavor, the study of people, how they use technology, and how technology

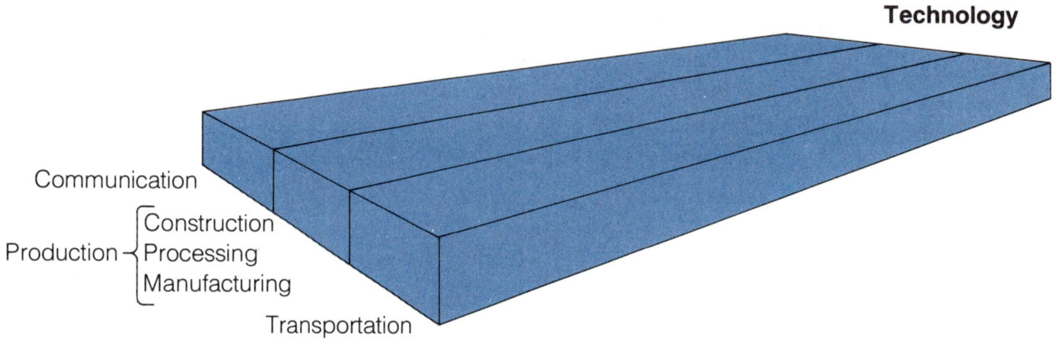

This model illustrates the three major technological systems and the subsystems of production.

ultimately affects them becomes an integral part of the study of technology.

The everyday needs of society may be clustered together into groups, each forming a system. Each system, working in cooperation with the others, serves to satisfy the needs of the total system. When the total system is analyzed the individual groups actually function as subsystems.

Components of the Systems

Within each of the systems are many elements necessary for the systems to function effectively. The model that follows shows these elements cutting across all three major systems,

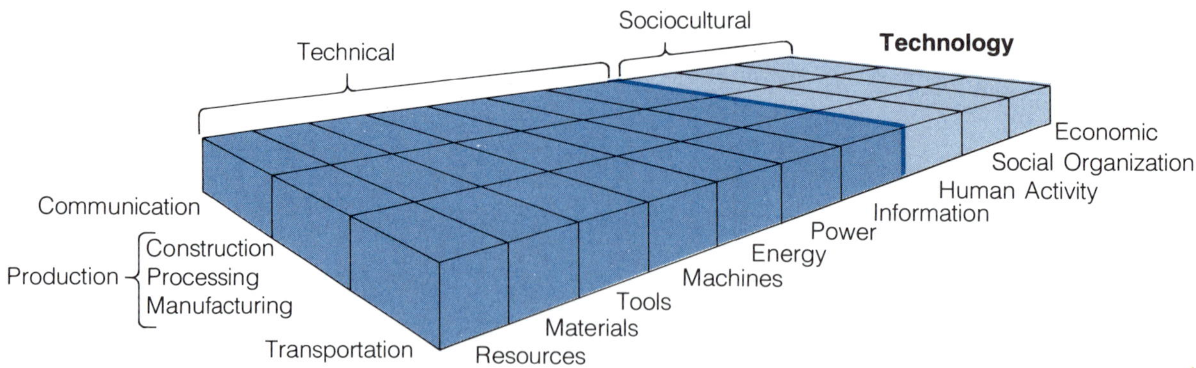

The addition of technical and sociocultural elements to the model indicates they are a part of all three systems.

indicating that the same components function in each. Components also fall into both technical and sociocultural categories. In many cases, it would be difficult to separate these components into strictly technical or sociocultural groups, because all technical activity evolves around sociocultural needs.

Complexity of the Systems

Technology had its beginnings with the first humans who used primitive tools and techniques to help satisfy basic needs. This same process has continued through history to today. The primary difference between the primitive and the modern is in the complexity of the tools and processes. The evolution of technology has passed through three distinct periods: handcraft, mechanization, and cybernation.

Adding levels of complexity shows movement toward higher technology. In handcraft technology, tools were made by hand, using only muscle power for operation. Mechanization used mechanical power to both make and operate tools

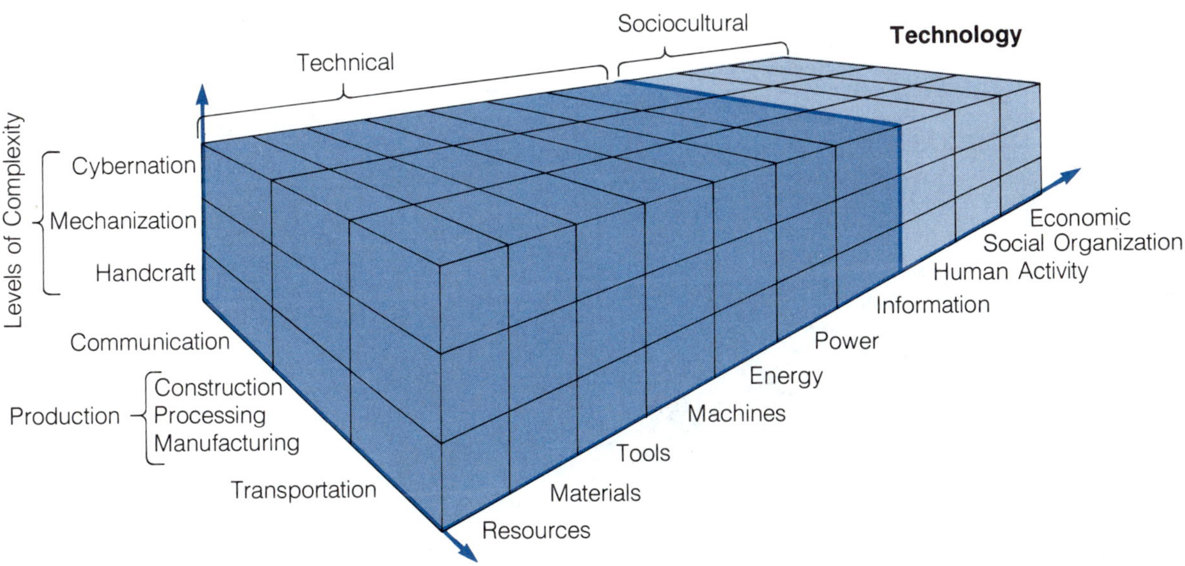

The addition of the levels of complexity to the model illustrates the movement toward higher technology. The complete model can now be read in three directions, providing information on each element in the systems at any level.

and machines. Cybernation involves automation (automatic functioning) coupled with cybernetics (machines controlled by other machines). This is the level of technology at which we are now functioning. In the illustration showing levels of complexity it should also be noted that at the handcraft level, some of the components would not logically apply, such as machines and power other than muscle power. Another important point to note is the use of arrows on the model indicating an open-end model. The future may very well add more elements in all three directions.

Environment of the Systems

The technology systems operate within our environmental system. Because they are part of it, the technology systems have a direct effect on the function of the environmental system. It is easy to relate examples where the effects have been positive or negative. Technology makes use of our natural resources, as

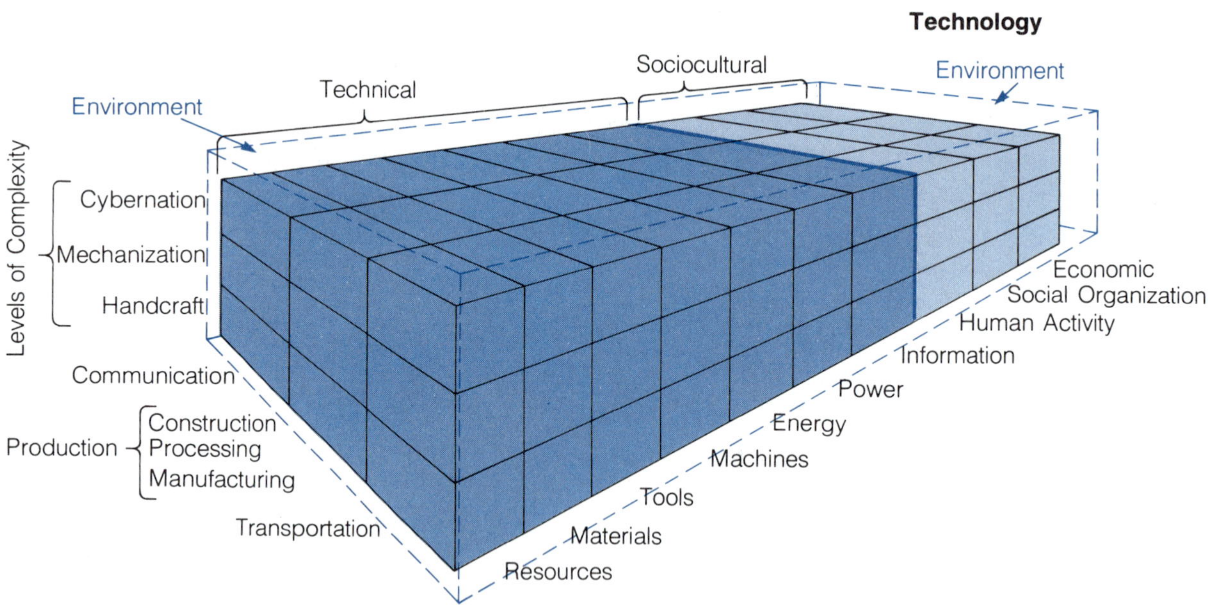

This model illustrates the fact that technological activity takes place in the natural and human-made environments, being both affected by and affecting the environment.

The interrelatedness of technology and environment can be seen in this photo of a commercial apple orchard. This land has been reclaimed after strip mining of coal. In many cases these trees grow better and are more resistant to disease because of the fresh subsurface elements brought to the surface.

well as contributing to their conservation and replenishment. This part of the model is extremely important because of the delicate balance between the environmental system and the technology systems.

The environmental element includes not only earth but space as well. It is important to include this in the model, because the future will most certainly include the use of our space environment in all three major systems.

SUMMARY QUESTIONS

1. Define the term *system.*
2. Identify the three characteristics of a subsystem.
3. Design a graphic model which would show how various types of audiovisual materials and hardware used in the classroom work together to make up a system. (Note that this requires analyzing how they function in order to place them in the correct subsystem.)
4. Name a component of the technology system and identify how it functions in each of the three major systems of transportation, communication, and production.

5. Explain the difference between mechanization and cybernation.
6. In analyzing the technology system, what should be considered as the environment in which technology functions?

CHAPTER 7

The Transportation System

In this chapter you will learn:

1. How the transportation system is organized.
2. How the subsystems of land, water, air, and space evolved.
3. How many of the transportation vehicles function.
4. Why some transportation subsystems have problems.
5. What the future holds for transportation subsystems.

ANALYZING THE TRANSPORTATION SYSTEM

Imagine for a moment what the world would be like if suddenly all mechanical transportation of goods and people stopped. How would it affect you? How would it affect society? The result would be so devastating that the entire world would feel the effects immediately. Not only our own society, but all societies around the world are dependent upon the movement of people and goods to satisfy basic needs and more. Transportation plays a crucial role in the distribution of food, shelter, and health care, for example. Thanks to the transportation system, we are able to move energy to where its need is felt most strongly. It provides millions of jobs, not only for people who build and operate transportation vehicles, but for people whose work is in seemingly unrelated fields.

The systems diagram illustrates the interrelatedness of transportation, communication, and production. The communication system, for example, is used extensively in the transportation system. Busy airports could not function without a control tower, spacecraft could not reach the moon and return without radio contact, and highways certainly could not function without the electronic communication system of traffic lights. The overlapping characteristics of each system are the use and generation of energy in the performance of its tasks. The production of energy also provides many jobs, as does the transporting of energy and the use of energy to power transportation vehicles.

In the previous chapter we showed that a system can be analyzed in many ways, depending upon the information de-

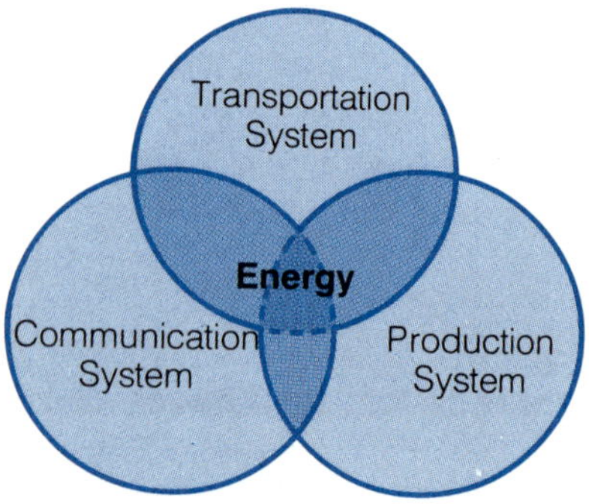

Energy is common to all three major systems. Without the cross-application of energy, none of the three could exist.

All transportation could logically be divided according to this model. The broken-line boxes indicate that additional transportation methods are possible.

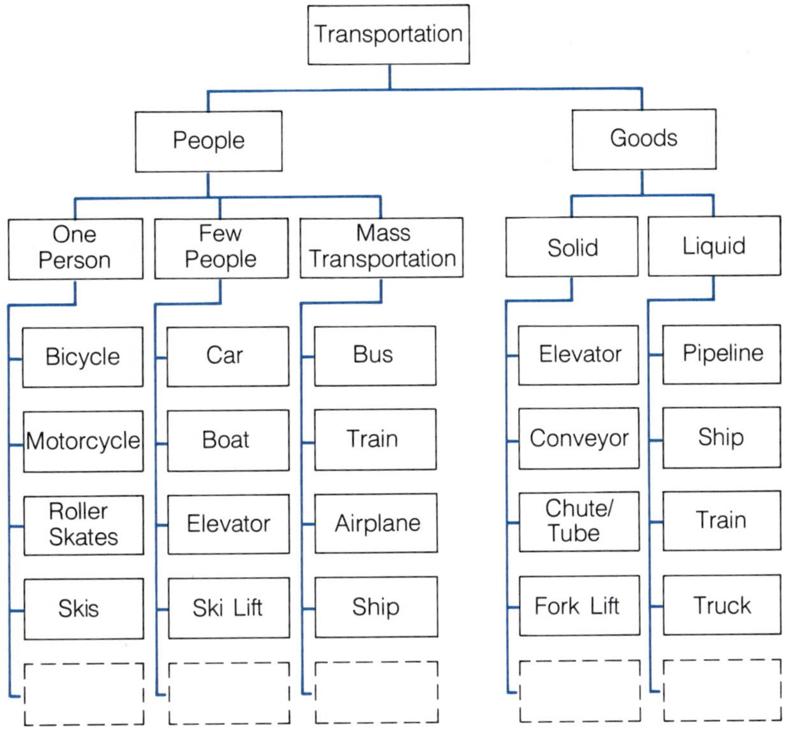

sired. We shall analyze the transportation system by using two different, yet related, models.

The transportation system can be broken down into two subsystems, according to what is transported. Goods refer to anything other than people and can be broken into two further categories of either solid or liquid. In the case of a train, for example, the same type of vehicle may be used for both kinds of goods as well as for people. The design of the specific carrier, however, would be different: a passenger car, a boxcar, and a tank car have little in common other than that they are used on tracks. Of course, the list of subelements is by no means inclusive.

The subsystem according to people is broken into three further categories, defined by the number of people a carrier may accommodate. This division shows a breakdown in the size of carrier as well as in the possible use. The model shows clearly that transportation involves more than independently moving vehicles. Elevators, conveyors, escalators, and pipelines are also important parts of the transportation system. By

Each transportation subsystem has to have a specified route of movement to prevent interference with other subsystems.

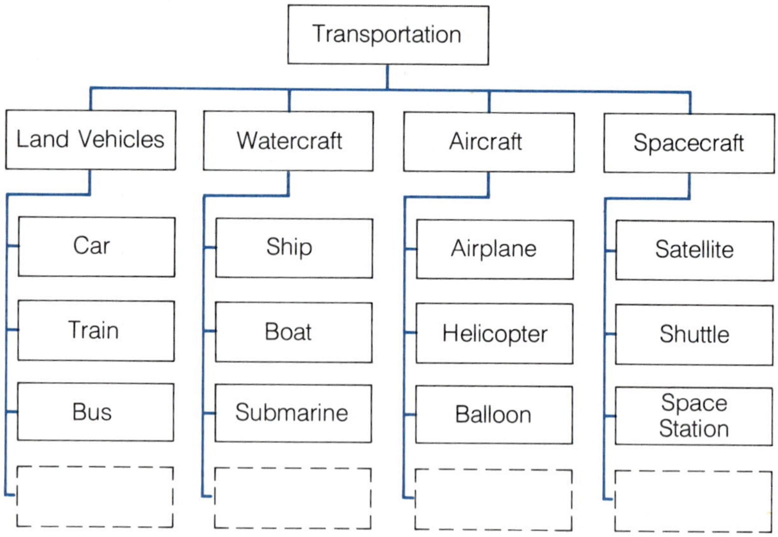

categorizing the system, we are able to show where transportation takes place. This categorization has the virtue of accounting for all vehicles. The vehicles under each heading differ from those in the other subsystems, because the environments in which they are used are different. However, despite the differences in environments where the vehicles are used, humans have learned to alter physical obstacles. We have canals, tunnels, bridges, and paved roads on which our vehicles operate. We even land aircraft on portable landing fields in the ocean (aircraft carriers). For the most part, rocks, rivers, trees, and valleys no longer slow or prevent movement as they did in early days when transportation was crude and ponderous.

When each subsystem is examined individually, it (land, water, air, and space) may be called a system. The vehicles within each system are designed to function in the natural and artificial environments of that system. In most cases, a network of transportation routes allows regular movement without interference from other transportation systems. For example, cars and trucks follow the highway system, trains follow a track system, ships follow a shipping lane, and airplanes follow flight routes. They all are able to function simultaneously, because they are organized into a system.

In designing most vehicles, seven factors are taken into consideration:

- surfaces on which it will travel
- environment (heat, cold, moisture, and so forth)
- weight it will carry
- type of cargo (people, goods)
- speeds it will travel
- acceleration (starting and stopping)
- costs of the vehicle and of the network on which it will run.

Nearly every design is a compromise because of the interplay of these factors. For example, a barge is not able to travel at even moderate speeds, and it cannot be used in open water where the sea is rough, but it can carry large loads and is inexpensive to operate. On the other hand, an air-cushion vehicle can travel anywhere at high speeds and carry moderate loads, but travel at high speeds makes much more noise and is more expensive to build and operate than the barge. The design of both vehicles is a compromise, because it is not practical to incorporate all the characteristics into one vehicle.

Until the last couple of decades, the primary trend in vehicle design in all the systems was to build bigger and faster vehicles. There are three reasons we are taking a second look at what has been done in the past.

The first involves rising costs of building and operating bigger and faster vehicles, which may now outweigh the benefits. It takes more energy to move at faster speeds, and with the high price of fuel, getting there faster is not necessarily cost effective.

Second, the speeds which we have reached today are near the limit of efficiency. An example has to do with airplane travel; it is not uncommon for an airplane flight of several hundred miles (km) to take less time than it takes to drive to the airport through highway traffic congestion. We are also reaching the peak of speed in terms of safety. Higher speeds would be acceptable if the transportation networks would be redesigned (highways, railroads, etc.). However, the tremendous cost involved far outweighs the benefits of higher speed.

Third, the loading and distribution centers are not able to handle the volume of increased loads. Whether for people or goods, the facilities would have to be redefined. If each airplane carried from 400 to 500 people, they would have to spend considerable time at the airport waiting for other flights. As it

Commercial tankers such as this are very efficient, but the larger super tankers, which may reach nearly a quarter of a mile in length, present problems. They can cause shore damage from the water movement and there is always the danger of an oil spill which could damage hundreds of miles (km) of water and shore for years to come.

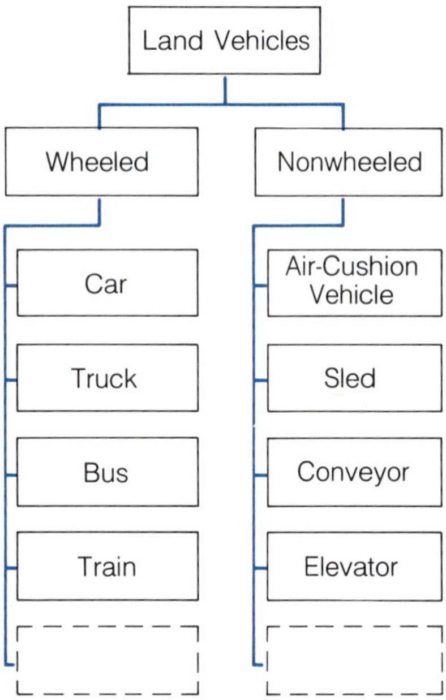

The term vehicle as used in this model refers to any complete mechanical system for moving people and goods.

is now, there are more scheduled flights in and out of the airports, which distribute the number of people more evenly throughout the day.

The concern now is beginning to be redirected toward efficiency. When the ratio of cost to speed and to load capacity is analyzed, we may see a number of our vehicles (both private and commercial) redesigned to a smaller, more efficient size.

LAND VEHICLES

We will take a look at land vehicles as a complete subsystem. Land vehicles can be separated into two groups; wheeled vehicles and nonwheeled vehicles. Although most carriers have some form of wheels either inside or outside, for our purposes we will look at carriers with wheels that provide contact with the ground.

The wheel is probably the most significant invention of humankind. From its beginning in the eastern Mediterranean region about 3000 B.C., the wheel has helped to advance technology. Before the wheel, animal power provided transportation. The domestication of animals has been dated back to 10,000 B.C. Between that time and the advent of the wheel, vehicles probably were a sled type, or much like the *travois* of the Plains Indian.

The transportation networks developed according to where people needed to travel. Footpaths became trails, and trails eventually became roads. Many modern roads today follow the same route as the footpaths of thousands of years ago. Cities grew up along these paths and have continued to grow. It is interesting to note that these paths were usually near water. Even today, many major cities are near the water.

Although there were many refinements in the design of wheels and axles, there was very little change in land vehicles

Population centers near the water have definite advantages. The high demand for goods and raw materials requires an efficient system of transportation. The water network does not interfere with necessary land transportation and, in many cases, provides a direct link with other population centers on the water. The link shown here is a unique watercraft called a hydrofoil.

Early car owners found that tires had to be changed quite frequently. This car carries two spare tires on the left side in case of such an emergency. The horseless carriage was at first a status symbol, but its practicality soon became evident.

for nearly forty-eight hundred years, other than adapting them in various ways to animal power. Until the middle 1800's wheels were made of wood with a heavy metal band for a tire. In the mid-nineteenth century, solid rubber wheels appeared, which gave a smoother ride. In 1888, John B. Dunlop, a Scottish veterinarian, improved on earlier tire inventions by developing an air-filled rubber tube. This became the forerunner of our pneumatic tires and appeared on some automobiles as early as 1895.

This tube would not hold air over a period of time, and in the early 1900s a new type was developed. This tire had two parts; an outer casing and an innertube. This provided a longer lasting, smooth "air" ride. In 1948, the tubeless tire was introduced. It has a soft rubber liner which allows it to seal the air in, as well as allowing it to seal itself in case of puncture. Early tires lasted only a few hundred miles (km), while many of today's tires last over 40,000 miles (64,360 km), partly because reinforcement is molded into the rubber. A number of materials are used for this purpose, such as plastic (rayon or nylon), fiberglass, and even steel.

At the time wheels were changing, land vehicles were also undergoing change. It is not a coincidence that the first major change in land vehicles came about during the time of the Industrial Revolution when new forms of power were being developed.

AUTOMOBILES

Because of the increase in knowledge about mechanical power during the Industrial Revolution, the dream of many inventors at that time was to apply mechanical power to horse-drawn carriages. Most inventions and major developments in technology may be attributed to a single inventor or to a group of people working together, but this is not the case with the automobile. The first internal combustion engine was built around 1878 by a German inventor, Carl Benz. By 1883, he had refined the engine and successfully applied it to the movement of a carriage. Early experiments were not just confined to the carriage. In 1885, Gottleib Daimler, another German, built the first successful motorcycle. He later produced a very successful automobile.

American inventors were behind the Europeans in the development of power-driven vehicles, because the Industrial Revolution started earlier in Europe than in the United States. The first American automobile was invented in Springfield, Massachusetts, in 1892 by Charles Duryea. However, once the evolution of the "horseless carriage" started, American inventors were soon far ahead in the development of new designs and applications. For companies already building carriages, it was a natural move to produce the horseless carriage. By 1900, there were over eighty registered motor car companies in the United States.

It must be remembered, however, these were custom-made cars; that is, they were produced one at a time. There were many companies that did not survive. Among some of the more recognizable names of pioneers who did survive, were Louis Chevrolet, David Buick, the Dodge brothers, the Studebaker brothers, R. E. Olds, and Henry Ford.

There are four primary reasons why the automobile industry grew so rapidly.

First, each company made extensive use of the developments and refinements of others. There was also much trading of designers and engineers. Ford started as an engineer with the Detroit Automobile Company, which soon became the

As the car came into greater use, the demand intensified for better roads and controls.

Cadillac Automobile Company. The Dodge brothers, John and Horace, worked for Ford before forming their own company. William Knudsen, Ford's top production man, went to work for Chevrolet and later became president of General Motors. Because of the cross-fertilization of ideas, many companies expanded, while those who could not keep up with the pace of development soon fell behind.

Second, growth of the necessary parallel industries allowed the car companies to expand their engineering developments rapidly. The petroleum industries, in response to the need for more fuel and lubricants, began growing. Rubber industries improved their products and expanded production. The steel industries also grew because of the increasing demands. It should be noted, however, that the rapid rise in production of these industries was not due solely to the automobile industry. It was due to the simultaneous expansion and resulting needs of an industrializing society.

Third, as technology moved ahead quickly, the needs of society changed. More people were moving to the cities because of the jobs available in expanding industries. As the population became more mobile, there was a need for more efficient transportation. The car met this need, allowing people to travel nearly anywhere at anytime. As jobs provided a higher level of affluence, more were able to afford the car, which in turn produced a greater demand for the product.

Fourth, new production methods were probably more

This photograph, taken about 1913, shows how the parts were moved to the correct point in the assembly line for installation.

responsible than any other single factor for the rapid growth rate. This is the area where Henry Ford made his greatest contribution to industry. Ford combined the principle of interchangeable parts with machine tool operations. Contrary to popular belief, he was not the first to introduce mass-produced cars. Many automobile companies were using mass production techniques, turning out several cars a day, but those cars were basically custom-made, the pieces being fitted together one by one. Ford applied an assembly technique where parts moved past the workers on a system of conveyors. The beginning of line assembly soon changed the method of production in nearly all industries.

The results of Ford's production technique were readily apparent. Until this time, cars were a luxury for only the wealthy. His dream was to produce a car which everyone could afford. At the time, most cars were priced between $1,000 and $2,000, which placed them out of reach for common people. In 1908, the first Model T came off the assembly line. It was this car which was to "put America on wheels." By 1916, a person could buy a new car for as little as $345. Between 1908 and 1927, over fifteen million Model T cars were produced. In some years during this time, Ford produced more cars than all the rest of the industry combined. The results of increased production efficiency were also reflected in salaries paid to workers. In 1914, Ford announced that qualified employees would be paid five dollars for an eight-hour day. This high rate made headlines around the world.

Railroads

Railroads were in use as early as five hundred years ago; wagons were run on tracks in German mines about 1465. What we would consider a true railroad, however, only came about during the 1800s. A definite correlation can be seen between the development of the railroads and the development of the automobile. Railroads were the practical application of developments in steam power, just as the car applied the technology of the internal combustion engine. European inventors had developed steam engines before American inventors, but, as with the automobile, once the movement started, rapid evolution helped change the entire country.

The first rail systems were designed for horse-drawn cars. By 1829, the first American-built steam locomotive was operating on a common carrier railroad. By 1834, horse-drawn trains were running from Columbia, Pennsylvania, to Philadelphia. Later, this was to become the Pennsylvania Railroad.

From that time on, there were many new developments. Many short lines had been laid, but in 1864, Congress passed legislation that paved the way for a transcontinental railway. Building began in both the East and the West, and the line joined at Promontory, Utah, in 1867. This was a giant step forward. The only means of transcontinental transportation up to this point was wagon trains, which had definite disadvantages. Wagons could not carry heavy loads, they were limited to cer-

This early steam engine carried its own wood or coal. The "cowcatcher" on the front kept anything on the track from going under the wheels.

tain times of the year because of weather, and they were often in danger of attack by hostile Indians. Heavy equipment and supplies had to be shipped around Cape Horn or carried across Panama, because there was no way of transporting them over the mountains until the rail system was built.

With the completion of the transcontinental railroad, the West expanded and the country became more unified. At the same time the railroad was built, telegraph lines were strung. The country was connected not only by transportation, but by communication as well. This is an example of how one technological system may speed another.

During the coming years, the railroads continued to grow, carrying both goods and passengers. By 1916, the railroads had reached 254,000 miles (408,686 km) of track. The capacity continued to grow as parallel tracks were developed. Several trains could then run simultaneously, using a system of main line tracks, secondary tracks, and sidings. The railroads carried a major portion of this country's transportation because they were able to carry heavy loads, such as coal, cattle, and manufactured goods. They were not affected by weather to the same extent as dirt roads, which became impassable with mud; they connected most cities, making stops in small towns along the way, and they were inexpensive for travelers because one could eat and sleep on the train.

The railroads continued to be an important element of the total transportation system throughout the first half of the twentieth century. During the 1960s, however, they found themselves in serious trouble for a number of reasons, one of which was competition with other transportation systems. The highways allowed people more freedom of movement for shorter trips. Also, goods were transported much more easily by truck to specific destinations away from the rail routes.

Another reason was heavy government restrictions. While the rail system paid for its own expansion, both the trucking and airline systems received federal help. In addition, there were provisions requiring the railroads to carry federal freight, such as troops, equipment, and mail, at reduced rates even after World War II.

A third reason was that poor management allowed the railroads to fall behind other transportation systems. Management failed to recognize changes taking place in society and in other transportation systems. Little money was put back into the railroads for upgrading equipment, scheduling, and service, and as a result, more efficient systems gradually took over.

Although the railroads are in a precarious position at present, there are indications they may be able to come back as a strong transportation competitor. To do so, however will require major changes, a tremendous amount of money, and a plan for articulation with other transportation systems.

Other Road Vehicles

There are also other types of vehicles which move through the transportation network. Cars and railroads have been discussed as subsystems because they represent the beginnings of two major forms of land transportation, but they have also led to many other subsystems. The first trucks were simply cars with a bed on the back, or a wagon that had been motorized. They became larger as more power became available, and they became a major part of the transportation system as the network of roads improved. These two elements, trucks and roads, contributed to an increase in the volume of goods that could be transported to nearly any destination. Today, a range of vehicles from light delivery vans to huge double-bed tractor-trailors carry our goods from one end of the country to another.

Vehicles to carry many people were also designed. Large busses now carry people over our highways from coast to

Early trucks were not produced by car manufacturers, rather, independent companies built special cabs and beds to be made into trucks and other specialty vehicles.

A track vehicle turns by stopping the movement of the track on one side while the track on the other side continues to move. This is accomplished by using two levers rather than a steering wheel.

coast. They also carry millions of people to and from work each day. They range from a basic box design with seats to a luxurious coach with restrooms and a kitchen for passenger comfort.

Another category of vehicle, which is usually called an off-road vehicle, combines the use of wheels and a track which are actually part of the unit. Examples would be bulldozers, cranes, and various kinds of construction equipment. The track moves as the wheels move, providing a portable surface on which to ride. There are three main advantages this system of wheels has over conventional systems: they can travel over very soft terrain, because the total contact surface with the ground is much greater; they can move over obstacles and rough surfaces, because the track can climb steep angled surfaces and span depressions; and because they have a large contact surface with the ground, the vehicle can be much heavier without sinking into the soft soil. As a result, they can be made more powerful by using much larger power plants (engines).

There are many transportation vehicles which are used every day but do not fall into the same category as cars, trains, and trucks. Many of them may be classified as work or recreation vehicles. Two-wheeled vehicles many times are used for both: bicycles and motorcycles are often used in this way. Examples of light vehicles of transportation might be roller skates, skate boards, and toy wagons.

Nonwheeled vehicles also play a part in our transportation system. As with wheeled vehicles, some are used for work and some for recreation. One example of a nonwheeled vehicle which has become popular is the snowmobile. The steering is done with skis, and it is driven through the snow by means of a wide moving belt under the machine. Snow skis could probably be classified as one of the simplest means of land transportation, because there are no moving parts. Both the skis and the poles would be classified as simple machines.

Nonvehicular Transportation

Many transportation subsystems could not be called vehicles. They are more closely related to a system of movements within a short distance, involving a mechanism which repeats itself.

This fork lift is a multimovement transporter, moving goods both horizontally and vertically.

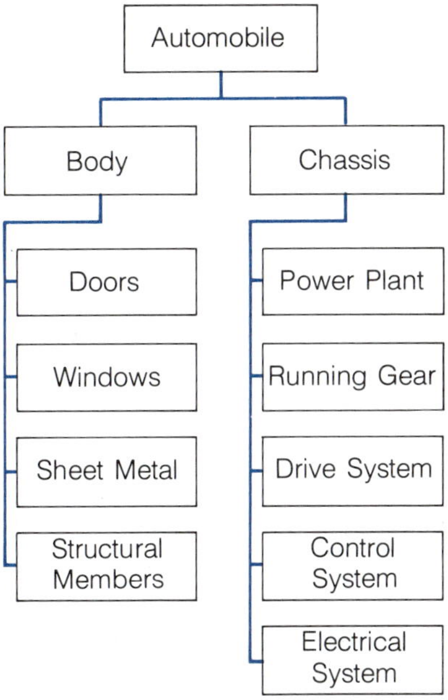

This model illustrates how parts of an automobile would be placed under the two subsystems of body and chassis.

Examples might be an elevator or escalator (upward and downward movement), conveyor (horizontal movement), and chute (downward movement).

Pipelines are used to transport huge quantities of liquids and gases over vast distances. Many pipelines carry combustible materials more safely than other means of transportation. Because of continuous flow through the pipe, they are usually less costly. The flow can be controlled by sending it in different directions through the same system of pipes and valves. Solids may also be transported by pipeline, by mixing them with water or other liquids. They are then separated at the other end by straining or evaporation.

Another transportation method is the combining of non-vehicular systems with vehicles. For example, a crane is a vehicle, but the mechanism which transports materials usually involves a boom and cable with a series of pulleys. A smaller unit called a fork lift also combines a vehicle with the capability of moving goods vertically.

Land Vehicle Design

Most land vehicles are built with two major parts. Each one becomes a specific subsystem because each has a specific function.

Body. The subsystem known as body varies greatly, depending upon the function of the vehicle. Some, such as busses, have many windows, and some, such as heavy equipment, have none. Variation also occurs in the number and shapes of

Unibody ***means the body is essentially one piece after it is welded together. The aperture panels contain the openings for doors and windows, and are the first pieces to be welded to the floor as assembly is started.***

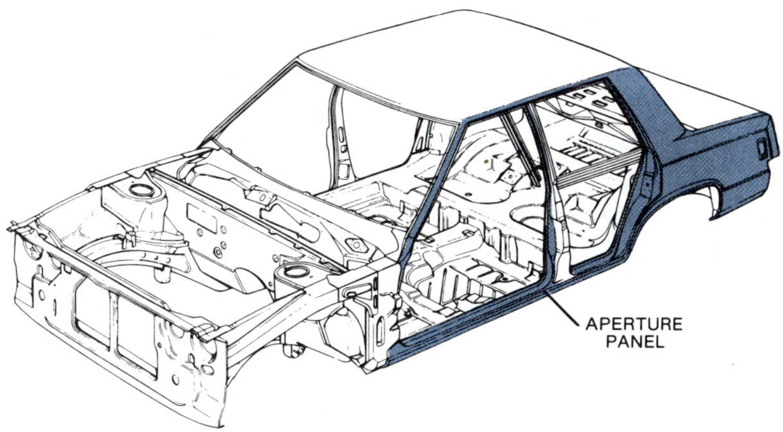

UNIBODY

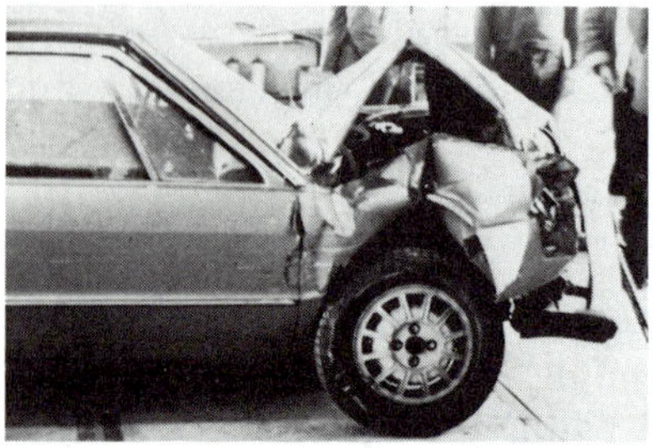

The entire front end of this car was crumpled upon impact, absorbing the energy. The passenger compartment was not compressed; the windshield is not even broken. Passengers are kept safe by making the compartment extra strong, preventing the space from becoming smaller.

doors and in other major features. One of the most important aspects of vehicle design is safety. The body of cars, as well as most other vehicles, contains much more than just the thin sheet metal which we see. Under the sheet metal is an entire system of steel structural members designed to protect the occupants if an impact occurs.

You might ask, "If this is true, why then does a car receive heavy damage when it is involved in an accident?" The answer is simple. A car is designed to crumple under impact. Each crumpling movement absorbs energy from the impact, thus, the more distortion of metal, the more energy absorbed and the less the amount of force on the occupants. Other safety features are bumpers, windshield wipers, and defrosters. There is an additional safety feature which is very important in the design of heavy equipment vehicles. This is a system of roll bars. These heavy bars provide protection for the occupants in case the vehicle rolls over or something heavy lands on top of it.

An additional function of the body subsystem in some vehicles is to control the air passing over, under, and around the vehicle as it moves. This is a consideration for vehicles that move at high speeds, such as passenger cars, race cars, and some trains. This streamlining, which reduces air turbulence, allows air to flow smoothly past the surface of the vehicle, causing less resistance. A flat vertical surface increases air turbulence and causes more resistance, slowing the vehicle and even making control difficult at high speeds.

Chassis. A chassis is composed of several parts. The power plant is commonly called the engine. Land vehicles usually have an internal combustion engine. This type of engine burns a fuel inside the housing. Pressure is created, causing a piston to move up and down. A system of mechanical linkages causes

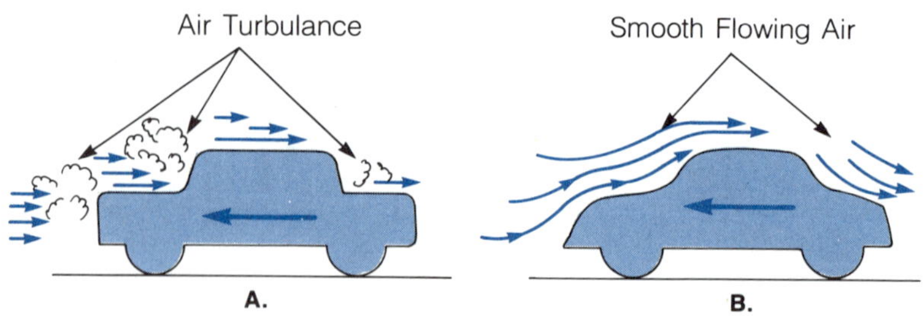

Air turbulance is somewhat like an invisible wall. It creates resistance, causing vehicle vibration and requiring more energy.

This illustration shows the complete chassis, which includes all components except the body.

the wheels to rotate in response to this movement. A motor, on the other hand, is usually associated with electrical energy.

The electric car has a number of advantages over the internal combustion engine car. There is no noise pollution, no atmospheric emissions, and no requirement for fossil fuel. However, there are also disadvantages. Electric cars are small, have a relatively short range of about sixty miles (95.5 km), have batteries that have to be recharged, and are relatively expensive. There is considerable research being conducted on this type of power plant, and it looks promising for the small, short-distance vehicle. It would be particularly appropriate for urban use.

Until there is a major breakthrough in battery technology, considerable room is necessary for the storage units. The size of the battery tunnel in this illustration shows one reason why electric cars are still in the research stage.

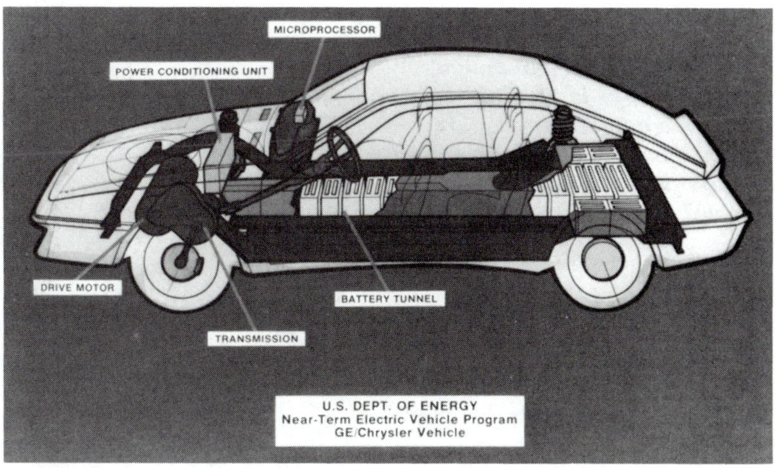

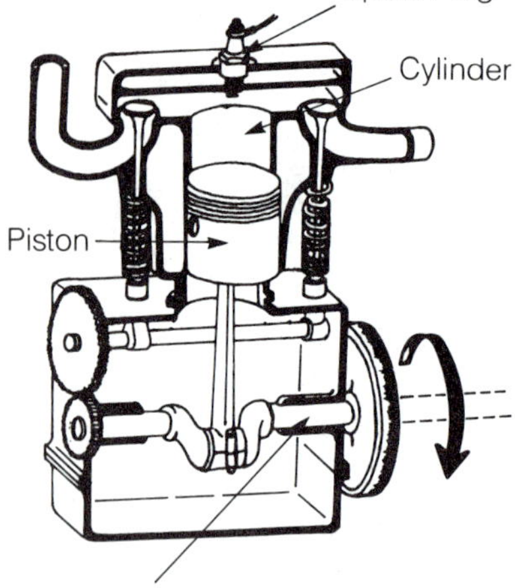

This cutaway view of a one cylinder engine illustrates how the internal combustion engine functions. When the spark plug fires, the gasoline in the cylinder ignites, making a compression force pushing the piston down. This turns the crankshaft, which transmits the energy to a drive mechanism, causing the wheels to turn.

Most power plants work in the same basic way. The amount of energy used changes according to the needs. The application of energy, called power, may be measured in horsepower. This may vary from as small as a ten horsepower (7.6 kw) garden tractor to a 2400 horsepower (1790.4 kw) diesel locomotive. Several things can determine horsepower, including engine size, weight, and piston size.

The running gear includes the tires, wheels, springs, and other mechanisms related to moving the vehicle. Tires vary according to the vehicle and use. Tires can be made in very large sizes. Large tires are designed to carry heavy loads on soft surfaces. The large contact area keeps them from sinking into the ground.

The drive system involves mechanisms that transmit the power from the engine to the wheels. One of the most important elements is the transmission. The gears that are part of the transmission system determine the speed of the vehicle and whether it moves forward or backward. Other elements include the clutch, which engages and disengages the transmission gears with the engine, and the gearshift, which selects the correct gears.

The control system provides the driver with control of the vehicle. The brakes slow and stop the vehicle. When the brake pedal is depressed, it tightens metal bands around the wheels, causing them to slow down. It is even possible to lock the brake bands so tightly that the wheels stop turning, even though the vehicle is still moving.

The tires on this earth mover stand well over six feet (1.8 m) tall. Many times large tires such as these are filled with liquid to reduce the amount of air required and to provide more weight to the vehicle.

The coil springs and the shock absorbers work together to make a smooth ride. The shock absorbers use a telescoping action to shorten under pressure, absorbing the energy of a bump. Some cars may have flat leaf springs rather than coil springs.

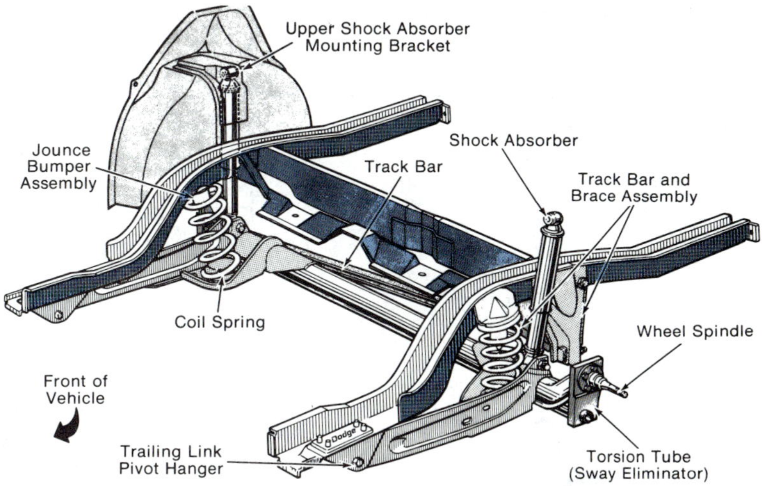

The steering mechanism allows the vehicle to change directions. The steering wheel causes the front wheels to turn in the same direction as the steering wheel is turned.

The electrical system serves several functions. It allows the energy in the battery to initially start the engine. After it is started, an alternator (or generator, depending upon the engine) generates electrical current which keeps the battery charged. It also provides power for lights, electric windows and seats, and any other electrical accessories.

Earlier vehicles were manufactured with the chassis and the body in separate units. During the 1950s, cars began to be built with the two together. This method, called unitary construction, is used in most cars today. They are much lighter and safer, and the design lends itself well to efficient production techniques.

Most vans and light trucks, as well as some buses, also have the unitary construction, while heavier vehicles, like commercial trucks, are still built with separate units. One reason for this is that a variety of bodies may be used on the same chassis, providing a wide range of design for commercial uses.

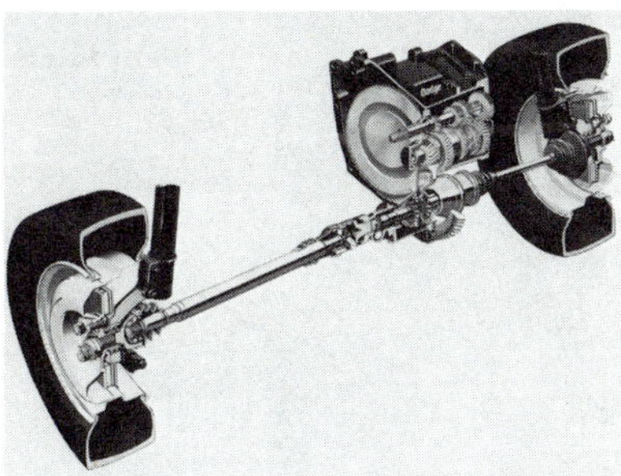

This cutaway view of a transmission shows the different sizes of gears needed for various speeds and reverse. This is a front-wheel-drive vehicle (the front wheels pull the car rather than the rear wheels pushing it) showing the front axle. The engine sets in the center of the axle beside the transmission.

Nonconventional Land Vehicles

These vehicles are in operation at the present, but they may be considered either somewhat experimental or rare because of their limited use.

People Movers. This term is given to vehicles used in a transportation system called Automated Guideway Transit (AGT).

Gears transmit energy, and also, by their size, regulate it. Large gears increase the speed of rotation by transmitting energy to smaller gears, and vice versa. RPM, which identifies gear speed, stands for revolutions per minute.

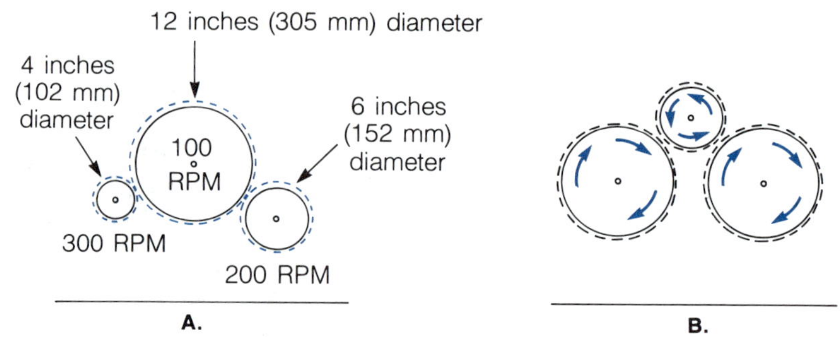

In the past, most standard cars had only rear brakes. This car has brakes on all four wheels, making it much safer. It has a split system linking the right-front and left-rear wheels, so if one side fails there is balanced stopping.

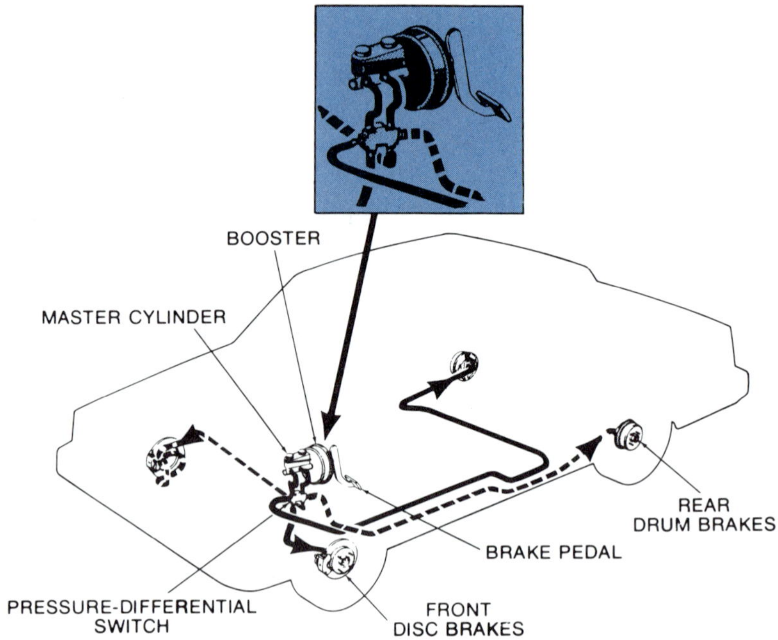

A simple gear box contains gears which rotate right or left as the steering wheel is turned. The wheels then pivot on the center points at each end of the axle. Today's cars have power steering, which uses hydraulics (oil pressure) to help transmit the energy required to turn the wheels.

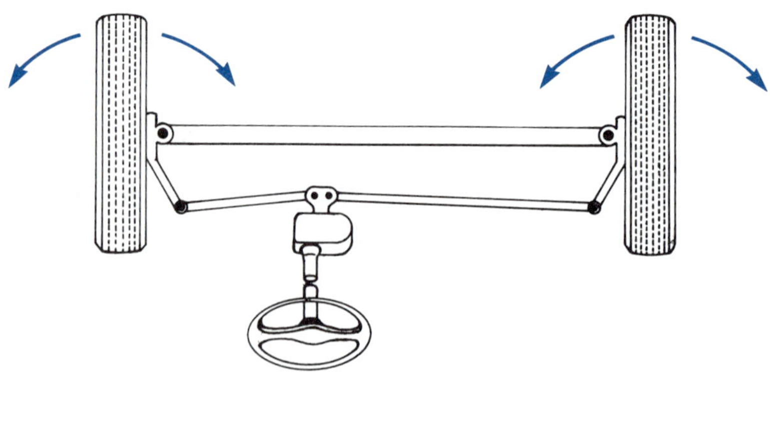

As cars move through the assembly line today each piece is set in place and permanently attached by welding, fasteners (screws, bolts, and rivets), or adhesives. This provides safety-design features which can not be easily changed or removed after assembly.

The system utilizes its own guideway, or traveling surface, and provides driverless, automated vehicles. The guideways may be located below ground level or elevated above the street-level traffic. There are three basic types of people movers.

The shuttle and loop transit (SLT) is the simplest of the three. The shuttle consists of vehicles which travel back and forth on a single guideway (shuttle). The closed loop system involves a vehicle that only goes forward. They may vary in size, and they may travel as separate vehicles or attached together in multiple units. The design allows accommodations for a limited number of people, and they travel on a continuous basis, providing immediate transportation at any time.

The group rapid transit (GRT) system differs from the

This people-mover operates on a single guideway, carrying people from a large hotel to a busy shopping center. It runs automatically as it is called for, much like pressing the button for an elevator.

This modern, efficient subway moves thousands of people daily; it is completely independent of other surface systems.

SLT in that it has a more sophisticated network of guideways and operations. It usually accommodates larger groups of people and provides more flexibility of movement because of branching routes. Off-line stations are also used so vehicles on the main line are not delayed by those at the stations loading and unloading passengers. The GRT usually moves at a higher rate of speed than the SLT. Although the traditional subways and elevated trains may be considered part of this system, there are also new designs in use and in the planning stages.

This monorail of the future is propelled by super-cooled magnets, which also suspend the vehicle above the track so there is no friction between surfaces. This system requires much less energy for movement.

The vacuum-tube train would be a highly efficient transportation system.

There are several monorail trains now in operation which operate either above or below a single track. The structural system for the track raises it above the normal flow of traffic. A unique concept in subway design is being tested but has not yet been attempted on a full scale. It consists of a high-speed

This all-terrain vehicle uses a combination of wide rubber wheels and a track which operates much like a bulldozer.

vehicle moving through an underground tube. Considerations under study include magnetic suspension rather than wheels, using air to drive the vehicle and the creation of an air vacuum to cut down resistance.

Personal rapid transit (RPT) is still in the conceptual stage. It appears to complement the other two systems extremely well and promises to provide another dimension to transportation in the future. It is characterized by smaller vehicles, designed to carry fewer than six people. The time schedule provides for only a few seconds wait, making the efficiency and convenience as nearly like the automobile as possible. Problems yet to be solved include cost, policies of use, and the development of sophisticated controls.

All-Terrain Vehicles. These vehicles, called ATVs, are able to travel nearly anyplace on land. They may be used for either work or recreation. Most have very wide heavy tires to prevent sinking into soft surfaces. If they have wheels, they usually have several, so that if some are not touching the ground for support or traction, others will be.

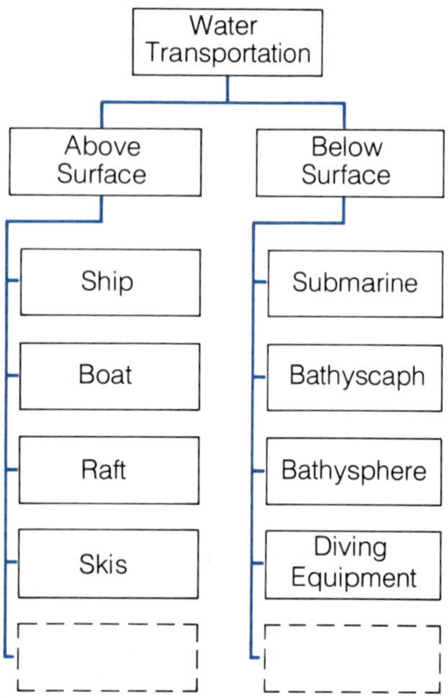

Water transportation vehicles and equipment can differ greatly because of the totally different environments found above and below the surface.

WATERCRAFT

The transportation of people and goods by water plays an extremely important part in our society today. Not only can we move over the earth's entire surface, but we can use the resources which are found in and around the water.

The many types of water vehicles vary most in size, ranging from a one-person boat, such as a dinghy, to huge tankers nearly a quarter of a mile (402 m) long, although they also vary in another more dramatic way. Watercraft are divided into those able to navigate on the surface and those able to navigate under the surface. Through the use of technology, humans have been able to create an artificial environment within a hostile environment, which is compatible with our biological requirements for life. All subsurface vehicles have to artificially recreate an environment that provides air, temperature, atmospheric pressure, and lights, but they do differ in shape depending on the marine environment in which they will operate, their purpose, and their size.

Above-Surface Craft

Early attempts at building a water vehicle were very unsophisticated because little was known of the physics involved. There were probably many types of small boats and rafts. The design depended upon the natural resources available. For example, logs were tied together to make rafts, animal skins were sewn together to make air-filled bags, clay pots were tied together to provide flotation, and water reeds were lashed together into bundles to provide a lightweight material. Each of these flotation methods are from a different part of the world, but the primary objective was the same: to extend the possibility of travel beyond the physical boundaries imposed by water.

Little progress was made in building boats until a method was discovered that could harness the natural energy of the wind. Sails were first used around 3200 B.C. Because of the increased power, larger boats could be built. About 3000 B.C., the Egyptians made a second major breakthrough when they discovered how to make boats with wood planks. This led to the building of ships, which were much larger vessels.

By the time of Christ, ships had become important for travel, warfare, and trade. The Romans built the largest fleet of ships in ancient times. Their largest ships could carry more than 1000 tons (907 t) of cargo and as many as 1000 passengers.

Around A.D. 1000, ships were traveling across the Atlantic, exploring new lands, and plying the Mediterranean and English Channel, visiting neighboring countries. There were four important results of this expansion in world transportation. First, new lands were settled, allowing people the freedom to start a new life. This helped spread technical knowledge. Second, more plentiful deposits of rare resources were discovered, making world trade an important element in the economy of countries. Third, technical knowledge was traded, which served to raise the overall level of knowledge and contribute to new applications. Fourth, more was learned about world geography; the earth's lands, climates, and peoples. This knowledge helped in charting voyages and scheduling timetables for travel.

There was little innovation in ship design until after the 1400s, when the full-rigged system of sails came into being. Previously, ships employed one or two large single sails, which were usually rectangular. A full-rigged ship had several sails which worked together as a system for more efficiency.

The Mississippi River steamboat became a part of our American folklore, carrying both passengers and goods up and down the river. From major points, such as Saint Louis, land transportation went both east and west.

For the next three hundred years, ship design did not change greatly. The next major breakthrough came in the latter part of the 1700s, when a number of people began experimenting with steam power. In 1769, James Watt developed a steam engine that could be applied to many kinds of work, and some began experimenting with steam power in boats. Contrary to popular belief, Robert Fulton was not the first American to build a steamboat. John Fitch, an inventor, built the first in 1786, utilizing a series of paddles on each side of the boat. Robert Fulton, however, built the first commercially successful steamboat in 1807. He later designed twenty other steamboats.

Boats and ships have figured in military affairs since their earliest beginnings. It could be said they have played an important part in shaping the world. Their superiority, or inferiority, determined the outcome of many naval battles, which may have held at stake ownership or dominion of lands.

Ships and their military roles, provide an interesting example of how technological developments in one field affected other fields. Until the early 1800s, large ships were made of wood and powered by sails. Even in the mid 1800s, large clipper ships with as many as thirty-five sails were used, because they were fast for ships at that time (they could travel up to twenty knots). Their hulls were built long and narrow to allow high speeds. Modern ships came about during this time in response to developments in military technology, the iron and steel industries, and steam power. Major advances in artillery

The U.S.S. Constitution was built about 1795. Nicknamed, "Old Ironside," it did not have sides of iron, but its design and the weight of the heavy oak construction allowed it to survive many battles. It has been reconditioned several times over the years and can be seen today at the Boston Naval Shipyard.

ammunition, including explosives and time fuses, made wooden ships very vulnerable and metal-plated ships more desirable. The first armor-plated ship was built by the French in 1859. It had a wooden hull covered with iron 4.75 inches (121 mm) thick, which made the ship, and others like it, very heavy. A more effective power source was needed, and developments in the application of steam power provided the answer. Refinements to this new power source placed it below the waterline, out of danger from bursting shells.

The next step was to build a ship made completely of iron and steel, rather than partly of wood. Such a ship would have many advantages, including freedom from rot, lower maintenance requirements, more strength, and higher fire resistance. Because of the change in the source of power, the shape and size of both military ships changed rapidly. Major changes in design included the use of steel hulls rather than wood and of propellers rather than large paddles. Even today steam continues to be the major power source in large ships.

The golden age of the ocean liner took place during the early part of this century. Luxury liners crossed the ocean in a little over four days, providing not only regular world travel but also mail service. This glamorous form of transportation,

This is one of the world's first liquefied natural gas carriers. One shipload will provide enough gas to heat 2.5 million homes on a cold day.

with every comfort of home, provided the first opportunity for the general population to travel any great distance.

After the Second World War, popular passenger service by ship decreased greatly because of increased air travel. The cost of air travel was much less than by ship, and the trip could be made in a few hours rather than a few days. Today, many goods are also transported by air, while the shipping networks generally transport goods that are not appropriate for air travel, such as crude oil and grain.

Subsurface Craft

Subsurface vehicles, which allow movement under water, are used mainly for military and research purposes. The first submarine was developed during the Revolutionary War by an American named David Bushnell. It was made of wood and used a pump to force water out of the hull to allow it to surface. Although submarines made a brief appearance during the Civil War, it was not until World War I that they were used in a systematic, large-scale way. Early twentieth-century submarines were powered by two engines: oil burners were used for surface cruising, while electric motors were used for submerged travel. Later, steam engines were substituted for surface travel. These were used to power the electric motors that propelled the craft under the surface. Electric motors had to be used because a combustion engine would have required oxygen for burning and an exhaust system, neither of which was possible under water. Today, diesel engines and electric motors are used on non nuclear submarines.

The first nuclear-powered submarine, the USS Nautilus, was launched in 1955. It is now common for submarines to be nuclear-powered. A nuclear reactor produces steam, which, under pressure, drives large wheels called turbines. These turbines generate electricity, which powers the sub. The exhaust steam is condensed back into water and is returned to the reactor for reheating. Thus, it is a closed system, meaning there is no loss of exhaust material. Since a nuclear reactor can run continuously, it is possible for a submarine to stay submerged for many months at a time.

The oceanographic submarines, called submarinos, are

This is the bathyscaph Trieste in the Pacific Ocean. This craft is used for underwater research and rescue operations.

much smaller vehicles. They usually carry only a couple of people, and their primary purpose is research and exploration. The *bathyscaphe* is a thick-walled steel sphere with windows, which is attached under a cigar-shaped larger hull. The researchers are in the sphere, while the hull is full of gasoline. Because gasoline is lighter than water, it provides buoyancy. To descend, the gasoline is let out and water is let in. This makes the craft heavier, causing it to go down. To come up, weights, called ballast, are released from the sides, making the vehicle lighter. It is driven by two small battery-powered motors. Bathyscaphes are able to dive nearly three quarters of a mile (1207 m) deep.

The *bathysphere,* the forerunner of the bathyscaphe, was first developed in 1930 and is still used for certain kinds of research today. It consists of a sphere connected to a surface vessel by cables, so it has little freedom of movement. The spherical shape helps to resist the tremendous underwater pressure. At the surface, pressure is about 14.7 pounds per square inch (101.4 kPa), while at 4000 feet (1219 m) under the surface, the pressure is nearly 1500 pounds per square inch (10342.5 kPa). At that pressure the craft would be crushed if the vessel could not withstand the tremendous forces (pressures) from all directions. A sphere withstands pressure in much the same way that blocks within an arch distribute the stress borne by the arch. The pressure inside the vehicles must be kept close to surface pressure, regardless of the depth of a vehicle. This is accomplished by onboard systems in the case of the bathyscaphe and by hoses attached to the surface vessel in the case of the bathysphere.

Watercraft Design

An object floats because of a condition called *buoyancy,* which is achieved when the weight of an object is less than the weight of the water it displaces. If a large object displaces enough water, that object will float. As ships became bigger, they had to be made wider and deeper, so that the volume of water displaced by the ship always weighed more than that of the craft.

In ship design, it is important to calculate the weight and volume, particularly in terms of their effects on the center of gravity and the center of buoyancy of a ship. The center of gravity pushes down, while the center of buoyancy pushes up.

The designing of ships is highly specialized, requiring the calculation of many types of forces.

The forces of buoyancy can be considered a single force pushing upward at the centerline of the ship. This is called the *center of buoyancy*.	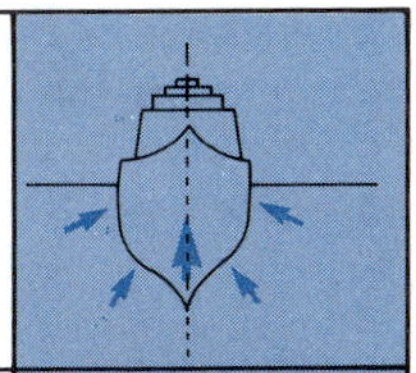
The forces of gravity, which is the total weight of the ship, can be considered a single force pushing downward at the centerline of the ship. This is called the *center of gravity*.	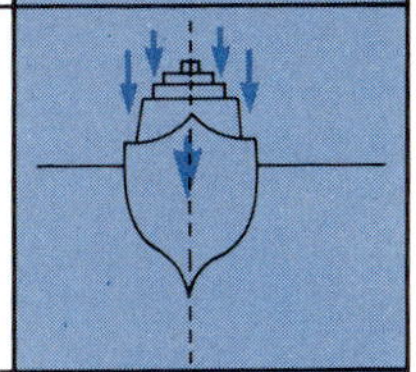
A ship is designed so both centers are in line with each other. This keeps the ship in a vertical position.	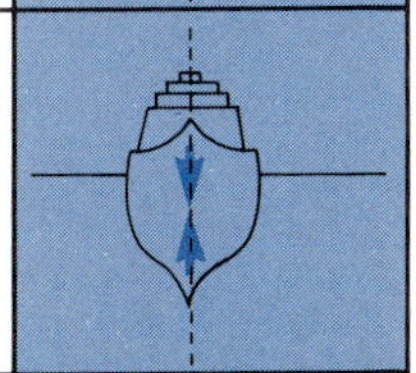
A stable ship will be forced back to its upright position if it leans because of high winds or waves. The force of gravity remains stationary while the force of buoyancy moves to the side of the ship which is lower in the water. The two forces work together to bring it back to a vertical position.	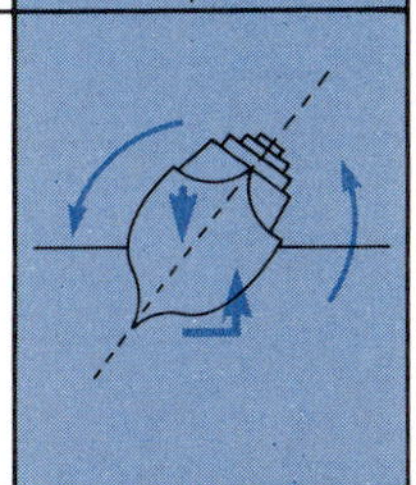
The more water that is displaced, the lower the center of gravity. A ship is unstable when the center of gravity is too high because there is too much weight at the top or when a hull design does not displace enough water. As the ship leans, the center of gravity moves with the top of the ship past the center of buoyancy on the low side of the ship. As a result, the two forces work together to turn it over.	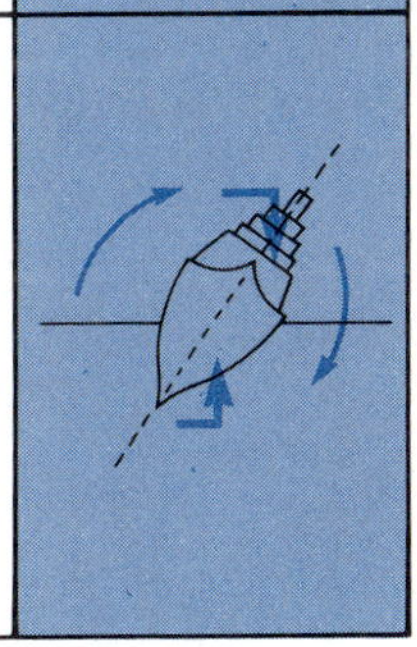

When the ship tips to one side, the forces are not lined up in a vertical center line. The forces shift to align again, thus bringing the ship back to a vertical position. If the two forces are too far apart the craft rolls over. Also important to consider in designing a ship is the type of cargo and the weight that the ship will carry. As a ship is loaded, it increases in weight, making it set lower in the water. The two forces of gravity and buoyancy must be balanced at all times.

The two forces of gravity and buoyancy create a constantly moving area of stress as waves pass the ship.

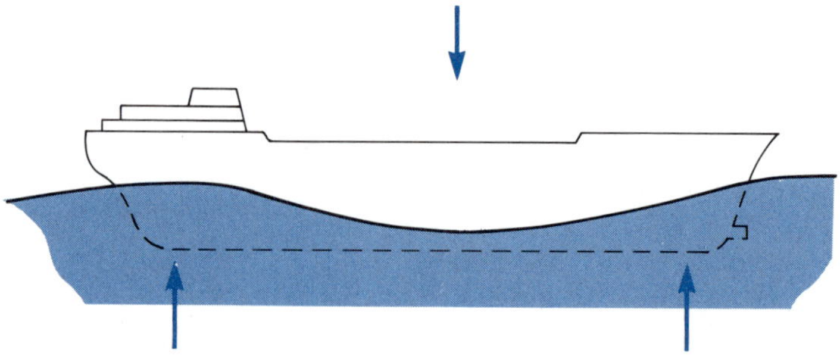

Force attempts to bend the ship upward at the ends ( ).

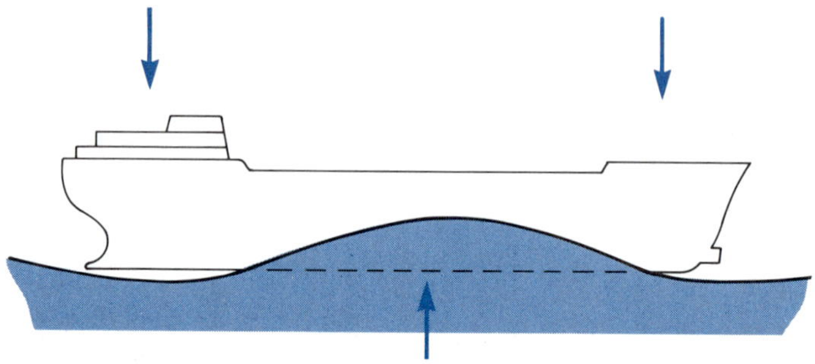

Force attempts to bend the ship downward at the ends ( ).

Hull strength is also very important. Waves place many different stresses on the hull of a ship. Because the waves are always moving, the stress is always shifting. Most ships are built in sections, which allows for a slight amount of bending at the areas of connection (areas called *expansion joints*). However, the main areas of stress have to be reinforced with more supportive, heavier material. A ship is made of steel girders (beams) and ribs which are covered with steel plates on the inside and outside, making a double wall. This provides more strength, as well as storage space for fresh water, fuel, and oil. A ship is divided into watertight compartments called *bulkheads*. If a ship takes on water, the bulkhead where the leak is located can be sealed off, preventing the ship from sinking.

Most smaller ships and boats use diesel engines, while large ships use steam turbines. These engines are among the most powerful machines known. Unlike the nuclear submarines, the turbines do not convert the energy to electricity; rather, they use a series of gears to turn the propeller. Most

The propellers and shafts on watercraft are sometimes called screws. This ship has twin screws. Notice the size of the propellers in relation to the man under them.

ships have only one propeller. The world's largest propeller weighs 72 tons (65.3 t), measures 31 feet (9.5 m) across, and is nearly as tall as a three-story building.

Smaller boats, such as those used mainly for recreation, may have one or two propellers, depending upon size. A pleasure craft usually has one of three basic types of engines and propellers. An *inboard* has an engine inside the boat, with the shaft and propeller extending through the bottom of the hull. This is found on the larger pleasure craft. An *outboard* has the engine and propeller built in one unit. It is usually considered portable, because it can be removed from one boat and attached to another. An *inboard-outboard* is a system whereby an engine inside the boat is attached to a unit with a propeller on the transom, or rear, of the boat. This unit can be raised and lowered much like an outboard engine. It combines the best features of the other types.

Water Networks

Although one would not expect the same type of traffic movement system on water as on land, there is still a system to water-borne traffic. Ships travel in specified shipping lanes across the ocean or on smaller bodies of water. A shipping lane may be several miles (km) wide, as opposed to the several feet of a highway, but it is just as important for ships to stay in their lanes as it is for cars to stay in theirs. There are also inland waterways for smaller craft, which are, like cars, subject to movement regulations. Canals provide easy movement between formerly separate bodies of water, and where water levels are different, locks have been built. The locks consist of a channel with gates at both ends. When a ship or boat enters the channel, the gates are closed behind it. The water level is then allowed to either rise or fall until it is equal to the level on the other side of the forward gate. The front gates are then opened and the vessel proceeds on its way. The Panama Canal, one of the most important achievements in construction, is a prime example of this kind of technology. The locks allow ships to pass through the narrow neck of Central America rather than having to go around the southern tip of South America, which saves more than 5,200 miles (8,320 km).

Nonconventional Watercraft

Nonconventional watercraft, which represent new directions in water transportation, may be divided into two categories: freight carriers and passenger carriers.

Freight Carriers. The last decade has seen the deployment of the megaton supertanker and the containerized freight ship (where the freight is shipped in huge containers that are loaded and unloaded by cranes). The supertankers which transport crude oil, displace so much water they cannot pass through some waterways that are commonly used by other ships. Despite the longer route, it is more economical for the vast supertankers to carry oil 11,300 miles (18,182 km) from the Persian Gulf to Britain via the Cape of Good Hope at the tip of Africa,

The hovercraft can navigate in many environments where other vehicles cannot. Underwater ice, which would be a threat to other watercraft, presents no problem.

than it would be for smaller ships to go the shorter route through the Suez Canal.

Passenger Carriers. There are two types of short-distance carriers that are designed for speed and for regular shuttle service and other specific purposes, such as shore patrol. The hovercraft rides on a cushion of air produced by large fans. The air trapped under the vehicle lifts it a few inches (mm) off the surface. Although they can operate on land, they are normally used on water, because it provides a smoother surface. They

The bottom of this Jetfoil is nearly twelve feet (3.6 m) above the water's surface. The reduced resistence from water provides a smooth, fast ride.

are usually pushed by propellers and guided by means of a fin and rudder much like those found on airplanes. They are used extensively in some countries for shuttle passenger service, as well as the ferrying of cargo.

The hydrofoil boat operates much the same way as an aircraft. Just as the wings provide lift for the aircraft, the hydrofoils provide lift for the boat as it moves through the water. Depending upon the type of foil, the entire boat may be lifted completely out of the water. The Boeing Jetfoil, a ninety-foot (27m) passenger boat, cruises smoothly in waves up to twelve feet (3.7 m) in height. Hydrofoils are usually powered by diesel engines or gas turbines, and travel up to fifty knots per hour.

AIRCRAFT

Probably since the beginning of time humans have wished to "fly through the air as free as the birds." Historical writings as far back as 400 B.C. abound with references to human flight, but most are from Greek mythology, so little is known about actual occurrences. The Chinese were flying kites by 200 B.C., so it is possible they attempted to construct a glider type of craft. In 200 B.C., Archimedes, the Greek mathematician and inventor, discovered the scientific reason why objects float on water, but it wasn't until about 1290 that Roger Bacon, an English monk, observed that air has much the same characteristics as water. The gradual accumulation of knowledge about the laws of nature over the next few hundred years laid the groundwork for flight.

Today, the world depends upon air travel. It has become a part of our everyday living, allowing rapid transportation of both people and goods. Ships made world transportation possible, but airplanes made it an everyday occurrence. In 1927, Charles Lindbergh made the first transatlantic flight in nearly a day and a half. At this time, ships took nearly five days to make the crossing. Today, the flight across the Atlantic can be made in less than three hours. The world has become, figuratively speaking, smaller, because travel time has been drastically reduced.

Aircraft may be categorized in several ways, but the most evident difference is between lighter-than-air and heavier-than-air craft. Heavier-than-air include prop planes,

Early aircraft, such as this glider, resembled large box kites. In this picture the pilot hangs on as the wind turns his glider over.

jets, gliders, rockets, helicopters, and parachutes. Lighter-than-air craft include balloons and dirigibles. Both types are used today, but heavier-than-air craft have so many more advantages than lighter-than-air craft that little use is made of the latter. Among the advantages of heavier-than-air craft are the following:

- They can carry heavy loads. The world's largest airplane can actually carry other aircraft and various kinds of transportation vehicles. The nose of the C-5A Galaxy opens wide enough for vehicles to be driven on board.
- They are not dependent on the weather, which makes scheduling more predictable. They have their own power source and instruments for control. With radar and computer controls, the aircraft can take off and land in almost any kind of weather.
- They can provide for the comfort of their passengers; light, temperature, and other elements of the natural environment are controlled within the aircraft.
- They are able to fly for extended periods of time; nonstop trips cut down on time and costs.
- They are easily maneuverable and can be used for many purposes because they are able to reach places otherwise inaccessible. They can, for example, dump water and chemicals on a forest fire, rescue flood victims, and lift building materials to the top of a skyscraper.

The history of flight demonstrates dramatically the concept of technological readiness. This phenomenon occurs when conditions are favorable, and the scientific knowledge is available to provide a major breakthrough in a field of technology. For hundreds of years people dreamed of flying, but efforts were usually in vain. The accumulation of knowledge concerning our natural environment by Archimedes, Bacon, da Vinci and others helped to launch us into the Air Age. Each bit of applied knowledge led to new developments. It is interesting to note however, like many other technological developments, discoveries about flight were occurring independently and simultaneously in various parts of the world.

The year 1783 was a milestone in the history of flight. That year two French brothers, who were papermakers, filled a cloth and paper balloon with hot air and sent it aloft with a

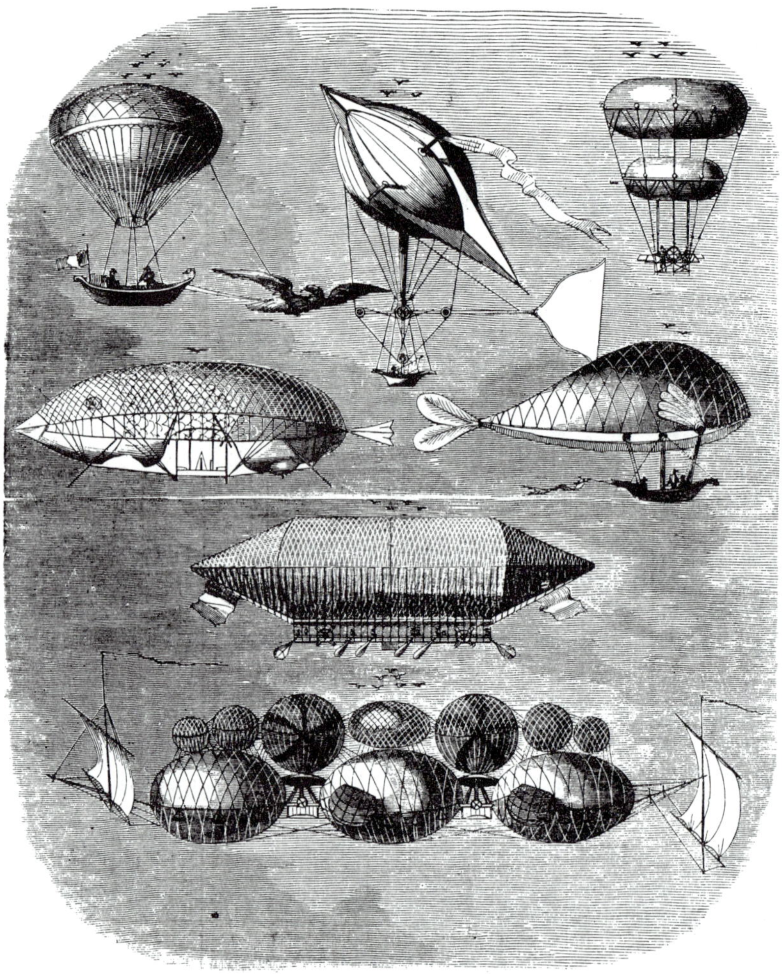

These drawings illustrate some of the early ideas for aircraft. Impractical ideas such as boats, fish, oars, and sails can be seen, but some evidence of scientific knowledge can also be seen, as in a network of reinforcement around the balloons and streamlined shapes.

rooster, a duck, and a sheep aboard in a basket. This flight of one and a half miles (2.4 km) was the first extended flight. Also, in 1783, two other French brothers sent up the first hydrogen balloon. Harvey Firestone, Jr., in his book *Man On The Move,* relates a conversation of two who observed this balloon as it ascended.

> In the crowd was the American diplomat, Ben Franklin. As the balloon rose through the drizzle, a French Army officer turned to him and remarked, "Interesting. But of what use is it?" "What use," Franklin replied succinctly, "is a newborn baby?" (Firestone, p. 218)

This, indeed, was the birth of aviation. And again in 1783, Jean-Francois Pilatere de Rozier, a French doctor, became the first human to fly. His balloon went nearly five miles (8 km). Ten days later another flight went twenty-seven miles (43.4 km). Also, in 1783, the first balloon flight in the United States took place. Over a period of years improvements in design were made and, by 1852, many experiments with elongated balloons and means of power were tried. Both steam engines and electric engines were tried, but they were found to be too heavy or lacking in power.

Winged Aircraft

By the mid-nineteenth century, many people were experimenting with the idea of heavier-than-air craft. Experimentation with gliders revealed the lack of a power source as a major drawback. The only real power source at this time was steam, but despite attempts by American and European inventors there was no instance of sustained flight.

Finally, in 1902, the Wright Brothers built and successfully flew the first powered airplane. The power was supplied by a gasoline engine. Their best distance was 852 feet (260 m). Following this, there were many firsts: first powered flight in Europe, in 1906; first public flight covering one mile (1.6 km) in

The Wright brothers' airplane, as it looked in 1908, had several modifications from the first one of 1903.

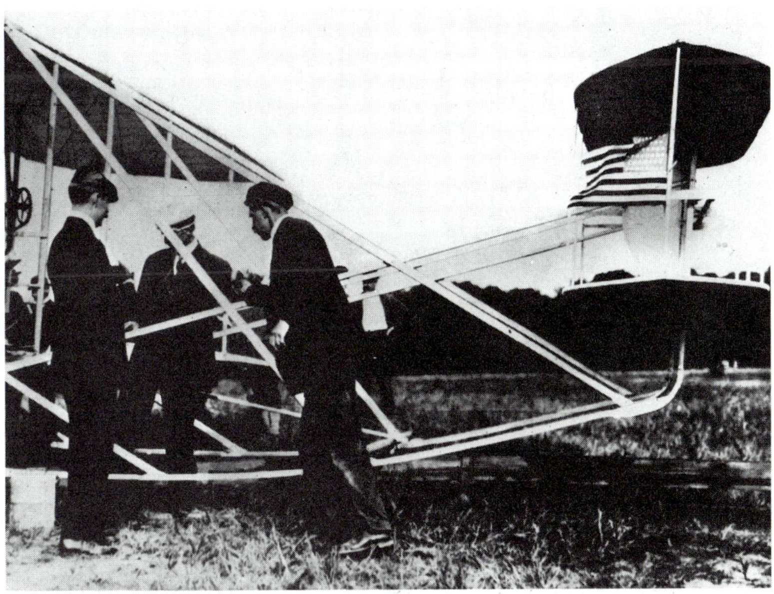

The Bell X-1 was the first vehicle in history to carry a human being faster than the speed of sound. This aircraft, which was a rocket plane (not a jet), made its first supersonic flight in 1947.

1908; first flight over the English Channel, 1909; and first coast-to-coast flight in the United States, 1911. World War I provided the impetus for many innovations and proved the value of air transportation. By 1919, there was regular airmail delivery, and in 1924, two US bombers circled the earth by air.

This amazing progress in little more than twenty years was primarily due to the development of the gasoline engine. The lightweight engine allowed continuous refinements to be made in aircraft design and enabled airplanes to become bigger and more powerful.

Although airplanes were used to deliver mail, they served few other major needs until World War II. It was the war that launched the true age of air. Aircraft were suddenly needed to transport service personnel and equipment and to serve in combat and reconnaissance. There were rapid changes in the design of airplanes; practically overnight, they became bigger, faster, and more efficient.

At the beginning of the war in 1941, the top speed was approximately 300 miles per hour (483 km/h). By 1947, two years after the war, planes could fly faster than the speed of sound. These new airplanes soon became a part of our technological society, being used for nearly every type of transportation.

As the discussions of various inventions and developments have shown, the discovery of new sources of power leads to technological innovations, and the progress of aircraft particularly exemplifies this fact. The tremendous increase in speed and power of aircraft during World War II resulted from the development of the jet engine. The increased speed called for new aerodynamic designs, which changed the looks of aircraft the world over. In the short period of forty-five years (1902–1947), human flight went from a "kite like" open vehicle to an enclosed metal aircraft that flew faster than the speed of sound.

On December 18, 1986, a modern record in the field of flight was set, which exemplifies the advancement of knowledge since the Wright Brothers' first flight. A lightweight two-passenger plane flew nonstop around the world without refueling. This only could have been accomplished by consolidating the latest scientific knowledge in aircraft design. What was learned will likely have a bearing on future designs of small planes. This feat is an example of how proven technology can be improved if new knowledge is applied to what is already known.

Nonwinged Aircraft

Prior to WW II, experiments were being conducted on another form of aircraft, one that had no fixed wings. The helicopter, like fixed-wing airplanes, went through phases of development that were not successful until the gasoline engine was a reality.

Also like fixed-winged airplanes, the helicopter appeared in the fifteenth-century notebooks of Leonardo da Vinci as sketches of an aircraft with rotors. It is not clear who made the first manned flight, but it was Igor I. Sikorsky, a Russian immigrant to America, who built the first practical helicopter. This flight took place in 1940, just at the beginning of World War II. They were immediately put into production and used extensively during the war to transport troops and equipment. They have proven invaluable since that time, for both military and peaceful uses. Some examples of helicopter uses are as follows:

- pickup and delivery of cargo without landing
- rescue of disaster victims
- wind from rotors dries fruit in critical growth stages, circulates air to prevent frost damage, and spreads fertilizers
- seeding crops and trees
- spraying farmlands and forests for insects and diseases

This was Sikorsky's first helicopter, the forerunner of the modern helicopter.

- delivery of heavy equipment and supplies to inaccessible construction sites
- air taxis and shuttle service
- traffic control and crime fighting
- cattle herding on large ranches
- mail delivery
- forest fire observation and fire fighting
- military

Although they are extremely slow compared with fixed-wing aircraft—top speed is only a little over 200 miles per hour (322 km/h)—they do have particular advantages. They are able to pick up and deliver heavy loads without touching the ground, which saves considerable turnaround time. This capability is also extremely valuable where landing is impossible, such as in mountain and sea rescue work. When they do land, they can set straight down and, thus, do not require a landing strip. Helicopters, sometimes called flying cranes, can transport cargo to and from areas where it is not possible to have an airfield, such as on ship decks and atop buildings. Because they take off and land in a vertical pattern, they are useful in rescue operations and for depositing huge air-conditioning units gently on top of a tall building.

In addition, helicopters can fly in any direction, including backwards. They also can remain stationary in the air. This

Helicopters can perform a variety of functions in nearly any type of environment. The illustration on Top shows a military helicopter landing in rough sea conditions. The other is an example of commuter travel.

makes them valuable for such uses as land surveys, farm crop dusting, and forestry.

How an Airplane Flies

Most first attempts at building a flying machine involved movable, flapping wings. This attempt to simulate birds' wings was unsuccessful because of the difference in weight. Even success with gliders was limited until basic laws of physics were applied to the design. Sir Isaac Newton's laws of motion, formulated in the seventeenth century, serve as the basis for aircraft design. It was not until the nineteenth century, however, that the principles were applied. Three basic laws of motion, developed by Newton, have a direct effect on how an aircraft flies.

- A force must be applied to an object before it will change position, speed, or direction. (This means there needs to be an energy source to provide a force.)
- A force acting on a mass will produce acceleration proportional to the applied force. (This means that more force will produce more speed and/or lift.)
- For every action there is an opposite and equal reaction. That is, if a force is produced in the air, where there is no ground resistance, the movement will be in the opposite direction. As a propeller turns, a force is created in the direction of the aircraft. The opposite reaction moves the craft forward. (The aircraft is really being pulled through the air as a result of this force.)

These laws of physics, coupled with another formulated in the eighteenth century by Daniel Bernoulli, a Swiss scientist, provide the basis of modern aircraft design. Bernoulli's principle states that as the speed of a fluid (liquid or gas) flows over a surface, the pressure on that surface is decreased; the faster the speed, the lower the pressure. If a wing is designed

Aircraft design is very functional. Every part is designed to serve a specific function in controlled flight.

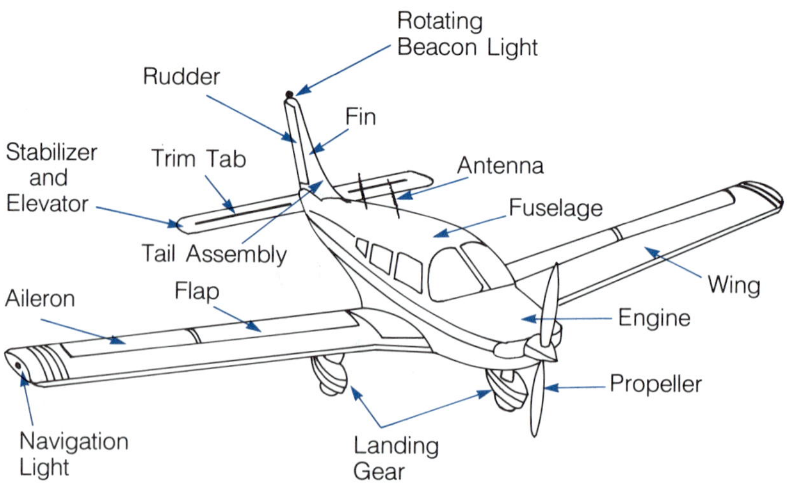

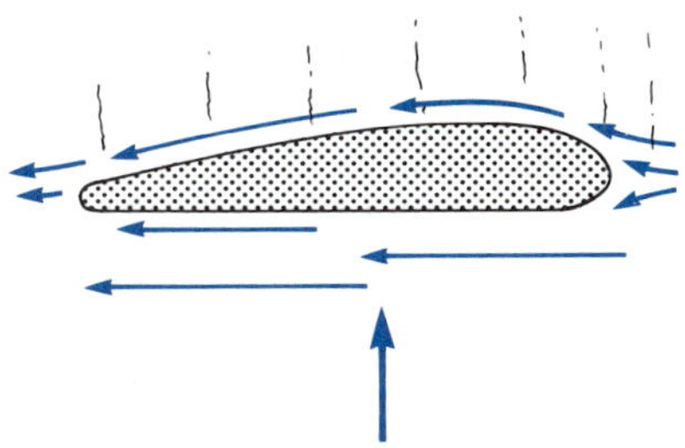

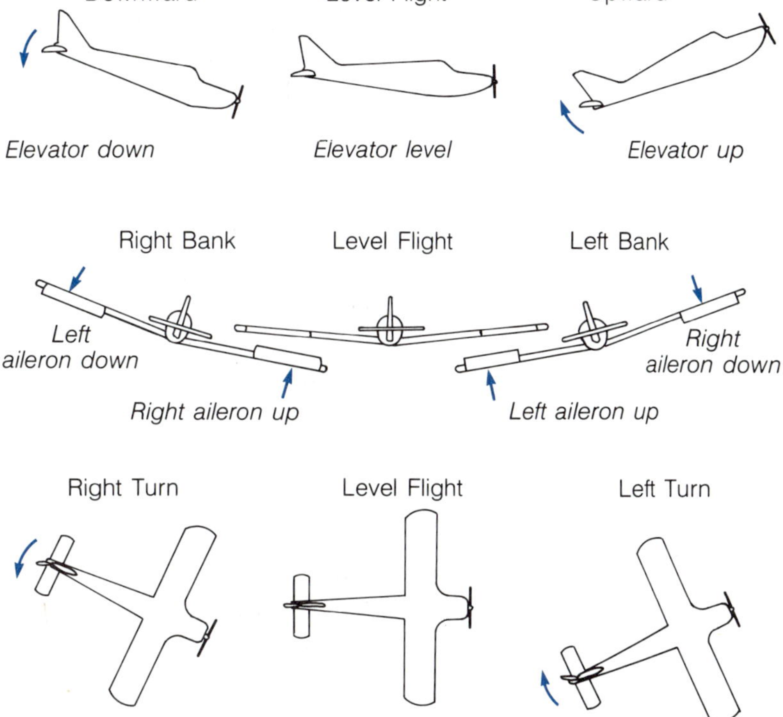

Lift is created when the air moves faster over the top of an airfoil. The use of the elevator, rudder, and ailerons changes the configuration of the airfoil and, thus, changes the speed passing over and under it, adjusting the amount of lift.

with a flat surface, the air rushes over and under it at the same speed and the pressure both above and below the wing is equal. If the top of the wing is curved, however, the air moving over it increases in speed. The result is less pressure on the top surface than the bottom, which causes the wing to rise. This is known as *lift* and is what makes an airplane fly. The engine provides the force that gets the air moving over the wing fast enough to create lift. This is why an airplane has to taxi down the runway to gain proper speed before it can leave the ground.

For an airplane to fly, the wings must be a particular shape, but for the flight of a plane to be controlled, the wings must have sections that are movable. The pilot controls the aircraft by moving these sections, called *ailerons,* up or down. Air speed is increased when the flaps on the trailing edge of the wing are down. This creates more lift, causing the airplane to rise. When the flaps are up, the opposite occurs. When the flap on one wing is up and the flap on the other wing is down, the plane goes into a bank (turn) because of uneven pressure.

More pressure under the wing pushes it upward. The tail of a plane also has flaps, called *elevators,* which function in the same manner as the ailerons. Elevators cause the tail to move up or down, while the *rudder* causes the tail to move to the left or right for turning. All of these surfaces, whether horizontal or vertical, are called *airfoils.*

Airplanes with propellers have a relatively straight airfoil (90° to the body of the plane, or *fuselage*), which gives maximum lift at slow speeds. Although propeller driven, planes are usually able to take off and land at slow speeds, they cannot cruise at high speeds because the straight-wing design creates too much air resistance, and the wing might be torn off. Jet planes, on the other hand, are able to travel at extremely high speeds because their wings are swept back to ease resistance. The necessary lift is still there, because the increased speed causes faster moving air, which in turn creates more lift. As a result, jet planes have to take off and land at much higher speeds.

Although the helicopter looks very different from a fixed-wing aircraft, the same principles of physics apply. The rotor, which may have two, three, or four blades, is a system of narrow wings. Each blade has an angle (called *pitch*) which can be adjusted. This makes each blade an airfoil. As the speed of the rotor increases, lift is generated. The smaller rotor at the rear controls the tail section and allows for turning.

The landing gear is the unit on which the aircraft sits while on the ground. Depending upon the purpose of the aircraft, there are four types of landing gear: wheels, skids, skis, and pontoons. On many aircraft these are interchangeable, allowing them to be used in different environments.

Air Networks

Aircraft also has a system of movement. The network of traffic control keeps them from coming too close together while flying, landing, or taking off. Basic to the system are runways, some of which are designated for landing and some for taking off. The control tower tells each pilot which runway to use. After take off, the control tower may also tell the pilot what direction to turn, how fast to climb, and at what height to level off. Each aircraft has to fly at a specified height. This part of the network has many vertical levels so the aircraft are able to cross paths safely. When a major airport is extremely busy, there may be many airplanes "stacked," or waiting to land. They fly in circles around the airport, each at a specified level and angle of turn. As one lands, each aircraft drops to a lower level. These "invisible roads" are very important, and have to be followed at all times.

Nonconventional Aircraft

Some aircraft are still in the experimental phase, and others are used on a limited basis.

Hang Gliders. Hang gliding has recently become popular. For centuries humans have attempted to fly, using wings as the sole power source; the hang glider comes close to this goal. It is really a large airfoil made of lightweight material. It can be controlled and stays aloft by hot air thermals (hot air rising from the ground). Height is usually gained by flying off the side of a mountain or hill. Although some of the design concepts are now being used in parachutes to provide more control, hang gliders are primarily used for recreational purposes.

VTOL Aircraft. One of the directions of research in the last few years is the design of an aircraft called a VTOL (vertical take off and landing). These have engines, either propeller or jet, which pivot vertically, pushing or pulling the airplane ver-

This VTOL aircraft achieves lift by directing the jet exhaust downward by means of the flaps in the wings. After proper height is attained, the angle of the flaps is changed to allow more forward thrust.

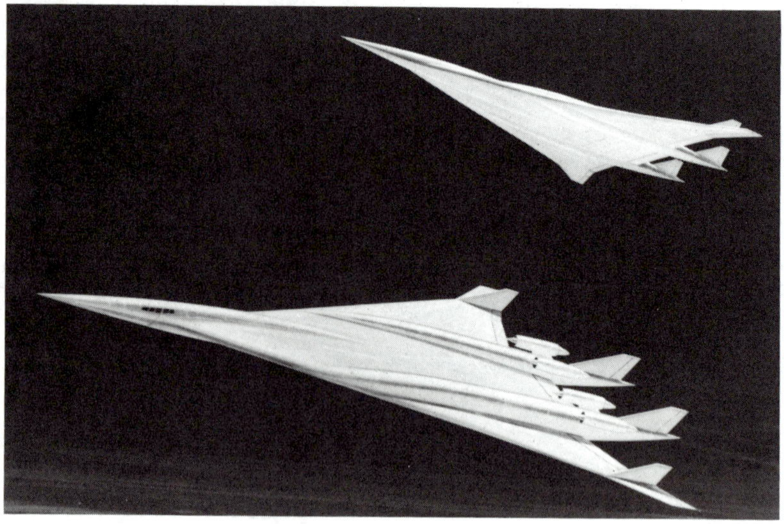

The supersonic transports of the future will have to be significantly different from those of today. Body design changes, advanced metal alloys, and a different fuel system would make them more efficient. Their range would be at least 4800 miles (7680 km) at a speed of over twice the speed of sound.

tically into the air. When the proper altitude is reached, they are returned to the normal position for horizontal flight. There have been many different successful designs, and major advances are regularly being made. The obvious advantage of this type of aircraft is that it eliminates the need for long runways.

Large Aircraft. The jumbo jet seems rather commonplace now, but the accomplishment of flying something that large is still mind boggling. The largest aircraft to date, the C-5A freight transport, is so large it is said that shaving only a few thousandths of an inch (mm) of paint off the craft would save 2000 pounds (907 kg) of weight. Its tail section stands as high as a six-story building.

SST Aircraft. Another term which has become well known is the SST (supersonic transport). These are large aircraft capable of flying faster than the speed of sound. A unique concept was designed into this aircraft. The nose section is movable so there is visual contact with the runway during take off and landing. The angle of flight and the height of the craft would make it difficult to see if conventional designs were used. There has been considerable controversy over the aircraft because of the noise and accompanying "sonic boom" when the sound barrier is broken. (The sonic boom is a loud noise cre-

The hypersonic transport would service cities which are over 6000 miles (9600 km) apart, traveling at a height of nearly 20 miles (32 km) at a speed of nearly four times the speed of sound. Our level of technology today has not yet reached the point which would make this concept of travel efficient.

ated by the building up of shock waves and the sudden release when the barrier is broken.) This is caused by the configuration of the wing, which is necessary for subsonic flight (below the speed of sound). Scientists are seeking a solution to this problem, possibly by using more movable parts like the nose section. In this case, the wings would be flared backward along the fuselage to cut down on drag, which causes the build up of shock waves. As the plane slows down to subsonic speed, they would be returned to their original position. This capability would allow the aircraft to have the best aerodynamic features of both an airplane and a rocket.

HST Aircraft. A new aircraft now under consideration is called the HST (hypersonic transport). It is still only a concept, but an interesting one. This craft would fly at 4000 to 5000 miles per hour (3437 to 8047 km/h) in the upper atmosphere. This was accomplished back in 1966, with X-15-2, which was a rocket aircraft. The speed would approach 18,000 miles per hour (28967 km/h), which is orbital speed. It is obvious that this type of airplane would not be used for daily flights; it may represent that intermediate step between aircraft and spacecraft.

SPACECRAFT

The rapid advances made in air transportation and related systems (controls, communications, and so forth) in the short time between the Wright brothers' first successful flight and the

Life in space will require living in a weightless environment. Both men and women are being trained for future trips to carry on research and other scientific activities.

modern plane pale in comparison to the tremendous advances made in the last few decades in space transportation. On October 4, 1957, Russia placed a satellite in orbit around the earth, launching us into the Space Age. The small satellite, not much larger than a softball, was the first of some 17,500 payloads (spacecraft and hardware) that were launched into space during the next several years. One of these carried the first humans to the moon. On July 20, 1969, only twelve years after the first unmanned spaceflight, American astronauts, Neil Armstrong and Edwin Aldrin, walked on the surface of the moon. Advanced technology made it possible, and today, we are still benefitting from the technological spin-offs of our early space endeavors. We did not have all the technology necessary for the "moon shot" when we started the project, because we were not even sure what was going to be required. Each new technological development was based on previous knowledge, starting a chain reaction which is still building.

Two questions might be asked about space travel: (1) Why did the Space Age come about at this particular time: (2) Why was the development so rapid? Both questions can be answered, in part, with the same answer. The advances made in the field of electronics provide sophisticated computers which were necessary for space research. Computers speeded up the development of the transportation system because they helped in the progress of hundreds of subsystems, such as communication, engineering, and medicine. Computers were used in space transportation primarily in many ways.

They were useful in simulating the chain of events if a certain situation occurred. This eliminated much of the trial and error that marked the development of the first airplanes. As a result, the hardware was correctly designed, different systems functioned together and alternate plans were outlined in case something went wrong during the flight. Many simulation programs were also available as the flights took place. This not only saved time and money, but very likely saved several human lives.

Computers were used in developing new materials and processes. Private industries from all over the country had contracts to develop the components to be used in the systems. In some cases, new materials with certain characteristics were required, and new processes had to be devised to produce them. In other cases, the materials existed, but they had to be integrated with other materials into a system. For example, the combining of plastics and metals into the same system presents problems, because at extreme temperatures of heat and cold they can have different expansion and contraction rates, causing them to break apart from each other. Also, some materials can withstand more vibration than others. Many hundreds of materials were used, and they all had to function together without failure.

Computers were important to an efficient communication system that could transmit both sound and images over many thousands of miles (km). Computers not only helped in development, they also were a part of the system. Voice and visual communications are transmitted through space as radio waves and are converted back to the original form so they can be understood at the receiving end. Computers are used in both the transmission and conversion processes.

Computers functioned as the control system. They were used in the guidance systems, which allowed the space vehicles to be controlled from the earth. They were also used in the tracking systems, which allowed us to know exactly where the spacecraft was at all times. This control also provided a very important element, a series of backup systems in case a primary system might fail. The vehicle could be controlled either by those in the spacecraft or those back on earth, with alternate systems available if needed.

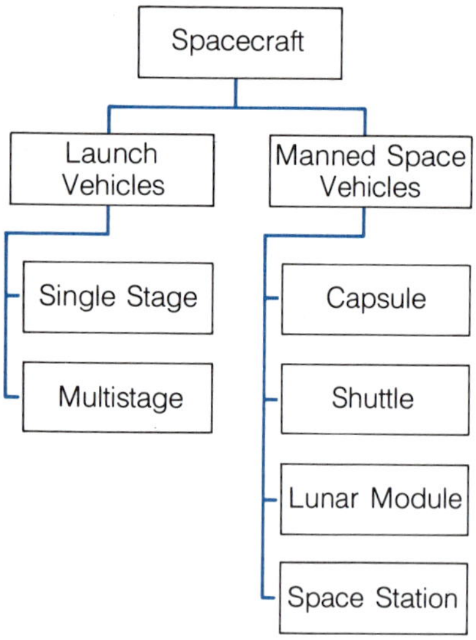

Launch vehicles are simply power generating systems, while manned space vehicles usually have many systems for different purposes.

This ancient Chinese woodcut depicts a flying chariot. The experiments of the Chinese with rockets and flight were the beginnings of our modern space travel.

Finally, computers were used in developing and monitoring life support systems. Because of the alien environment, a controlled artificial environment equipped with proper temperature, oxygen, water, and pressure levels had to be maintained at all times. This was necessary within the craft, as well as within the space suits. Medical data also had to be monitored to ensure good health.

Without the aid of computers, space travel would be impossible. They still play a major role today in space research and contribute to our understanding of the universe.

Spacecraft may be placed into two basic categories: launch vehicles and space vehicles. Each has a particular design and a very specific use in space travel.

Launch Vehicles

A launch vehicle is a rocket containing the fuel and engines to boost the payload into space. The payload may be either a manned or unmanned space vehicle. It sits at the front end of the launch vehicle, constituting the nose section. The first launch vehicles were called single stage because they had only one unit to boost the space vehicle into orbit. The Chinese used launch vehicles as early as the 1200s for warfare and fireworks displays. The "father of the modern rocket" was Robert H. Goddard, an American scientist who launched the first successful liquid propellant rocket in 1926. The use of the V-2 rocket by Germany during World War II was the first full-scale use of rockets for a specific purpose. After the war, the United States conducted high-altitude research with captured rockets. Single-stage rockets were also used on the first aircraft to reach the outer atmosphere. They were not used, however, to extend the aircraft into space, because at that time our knowledge and equipment capabilities would not allow it.

The single-stage rocket was not powerful enough to launch a payload into space. Too much fuel was consumed before the proper height and speed was attained. The multistage rocket was the answer. It consists of rockets attached end-to-end and designed to burn in stages. As the first stage burns out, the first rocket drops off and the second stage ignites. As it burns out, the third stage ignites. As each stage drops off, the vehicle becomes lighter, and the higher it goes, the less resistance it meets because the gravitational effect decreases. As a result, the speed steadily increases, taking the

This picture, taken in 1925, shows Dr. Robert Goddard with one of his advanced rockets. His contribution to the Space Age would be comparable to the Wright brothers' contribution to the Air Age.

vehicle into space. The spent rockets fall back toward earth, burning up because of the heat generated by friction with the atmosphere.

The rocket is a very powerful engine. It produces more energy than any engine known today. Its great power comes from the fact that it is able to burn fuel at an extremely rapid rate. The Saturn V rocket, which carried the first men to the moon, burned over 560,000 gallons (2,120,000 L) of liquid fuel in the first two-and-a-half minutes of flight.

Although jet engines work on the same principle of thrust (for every action, there is an opposite reaction), one major factor makes jet engines different from rockets. Combustion (burning) requires oxygen. A jet engine uses large fans, called turbines, to suck air in through a large hole in the front. The oxygen in the air is combined with fuel and burned inside the engine. The exhaust is then thrust out the rear under pressure produced by the combustion. Because space has no oxygen, jet planes cannot be taken into space.

Rocket engines, on the other hand, carry fuel and their own oxidizer (usually liquid oxygen), which provides onboard oxygen for combustion. Thus, they are able to function in an airless environment. The combination of fuel and oxidizer is called the *propellant.* A major portion of the nearly thirty-six-story-tall Saturn V was used for propellant storage.

Today, launch vehicles are used on the space shuttle as booster rockets. They also are used for unmanned launches of orbiting satellites and other deep space probes.

Space Vehicles

There are several types of space vehicles, each has a specific use, and may be manned or unmanned. The capsule was a bell-shaped vehicle which circled the earth while in orbit. It did not require powerful rockets. The capsule was able to orbit the earth, because it was under just enough gravitational pull from the earth to keep it steady. If there was no gravity, the capsule would speed off into space in a straight line. Gravity pulled on the craft, keeping it in a curvature around the earth. If it got too low and the gravity increased, it started to curve inward toward the earth. Small rockets on the craft maintained the right altitude. Because there was no atmosphere to cause resistance, only a small burst of energy from a rocket would make corrections. The small rockets turned the craft to any

The lunar module carried the astronauts to the moon's surface and up again to the command module, which remained in orbit. The rover still remains on the moon today.

angle desired. On the return to earth, rockets slowed down the craft, letting it fall into the increased gravitational pull at the correct speed. A heat shield protected the craft from burning upon reentry. At about 40,000 feet (12,192 m) parachutes opened, allowing the capsule to gently return to the earth's surface.

The lunar module was a special vehicle designed to allow astronauts to land on the moon's surface and return again to the command module, which remained in lunar orbit. Although built for landing, the module had no wings or parachutes, because the moon has no air to allow them to function. Instead, rockets were used to slow the descent to the surface. A second lunar vehicle was called the *rover*. This was a small battery-powered, four-wheeled vehicle with test instruments. The astronauts rode on the moon's surface for more than seventeen miles (27 km) on this vehicle.

Space stations are designed to orbit the earth continuously, allowing astronauts to travel back and forth to the vehicle. While in the station, they are able to conduct various kinds of research. The unit contains all the necessary requirements for sustaining everyday living. This vehicle has not been used yet but is under development. The space shuttle, a reusable manned spacecraft is now being used. Already astronauts have used it to travel into space and return. The shuttle takes off like

Even though the Shuttle Orbiter looks much like an aircraft, it is much more. It is a very sophisticated, high-technology, research laboratory. The scientific equipment on board can change in response to various research needs.

a rocket, lands like a plane, and is designed to replace both the launch vehicle and capsule of earlier manned space flight.

In the United States, development of the space shuttle was begun in 1972. The intent of the shuttle is to carry people and equipment into orbit and return to earth. The shuttles are used to launch, repair, and recover unmanned satellites and space probes. Also, the shuttle enables crews to carry supplies from earth to space. The supplies will be used to build and maintain space stations in earth orbit. The first shuttle took place in April 1981. Twenty-four successful flights followed, until disaster struck on 28 January 1986. On this date, the space shuttle *Challenger* exploded 73 seconds after take-off, killing the seven astronauts on board.

As a result of the accident, space travel was suspended for a period of time to allow more complete planning and the establishment of new procedures. This time was needed to allow the human element to "catch up" with the rapidly advancing technology, so that correct decisions could be made. As space exploration continues, there will be a greater control over the technology.

TRANSPORTATION PROBLEMS

Even as technology provides new and better means of transportation, problems arise. Some result from circumstances, but many result from poor planning and a failure to analyze needs.

The following illustrates a few problems that need to be addressed in the immediate future.

Some of the environmental pollution can be attributed to transportation systems. Burning fossil fuels in engines releases harmful hydrocarbons into the air. This may cause unclean air, acid rains which may kill plants, and changes in the atmosphere which may weaken our protection from the sun. There is also the possibility of large-scale environmental damage as a result of accidents. An oil tanker spill can spread oil over hundreds of miles (km) of water.

Space is a scarcity in large cities, and much of the area of any large metropolitan city is taken up with transportation networks. Streets, roads, interstate highways, beltways, and interchanges take up thousands of acres of valuable land. At one time, problems of traffic congestion were solved by building more roads. However, this practice cannot continue indefinitely.

There are greater distances between work and residence. During the last few years there has been a movement of population to the suburbs and even further out into the country. Because most of the jobs remained in the metropolitan areas, it has become necessary to drive longer distances to and from work. This has created tremendous traffic problems in many areas.

The population has increased at the same time population mobility has risen. The solution has been to sell more cars and add more flights to the airline schedule. We are rapidly approaching the saturation point in both of these instances.

Transportation costs are rapidly increasing. The high cost of energy and the increasing costs for operation demand that new approaches be considered. Efficiency has to be a key consideration.

Many of the network surfaces are deteriorating because of age and use. Many railroad tracks have deteriorated to a dangerous point, some bridges built during the early part of this century are called upon to support a volume of traffic not foreseen at the time of their construction, and many miles (km) of highway have not been kept in repair.

There have been many instances where decisions concerning design and application of various systems systems have proved to be unsafe. For example, at a time when nearly fifty thousand people die each year in car accidents, new cars

A master plan is necessary for improving our transportation system. A coordinated system of networks, including roads, rails, aircraft runways, and rapid transit guideways, will allow each to function at its potential. This illustration of an interface between various modes of transportation includes a hovercraft, moving walkways, people movers, and cars.

are being designed to travel almost 100 miles per hour (161 km/h). Also, the control tower equipment of many busy airports needs updating to make the increased flight traffic safer. The equipment is available, but it is too costly for many airports to install.

There has been no master plan for solving transportation problems. Because of the rapid rate of technological innovation and application, these problems and others were upon us before we realized it. Now that the problems are recognized, a plan is needed to help reconcile them. Rapid transit systems, new alternative systems, and integration of traditional systems are probably all needed to solve the problem, but there has to be a plan for them to work together efficiently.

TRANSPORTATION AND THE FUTURE

It is interesting to speculate what new methods of transporting people and goods might be used fifty to one hundred years from now. The vehicles will most certainly look different from what we now have. One of the major developments which allowed transportation vehicles to evolve during this century was the gasoline engine. It also has determined, to a certain

The return of the propeller is being studied as a partial solution to energy shortages. The newly designed "propfan" aircraft would use a propeller with eight to ten blades and, with the change in body design, could carry 120 passengers on medium-length routes about 500 miles (800 km) per hour.

extent, their design and use. Because fossil fuels were relatively inexpensive, there was little interest in exploring new sources of power. Most of the research was directed toward refining and improving the internal combustion engine.

Today, the higher cost of petroleum and the pollution caused by more and more vehicles require that new sources of power be explored. Perhaps some of the most fruitful research concerns the development of a more efficient electric-powered vehicle or of a gas turbine engine appropriate for small vehicles. But even if specific engineering problems are overcome, these two methods still have possible long-term drawbacks. The electricity needed to charge the batteries of an electric car still has to be generated. The problem of pollution may be solved, but as the supply of energy becomes more critical, the generation of electricity by conventional means may not be able to meet the demand. The turbine engine is most certainly more efficient than electric power, but it still contributes to some pollution and it operates on fossil fuels.

Most vehicle design today incorporates lighter materials into a smaller vehicle. The total reduction in weight produces better gas mileage, which cuts down on overall fuel consumption. This is, however, only a short-term solution to our increasing problems of pollution and energy shortages. Two

The hydrofoil boat has many energy-saving elements. This boat of the future will carry 50 cars and 400 passengers much faster and more efficiently than they could travel by land.

issues are vital in solving our transportation problems: energy and planning.

Energy

Developing new energy sources is critical. A number of corporations are seriously exploring long-term solutions, such as nuclear and solar energy. Either might be used directly or indirectly in powering vehicles. Indirect application might involve the generation of electricity, while direct application might involve an engine powered by a system of very efficient solar cells mounted on the car. With more research, either energy source might be made to operate efficiently; both are in abundant supply. The logical answer to our problems seems to be concerted research directed toward both short-term and long-term solutions. Research should not be limited to only a few possibilities. A number of somewhat less efficient methods of energy generation may have to be used as intermediate measures. Intermediate steps in development appear to be wise, from the standpoint of both engineering and public acceptance.

Planning

Long-range planning for use and control of transportation subsystems is essential. Planning enables the integration of the various subsystems now available, such as cars, rapid transit systems, and air travel systems. In the past, each subsystem had been considered as separate from the others, cutting down on the total efficiency. In planning for the future, each subsystem should be viewed as a component of the total transportation system.

Analysis of the various subsystems reveals basic differences in movement, speed, and weight-carrying capacity. A fixed-track vehicle, such as a subway, monorail, or dual-rail train, is restricted. It can move only forward or backward. It may be called a *unidirectional* vehicle. A car, on the other hand, is able to turn left and right. It might be called a *bidirectional* vehicle. An airplane is able to move vertically from the surface of the earth, making it a *multidirectional* vehicle. Using the same criterion, a submarine is also a multidirectional vehicle, because it is able to travel below the surface of the water. All transportation vehicles may be placed in one of these three categories. The bidirectional vehicles cause the major problems, because they are smaller, more prevalent, and harder to control.

Safety within each subsystem depends on separating its vehicles, which in turn depends upon the speed and range of movement. The separation of moving cars on a highway may be 50 feet (15 m), front and back, with side separation of 10 feet

This proposed transportation system is an automated highway. Cars are carried on pallets which are propelled by magnetic suspension. This makes the car a unidirectional vehicle which can interface more easily with other subsystems.

One possible solution to subsystem intergration is shown here, where passengers can board either the aircraft or the rapid transit system directly. This would eliminate the congestion experienced now at air terminals where nearly every passenger arrives in a car.

(3 m), while an aircraft may have 100 miles (161 km), front and back, with 1000 feet (305 m) in height separation. Each subsystem has a critical point of separation for safety. As more and more vehicles crowd their networks, the point of separation becomes more important.

The integration of subsystems requires an analysis of interfaces, those intersecting points of loading and distribution centers where passengers or freight change from one subsystem to another. These are locations where congestion usually becomes a problem, because the problem often arises where there is an interface of bidirectional vehicles and either unidirectional or multidirectional vehicles (such as parking problems at airports). A new rapid transit rail system is not efficient, even if it handles 10,000 people a day, if there is no provision for moving people efficiently in different directions at each end of the line. That factor of movement is important to users. This is the single most important reason for the popularity of the automobile; people are able to travel when they wish, where they please, and make as many stops as necessary. Any new plan has to be flexible enough to fit the needs of the people, or it will not be accepted.

Control of transportation vehicles does not mean the control by one person or a group of persons. It simply means that flow, speed, and volume have to be balanced in order to retain safety and efficiency. This requires a "total system approach."

SUMMARY QUESTIONS

1. Why did the development of the automobile have such a great impact on society?
2. What is meant by nonvehicular transportation?
3. Identify the names of the following abbreviations: AGT, SLT, GRT, PRT, ATV, VTOL, SST, and HST.
4. Name two reasons why space transportation is important to our future.
5. Identify four major transportation problems we are facing today.
6. Explain a hypothetical solution which might help solve one of the many transportation problems we may be facing in the future.

CHAPTER 8

The Communication System

In this chapter you will learn:

1. How the communication system is organized.
2. How the subsystems of visual and audio communication evolved.
3. How several of the communication subsystems function.
4. What problems are facing the communications systems as new technology develops.
5. How changing technology will affect methods of communication in the future.

ANALYZING THE COMMUNICATION SYSTEM

As we stated, technology serves to satisfy our needs. In addition to the basic needs of food, shelter, warmth, and clothing, the need to communicate is extremely important in everyday living. Early humans felt this need even though the technological level of communication was low compared with today's standards.

Communication enables the diffusion of information, and for this reason is a critical element of every society. The more complex the society, the more important communication becomes. A spiralling effect can be clearly seen; as communication improves so does technology, and as technology advances so does communication. As a matter of fact, communication, technology, and society are so closely interrelated that it becomes difficult to study each separately.

The communication system is somewhat special, compared with other technological systems, because it has so many different dimensions. All technological systems are characterized by their particular procedures and hardware, but the communication system is, in addition, continuous, universal, and inclusive of physical, psychological, and sociological factors. These attributes make human communication different from that of other living organisms. Humans not only have the ability to communicate, but the ability to develop new forms of communication and choose the appropriate forms for any given situation. Even these powers of discernment, however, cannot prevent problems from sometimes arising. Despite the diffusion of technical knowledge and advances in hardware, a

We depend upon the daily news to keep us informed. It is carefully planned and presented to make optimum use of the available time, by using both live reports and taped segments.

communication system will only be effective to the degree to which it engages the physical, psychological, and sociological aspects of human nature.

The advancement of technical knowledge can allow the development of hardware, and at the same time ignore the needs of people who use it. The effectiveness of any communication system depends upon the ability of the communicators to use it. The physical, psychological, and sociological functions have to be considered when new methods and hardware are developed; if they are not, problems of ineffective communication can develop.

The communication system is basic to the other technological systems. Communication, in one form or another, guides all our daily activities. There are three primary purposes of communication. First, communication conveys past happenings. This record of the past allows the advancement of technology because it adds to our knowledge and frees us from the necessity of having to rediscover scientific principles and procedures. Second, communication relates current events. The technology of television and communication satellites allows us to watch events all over the world and also in space, as they occur. Electronic communication places us in direct contact with major events at the flip of a switch. Third, communication alerts us to possible future occurrences. Early warnings of impending tornadoes, floods, and hurricanes have helped

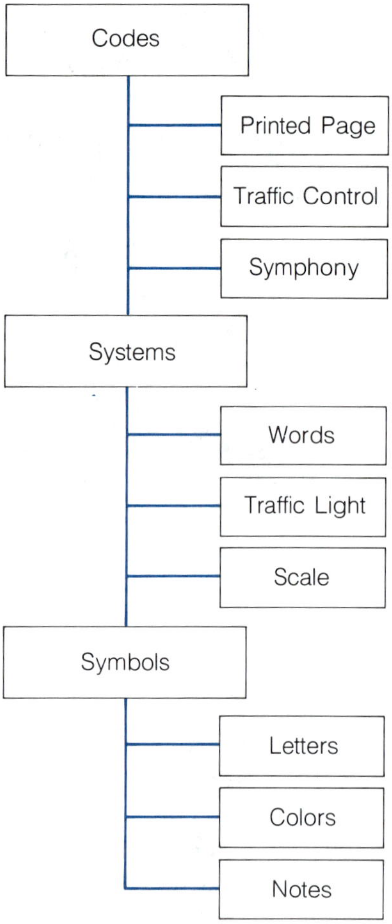

A code requires a system which places symbols together with a specific meaning.

save many lives. Dates for future events can be announced, as can last-minute changes in schedules and alternative programs. The communication systems can be analyzed in many ways. For our purposes, we will examine three models that will relate both the theory and methods of communication.

The first model analyzes communication as a set of codes that can be broken down into subsets of systems and further subsets of symbols. Communication is the transfer of thoughts and ideas. For such intangibles to be communicated, they must be coded in a form that allows transmission. Radio announcers, for example, provide words that convey meanings. We hear their sounds as words and understand their message. Radio announcers code their ideas and transmit them in a form that our sense of hearing enables us to pick up; our brain then decodes the transmission. The communication of ideas occurs as symbols and systems are combined into a code of common understanding.

The difference between a symbol and a system is largely one of complexity. A *symbol* can be considered an element that stands alone, such as the letter *h*, or a simple unit that conveys a single message, such as an arrow on a sign. A *system* is formed when symbols are joined with other symbols to create an organization or to convey a meaning of relative complexity. Individual letters, for example, are organized into a system of words, and flashing colors on a traffic light convey directions of when to stop, go, and turn. Flashing lights and strokes of black on a white page would mean nothing, however, if we could not bring our knowledge of these and other systems to bear in our interpretation of these recurring images. A combination of systems becomes *code* at the level where to see, to taste, to hear, to smell, or to touch is to understand. Codes have two major advantages: they are flexible, allowing for a combination of systems to provide different meanings, and they are capable of communicating highly complex ideas.

Ideas are communicated through our senses. Although our senses of hearing and sight are used the most in communicating, all five of our senses are used in various combinations with each other. The choice of communication methods, then, involves not only the hardware, but a basic decision as to which of the senses would be the most appropriate receiver. Technology today provides many choices of hardware, each appropriate for a specific situation.

Why are some types of communication effective and

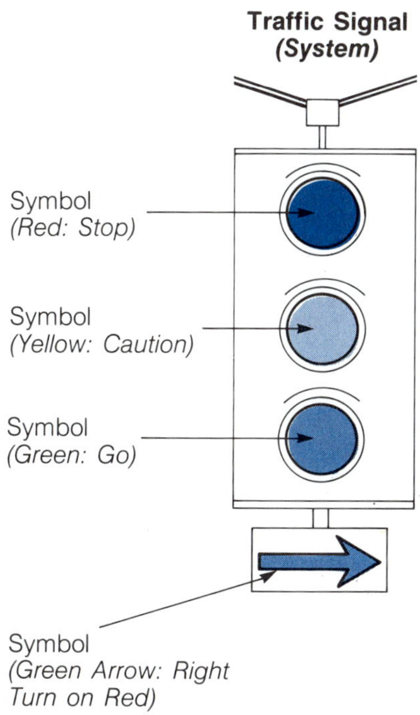

A traffic signal demonstrates an example of a system that combines the symbols of color and shape with on/off, sequence, and combination actions to constitute a code for traffic control.

others are not? The answer could be a combination of many reasons, as it relates to the process involved in transferring thoughts and ideas. This process, although complex in terms of our physiology, may be identified in a simple model.

The model shows the route of a typical communication. A principal posts a notice of a meeting on a crowded bulletin board. Although we commonly think of *noise* as sound, it refers in this case to anything that might interfere with the transmission, such as crowding by the other notices or a poorly lighted display area. *Feedback* tells the sender how effective the communication was. Once the loop is completed, the way is cleared for another cycle. The same system is appropriate for person-to-person, group-to-group, human-to-machine, and even machine-to-machine communications, although there may be changes in the terms, symbols, or codes.

Although we can communicate by touching, smelling, and tasting, most of the technology involved in communication can be broken into the two categories of visual or audio. Both contain the same three processes: reproduction, transmission, and storage and retrieval. Reproduction is especially important in a highly complex society where it is necessary to communicate with many people at one time. This requires multiple copies of the same information. Reproduction methods allow mass communication in the form of, for example, newspapers (visual) and stereo records and tapes (audio). Reproduction also allows editing of an original message to make it more effective. Transmission, as might be guessed, involves the sending and receiving of information. Storage and retrieval systems allow choice in timing communications. Microfilm, for example, stores printed materials, and video tapes store both pictures and sounds.

VISUAL COMMUNICATION

We will discuss the hardware of visual communication in terms of three categories: printing, photography, and electricity/electronics. Each area is different, but each relates one way or another to visual communication.

Printing

It is not known at what point early humans began using technology for communication. Obviously, hand signals and body language were used first, but the earliest evidence of technol-

This model does not attempt to illustrate the hundreds of physical and psychological variables which might occur in a highly complex communication cycle. Rather, it illustrates only the basic elements required if communication is to occur.

Communication Model

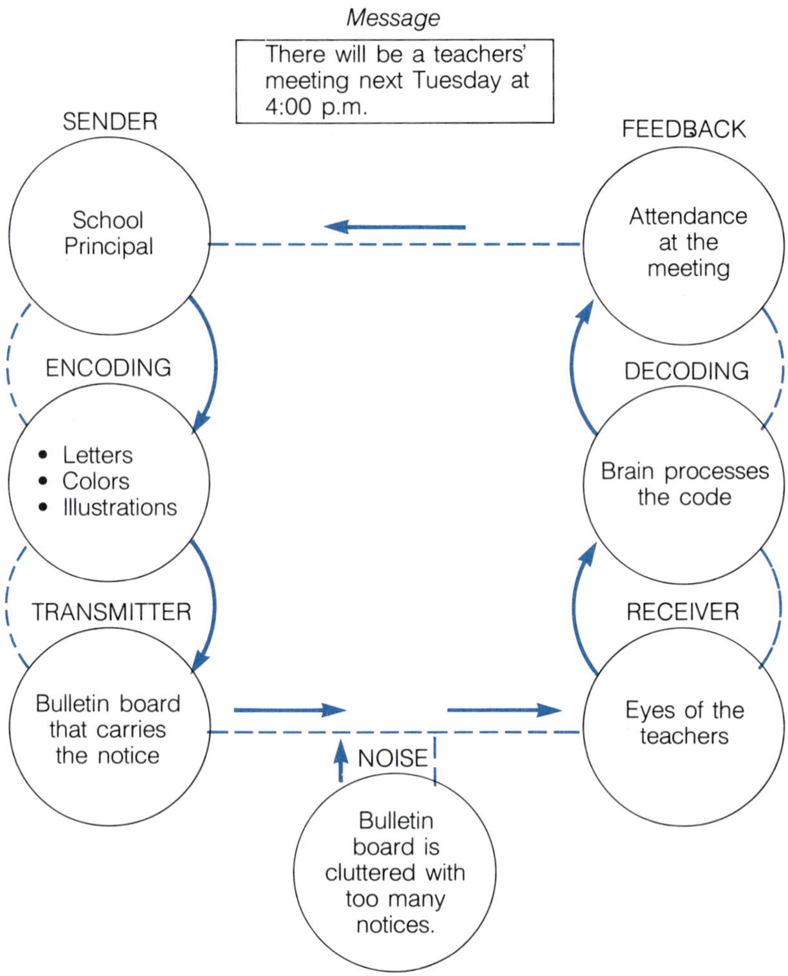

ogy used in visual communication would probably be cave drawings and carvings. Although the primary purpose of these drawings and carvings might have been art and/or religion, they were, in fact, the beginnings of our modern written language. Humans devised visual elements to record numbers such as notches in wood or knots tied in cords. In this way, past happenings could be related to others in the past or present.

As time passed, pictures were combined and placed in a sequence to convey a series of events, thus forming a picture story. Because detailed drawings took considerable time to

These are original cut stones with Egyptian hieroglyphics carved on them. Many symbols of birds and animals were used in this visual system of communication.

	Egyptians	Semites	Phoenicians	Greeks	Romans

Time Period	About 3000 B.C.	About 1500 B.C.	About 1000 B.C.	About 600 B.C.	About A.D. 100
Meaning	Ox	Aleph: Ox	Aleph: First letter of the alphabet	Alpha: First letter of the alphabet	Alpha: First letter of the alphabet

Time Period	About 3000 B.C.	About 1500 B.C.	About 1000 B.C.	About 600 B.C.	About A.D. 100
Meaning	House	Beth: House	Beth: Second letter of the alphabet	Beta: Second letter of the alphabet	Beta: Second letter of the alphabet

Although not all letters can be traced back to early picture writing, some do show a gradual evolution.

complete, pictures of specific objects became more abstract and symbolic. These developed into a true picture language using standard symbols. This system is called *pictography*. A pictography displays a sequence of pictorial ideas, which could be compared with our sentences. Once a society reached this stage, it achieved a true communication system that could easily be recorded and understood.

Early peoples from nearly all continents of the world used this form of writing. In some of the underdeveloped countries of the world today, pictography is still used.

The oldest forms of writing known are pictographic systems developed by the Sumerians (what is now Iraq, Syria, and Turkey) and the Egyptians (the area around the Nile River). These date back as far as five thousand years. Over a period of several hundred years since that time, a number of alphabets emerged where letters which had specific sounds took the place of pictures. The most commonly known, of course, is the Greek and Roman alphabets, which eventually evolved into the system of writing used today in most Western countries.

This engraving shows Gutenberg taking the first proof of a completed page—an event that changed the course of human history.

During the evolution of the various alphabets, the ability to read and write was limited to only a few persons, and all reading matter was original, that is, one of a kind. For nearly thirty-five hundred years writing was done by hand. A method of printing did come into being in the early Middle Ages, but it was slow and laborious and permitted only reproducing. The letters of a page were cut into a block of wood, the block was inked, and paper pressed to it. Then, about 1440, Johann Gutenberg, a German printer, developed a method of type printing that allowed multiple copies to be made in a very short time. Rather than carving a whole page into a block, Gutenberg used many small blocks made of metal, each with a single letter. This system of moveable type allowed each new page to be composed by rearranging the blocks. For the first time in history, exact duplicate copies were possible. Thus, printing became the first true mass production industry.

As with all new significant technologies, this development changed the entire course of human civilization. For the first time, the mass of people were able to read and write. Knowledge was no longer a monopoly of the ruling class and religious leaders. In the past, knowledge meant power and at times was used as a weapon to retain control of authority. When this was no longer possible, many changes took place in the religious, social, and political structures of societies.

The first book Gutenberg printed was the Bible. It contained 290 signs, 47 capital letters, and 243 small letters and punctuation marks. This illustration shows the type style used. The letters were created as if they were written by hand with a quill pen.

Dictionaries became necessary to aid in reading and writing. Dictionaries served to standardize languages and alphabets so that everyone would be using the same system. At the same time, there also was a negative effect produced by the dictionaries. The evolution of changes in alphabets virtually stopped, which prevented improvements in usage and comprehension. Some of the flaws which needed improvement are still prevalent in our systems today. Also, with the spread of literacy and learning from country to country, technologies of one part of the world began appearing in other parts. This, coupled with higher general knowledge, produced an upsurge in the level of technology in nearly all aspects of life.

Fof almost three and a half centuries following the development of moveable type, printing methods changed very little. During the 1800s, however, printing joined the Industrial Revolution. Gutenberg's press could print about three hundred sheets a day. In 1811, Friedrich Koening, a German, invented a steam-driven press that used a cylinder. It could print eleven hundred sheets per hour. In 1846, Richard Hoe, an American, invented the rotary press with two revolving cylinders. His first press could print eight thousand sheets per hour, and the number soon jumped to twenty thousand per hour.

Until 1884 the type was still set by hand, a slow process compared with printing time. In that year Ottmar Mergentha-

The linotype machine was a big, noisy machine. The copy was typed first and given to the operator to retype, making a line of raised metal letters.

ler, a printer living in the United States, invented the Linotype. This machine had a keyboard much like a typewriter, and it cast molten metal into a full line of type, as opposed to the single-letter pieces used previously. This machine was used until the middle of this century.

The increase in printing speed illustrates dramatically the effects of the Industrial Revolution on visual communication. Other developments in paper, inks, and photography have also contributed to the growth of the printing industry. In the United States today, printing ranks as the eighth largest industry and is one of the fastest growing.

Photography

It would be impractical to discuss the names and contributions of all the individuals involved in the development of photography. Many people contributed in many ways, and the fields of chemistry, optics, paper, journalism, research, and even recreational photography are represented among photography's pioneers. One reason for the large number of people was that photography did not require a specific industry for research and development, as did the steam engine, for example. Most of the early research was conducted by people who had some knowledge of chemistry and physics and considered this interest a hobby.

The first camera, called the *camera obscura* (dark chamber), appeared in the 1500s. It was a darkened box with a small hole to let in a pinpoint of light. The image produced on the

This illustration shows what the first camera was like. The picture had to be drawn by a person inside the darkened room.

Mathew Brady was a famous American photographer who recorded the Civil War on film. He used the wet process, so he had to carry his darkroom and equipment with him in wagons. This is a photograph taken during the Civil War, using the wet process.

inside wall of the box was large enough to hold a person. A picture was then drawn by tracing the image on the wall. During the 1600s a lens was placed in the hole to make a sharper image, and in the 1700s, the first permanent image was made.

It was discovered that silver salts are sensitive to light, causing the material to become dark. Joseph Nicéphore Niépce, a French physicist, coated a metal plate with the substance and then used an engraving process to "fix" the image, to make it permanent. He has been credited with producing the first photographs made with a camera. In the early 1800s, a British scientist produced a paper coated with salt and silver nitrate that could be used as a negative. This could then be exposed to light with a second paper to make a positive print.

During the last half of the nineteenth century, there were major advances in the chemistry and in the optical quality of lenses. The exposure times required to take pictures went from about ten minutes (which limited most subjects to landscapes) to about one-twenty-fifth of a second. The development of plates went from a wet process, which required the picture to be taken on a wet plate, to a dry process. Until the dry-plate process was discovered in 1871 by a British physician, the photographer had to carry the complete darkroom wherever pictures were to be taken, because the development had to be done immediately. This took a large amount of

George Eastman's first camera held one hundred pictures and sold for twenty-five dollars. The camera and exposed film were returned to the company where the film was developed, prints made, and the camera reloaded for ten dollars.

equipment and a specially equipped wagon to serve as the darkroom.

In 1888, George Eastman, an American dry-plate manufacturer, produced the first Kodak camera. This revolutionized the field of photography, making it a major communication tool. The new camera was very easy to use, so anyone could take photographs. It held a roll of film which was not removed for processing. Instead, the entire camera was sent back to the company for developing, then returned, loaded with new film, to the owner.

In the years that followed, cameras became smaller, electronic flashbulbs appeared, and more light-sensitive materials were developed for faster exposures (faster exposure meant that pictures of movement would not blur). In 1947, Edwin Land, an American, developed the Polaroid camera, which not only took the picture but developed it immediately. Thus, in about seventy-five years, photography went from a technology requiring special skills and a completely equipped darkroom, to a technology that requires no special skills and no darkroom at all for a successful picture.

During the 1800s, newspaper editors recognized the tremendous potential of photography to bring people closer to current events and news, allowing them to see faraway places which they would otherwise never see. Photography could not only capture a scene, it could capture character and emotion. Also, advertisers saw its potential as a tool to inform newspaper readers of available products. Photography became a powerful "equalizer," because people of all classes in all parts of the world could visually experience things that were not a part of their own lives.

This Polaroid camera contains a computer which automatically focuses the lens and sets the amount of light needed. A sonar system (high frequency sound waves) is used to determine the distance from camera to subject.

Electricity/Electronics

Compared with printing and photography, electricity and electronics are relatively new technologies. The role of electricity/electronics in direct visual communication began with Thomas Edison and his invention of the light bulb in his Menlo Park, New Jersey, laboratory in 1879. This discovery allowed small bulbs to be wired to a switch making light available at any time.

The idea of "on" and "off" is basic to the concept of communication by light. The variables of time, color, shape, and brightness must come into play because there are only two conditions of light: either it is on or it is off. The variables, in combinations with the "on/off" capabilities, provide a wide range of symbol possibilities.

Electrical flow is created when electrons are able to move through a complete circuit. A source is needed, such as a battery or electricity from an outlet, to get them moving and keep them moving. When light is generated, a situation exists which is much like a dam placed across a stream of water; some water continues to flow, but there is a buildup of water behind the dam. When a wire is too small to allow the electrons to flow freely, they build up, causing heat to occur. This heat energy produces light.

Electronics is a branch of the science of electricity. It deals primarily with the control of electrons, which determines variation in the direction and amount of flow. Combinations of these conditions make up small electronic signals. These signals may constitute images, such as are seen on a television screen or in the display panel of a pocket calculator.

Edison's bulb is referred to as an *incandescent lamp,* which means "glowing with heat." Oxygen is one of the elements necessary for burning. The reason the filament is not consumed by heat is because oxygen inside the bulb is replaced by a mixture of gases instead of air. The same basic

Although Thomas Edison is famous for the light bulb, he also conducted many other experiments with light and optics. This 1893 photograph shows him working on microphotography.

principle is still used today, although improvements have been made in materials and manufacturing processes which make bulbs more efficient.

There was no major change in bulbs until the 1920s, when the fluorescent lamp was developed. This light source, called a *gaseous-discharge lamp,* is a long tube containing mercury vapor. Electricity flowing through the tube causes the mercury to give off ultraviolet energy. This in turn causes an interior coating of phosphor to glow, lighting the entire tube.

Neon lamps, such as those used for advertising, contain a mixture of neon gas. This gas produces red light waves. When mixed with other gases, different colors can be achieved. These gases glow when they come in contact with electricity, so no coating is needed inside the bulb.

Another major development of light used for visual communication took place in the 1960s. This one is called *LED,* (light-emitting diodes). These are small red-colored lights which will burn almost indefinitely. They use direct current (where electricity only flows in one direction) such as produced by batteries, and they use very little electricity. They can be found in such items as digital watches, electronic calculators, and many other products which display small numbers and words.

Systems and codes of lights are all around us. Our traffic signals use both color and on/off switches, as do emergency vehicles, such as ambulances and police cars. Airports use warning lights around runways, and advertisers grab attention through large, colored, flashing signs. Watercraft communicate direction at night or in heavy fog by means of colored lights. A red light, which is always on the right side, indicates movement to the right; a green light, always on the left side, shows movement to the left; a red and a green light indicates movement toward the viewer.

The Hardware and Processes of Visual Communication

As previously mentioned in this chapter, visual as well as audio communication may be broken into three divisions: reproduction, transmission, and storage and retrieval.

Reproduction. Other than visual items produced for an artistic purpose, nearly all visual communication involves reproduction. It may be a repeating message, such as traffic lights and electronic calculators, or it may be multiple copies, such

This intersection has thirteen sets of computer-controlled traffic signals. The coding of lights changes constantly according to the traffic volume from any given direction, providing a smooth traffic flow.

as newspapers and books. Both have to be designed (coded) to convey the appropriate message.

In the electricity/electronics field, the coding is called *programming*. Most of these systems are able to store information which will be used at the right time in correct sequence. In some cases, the programmer has to know the correct language (or method of communicating with the hardware) in order to program it. In other cases, the process has been simplified, so a person has to know only the step-by-step instructions provided by the manufacturer.

Clip art is preprinted artwork which can be cut, trimmed and glued into the composition. Preprinted type may also be used in this way, or it may be in the form of transfer letters as seen here. When pressure is applied, the thin plastic letters are transferred from the clear plastic sheet to the paper.

The printing industry calls the coding process *composition*. The type style, color, pictures, and text are organized into a complete unit, so the message will be easily read and comprehended.

There are five basic methods of generating an image for printing composition. First, artwork involves such elements as drawings, clip art, and preprinted type. The use of this method probably allows the most individual creativity, but it also is the most time consuming.

Second, the strike-on method involves the use of typewriters and related kinds of machines, such as computers, that use a keyboard and letters which have interchangeable type styles and sizes. The first typing usually is done with no concern for margins, errors, or breaking words at the end of lines. As the manuscript is typed, the information is stored in memory or on a magnetic tape. When typing is completed, the machine will automatically retype the manuscript from the memory, using automatic spacing to correctly justify (straight vertical lines on each side of the column of type) right and left margins. These machines are usually called automatic typewriters.

Third, photocomposition uses photographic negatives containing the various type styles. Light-sensitive paper or film is exposed through the negatives and developed in the ma-

This complete system is composed of a keyboard and screen, a controller, a disk storage unit, and a type setter. The final copy is automatically typed out on paper with preprogrammed type style, size, margins, and spacing.

This offset press produces high-quality color printing. The stacked sheets under the worker's arms are fed into the press and through each of the four roller assemblies. Each one prints a different color on top of the one previously printed. The end result is full-color pictures produced at a rate of several thousand per hour.

chine. Multiple lines with interchangeable type styles may be used in a variety of ways. This process also uses a keyboard much like a typewriter.

Fourth, word processing came about as a corollary to data processing where numbers are processed. It consists of a keyboard, a computer memory system, and a CRT (cathode-ray tube). The CRT is really a television screen connected to a printer which prints what is typed. A variety of sizes and type styles are available, and the machine has the capability of moving words and groups of images any place on the screen. The typing is done much like the automatic typewriter; that is, there is no concern for margins, errors, or word breaks. The manuscript is stored in the electronic memory and reviewed on the screen. It also provides automatic justified right and left margins to the width of column desired. Any errors can be corrected word-by-word without any other typing being required. When all errors are corrected, it is possible for the entire manuscript to be automatically retyped faster than one can read it. Even then, the format and size of the entire manuscript can be changed. Hard copies (a complete manuscript on paper) can be easily obtained directly from the system at any time.

Fifth, continuous-tone photography consists of the traditional method of taking pictures with a camera. Continuous

Each time an image is transferred to another surface it is reversed. The offset process of printing introduces an extra transfer step, causing two reversals. The final image then looks like the original.

Offset Printing

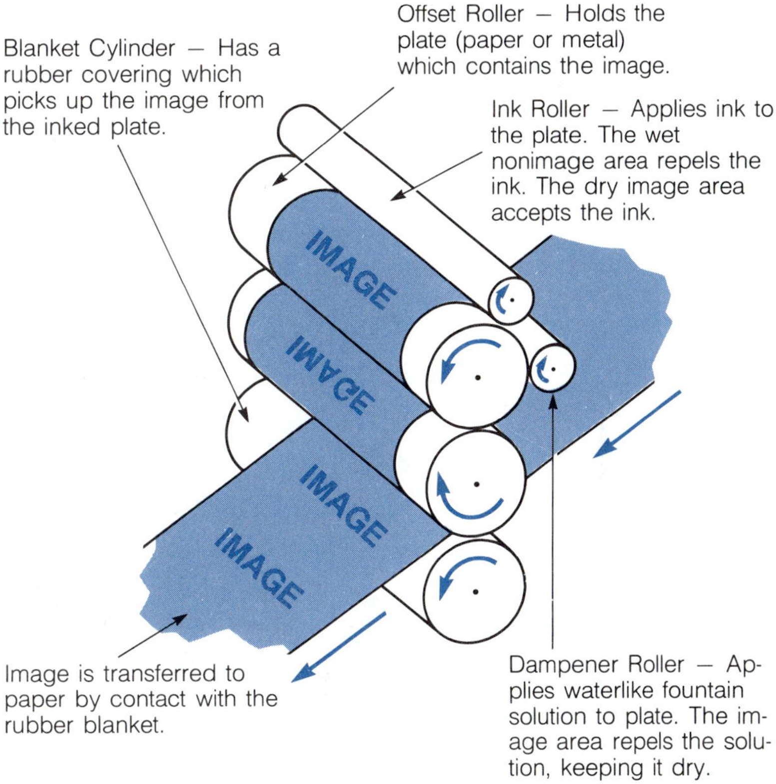

tone means there are varying degrees of gray tone ranging from black to white, providing shaded and light areas. The same is true with color photographs. Many things can be done to the picture to improve it, such as cropping unwanted portions (cutting the picture down), enlarging, and changing the tones to make them lighter or darker.

The printing process may be one of several types, each having particular advantages according to what is being printed. Probably the most common method today for printing on paper is the offset process. The press has a series of cylinders which the paper passes through. The offset roller has a plate which picks up the ink from the inking rollers. It then transfers the image to a blanket cylinder which comes in contact with the paper. The inked plate never comes in contact with the paper, thus the term *offset*. The primary advantage of offset printing is that it is a highspeed process. Most newspapers are printed using a web offset press. The term *web* refers to presses which use huge rolls of paper, rather than sheets. The rolls of paper run continuously through the press and are cut after printing.

A second printing process, which is used extensively for nonpaper products, is screen printing. This process, dating back many hundreds of years to Oriental cultures, is accomplished with a stencil. A fine-mesh fabric of plastic or metal is stretched on a frame and coated with a photosensitive solution. It is then exposed to light through a positive film (reverse of a negative), much like exposing and developing a photograph. When it is exposed, light hardens the material on the screen where it has passed through the clear areas of the film. Where it has not, the material remains soft and can be washed out. The result is a stencil. Ink or paint is forced through the screen during the printing process by passing a rubber squeegee over it with a slight pressure, producing the printed image. The primary advantages of this method are that most any shaped object may be printed, a very heavy coat of paint may be used if necessary, and a variation of inks and paints are available. Examples of some screened products would be cold drink bottles, T-shirts, clock dials and gauges, draperies, and trade names on household appliances.

Although not a high-speed, long-run printing process, copy machines and duplicators play an important role in our

This worker is operating a screen printing press. The screen automatically comes down and the squeegee passes over it forcing ink through the stenciled openings to the paper. When the screen raises, the sheet is removed and replaced with the next one to be printed.

everyday lives. Education, business, industry and others use duplicators and copiers daily when a large number of copies are not necessary. There are four basic types of processes used in duplication: Spirit duplication uses a desk top machine and a spirit master which can be typed or written on to create the image to be duplicated. The use of a waxy, dye-impregnated sheet creates the image much like using carbon paper. An alcohol-based developer dissolves a small amount of the image and transfers it to the paper as it passes through the machine. The spirit duplicator gets its name from the alcohol in the developing fluid. It is a relatively low-quality printing process, and it produces up to about three hundred red, blue, or green copies, depending upon the color of the dye in the master.

Offset duplication is much the same as larger presses used in printing books and other types of mass communication. The printing plate has a coating which prevents a grease-based ink from sticking to it when it is wet. The image may be typed, drawn, or made with a photographic process. It allows the ink to stick to it. When the ink and water are both applied with rollers, the ink sticks to the image and not to the background. The image is first transferred to the rubber blanket and then transferred to the paper. It can produce several thousand copies, depending upon the type of printing plate used. The color depends upon the color of ink.

Stencil duplication, sometimes called mimeograph, uses a paper printing plate which is a stencil cut with a special setting on the typewriter. Pictures may also be cut into the stencil using an electronic scanner, which is a photocopying process. While printing, the ink is forced through the stencil to create an image. The quality is good, and several hundred copies may be made from one stencil. Black is most commonly used, but other colors of ink are available.

Electrostatic copying is the principle behind several copying methods on the market. Most use static electricity, heat, and a photocopying process. The term *xerography* is used, which means "dry printing." An electrostatic charge picks up a fine graphite powder where the image is located on the original, and heat melts the powder to form the same image on the copy paper. There have been many major advances in electrostatic copying in the last few years, including color copying and the simultaneous reproduction of both sides of the paper.

Photography in reproduction is extremely important,

1. Surface of a plate coated with a photoconductive metal is electrically charged as it passes under wires.

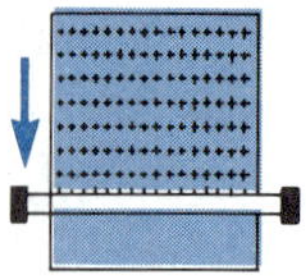

2. Plus marks represent positively charged plate.

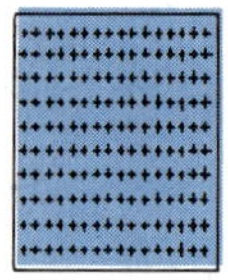

3. Original document is projected through a lens. Plus marks represent latent image retaining positive charge. Charge is drained in areas exposed to light.

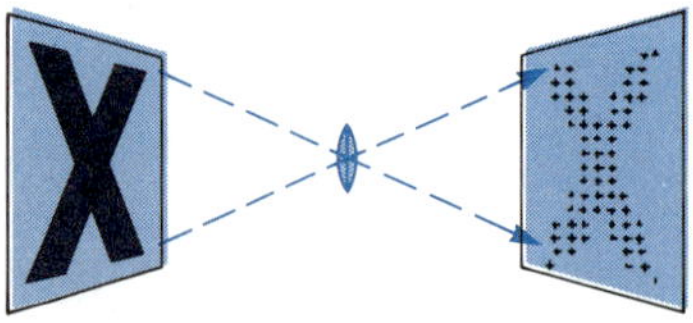

4. Negatively charged powder (toner or "dry ink") is applied to the latent image, which now becomes visible.

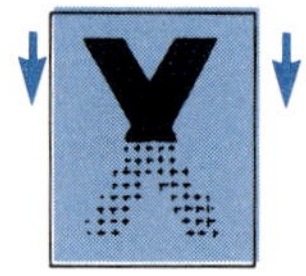

5. Paper (or other material) is placed over the plate and given a positive charge.

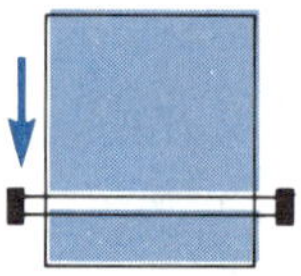

6. Positively charged paper attracts dry ink from plate, forming direct positive image.

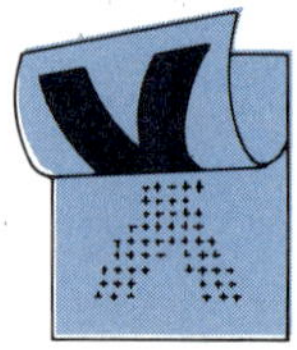

7. Image is fused into the surface of the paper or other material by heat for permanency.

The electrostatic process used in copiers can make a duplicate which is difficult to tell from the original.

because it provides a closer representation of an object than symbolic letters or numbers. The photograph begins as a continuous tone (various shades and tints) from an ordinary camera. This, however, cannot be printed on a printing press, because the press can only print one color at a time, just as a blue ballpoint pen (shades and tints of blue are not possible). As the paper passes through the press, either the single color is printed or no color is printed at all.

In order to print photographs, a negative has to be made by exposing light through a screen. The light reflected from the photograph passes through the small holes of the screen. The dark areas of the photograph do not reflect as much light while light areas reflect much more. As a result, different sizes of dots are produced, according to the amount of light passing through each small hole. The resulting negative is called a *half-tone negative.*

When a picture is printed on a press where a half-tone negative has been used, the various sizes of dots are printed. They are so close together that the human eye blends them. The reader sees a gradual variation of shades and tints, like a

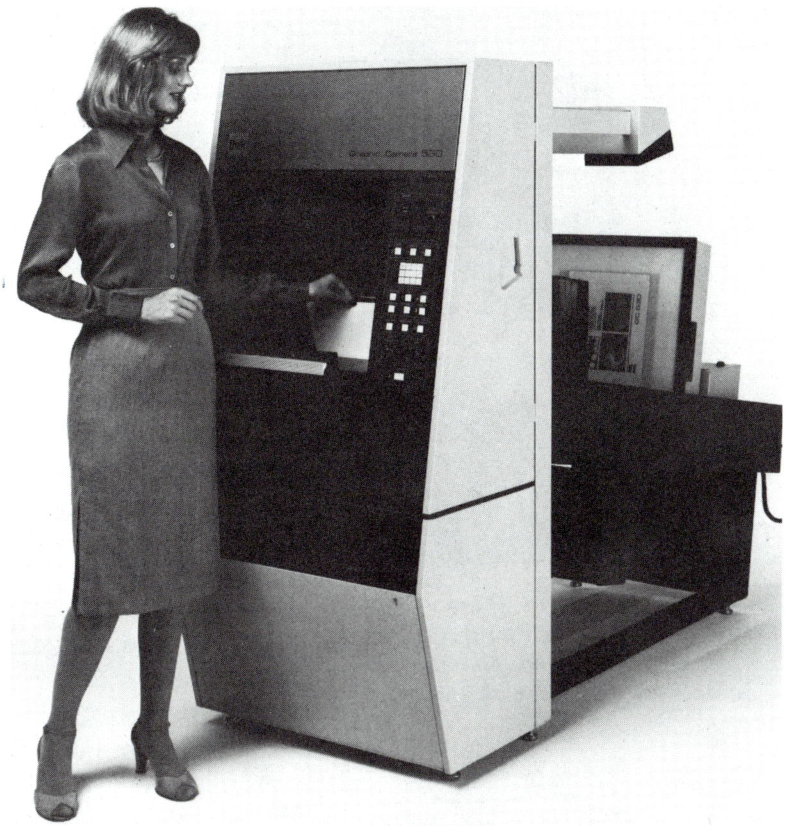

This process camera is a camera and darkroom combined. The original is placed on the copyboard and exposed. With the negative film and developing chemicals contained in the rear unit, there is no need for a darkroom.

continuous tone photograph. In reality, it is only one color being seen. If a newspaper photograph is examined under a magnifying glass, the variation in dot sizes is apparent.

Colored photographs, like those in magazines, go through the same process, with each color being printed over the next. Most large color presses are really four presses in one. Each part prints a separate color, usually magenta (red), cyan (blue), yellow, and black. Special rapidly drying inks are used so the paper is not smeared as it moves through the rollers from one color to another. These presses are able to print several thousand copies an hour.

Our eyes and a camera function in very much the same way in that the image in a camera is reproduced on paper and the image in the eye is reproduced in the brain. Our sight depends upon light because of the light-sensitive material in our eyes, called *rods* and *cones*. The rods allow us to see when

This is a close-up view of the Statue of Liberty taken from a newspaper photograph. The half-tone dots vary in both size and shape.

there is very low light, and the cones allow us to see colors and objects in very bright light. They act much like the light-sensitive film in a camera, picking up an image reflected through the lens (the eye and a camera both have a lens).

The type of photography with which we are the most familiar uses light waves to reproduce an image, because that is the way our eyes function.

Technology, however, is able to reproduce objects and conditions which can not be seen with the unaided human eye. The resulting photographs communicate to many people what only one person with the proper equipment could see. There are two processes which provide this capability: infrared photography and electronic microscopic photography.

Infrared photography creates an image by using heat rays rather than light rays. Heat rays are similar to light rays in that they can be reflected off an object (thrown back), refracted (bent such as a prism does with light), and absorbed. They cannot, however, be seen. Special film, highly sensitive to heat rays, picks them up from objects or areas that are warmer than others.

Infrared photography has been used extensively in environmental research. For example, pictures taken from an airplane over a river will produce vivid evidence of pollution entering the river because of the temperature difference. Photographs taken from satellites can help estimate the amount and type of food being grown throughout the world, because different crops reflect different heat waves. Medical science taps this procedure to diagnose illnesses in the body, because areas of pressure or infection have warmer temperatures than the rest of the body. Designers of tools or artificial limbs, can make use of infrared photography to identify areas of direct body contact and can change points of pressure. Because this method of photography needs no light, pictures can be taken in total darkness. Since World War II, this method has been used for military observation purposes by many countries, both in the air and on the ground.

Electronic microscopic photography has several variations. Each has specific advantages, but most function in the same basic manner. The instruments are able to magnify objects as much as two million times, enough to study the behavior of atoms. The electron microscope is used extensively today in industrial research, as well as in medical research and

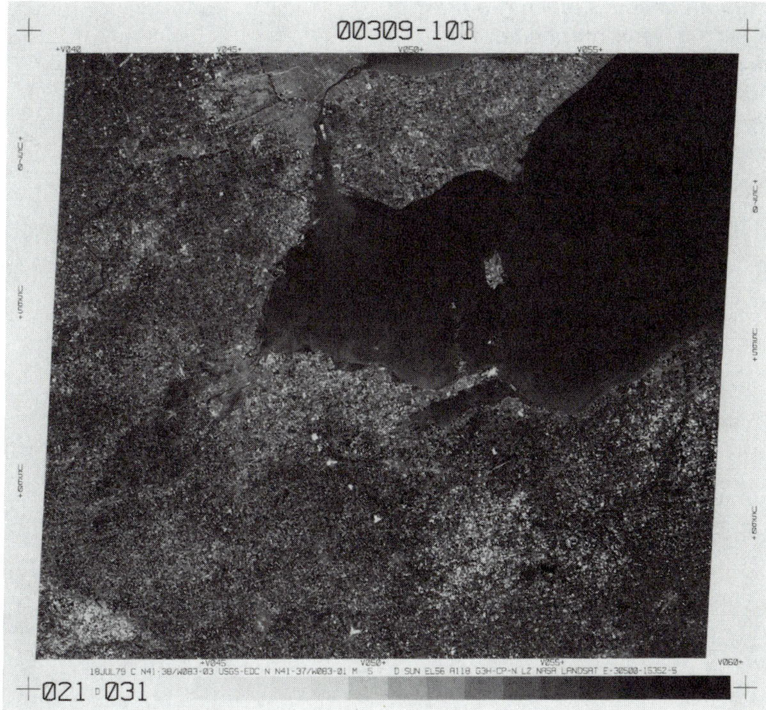

This infrared photograph of eastern Lake Erie showing Detroit and Toledo was taken from the Landsat satellite at a height of 570 miles (912 km). Note the pollution (lighter areas around the shore line) and the areas of the northern shore line where there is no pollution.

diagnosis. Rather than using light or heat rays, it produces an electron beam which, after hitting an object, is reflected on a phosphorescent screen. The scanning electron microscope works in the same way as a television screen. The beam "writes" a picture on the screen in lines; it is able to reflect each line in sixty-millionths of a second. A photographic plate may be introduced into the instrument, which will record a picture of an object much too small to be seen any other way.

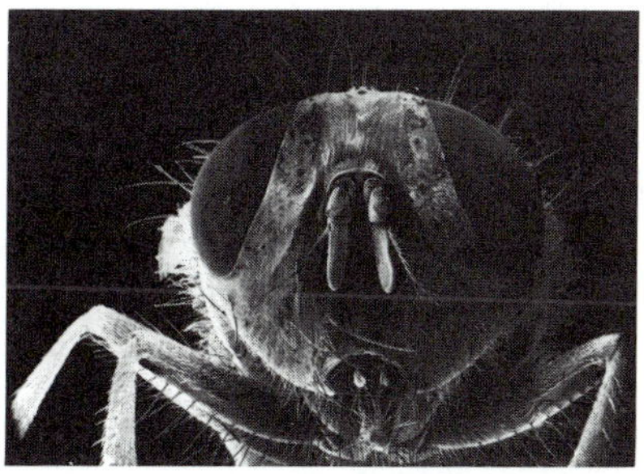

This is a scanning electron micrograph of the head of a common house fly. It was taken at approximately 110 times magnification. Note the detail of each hair, which would be impossible to obtain with normal light photography.

Another method of reproduction used extensively in the production industries is the diazo process. This system produces blueprints for manufacturing products, building roads and buildings, and recording geographic data. Although the term *blueprint* is sometimes used, the more common is simply *print*. The diazo process produces a whiteprint: (commonly called a print) it is a white background with a blue or black line. The drawing is made on tracing paper and run through the reproduction machine together with a specially coated paper that is sensitive to ultraviolet light. The machine exposes the reproduction paper where light passes through the tracing

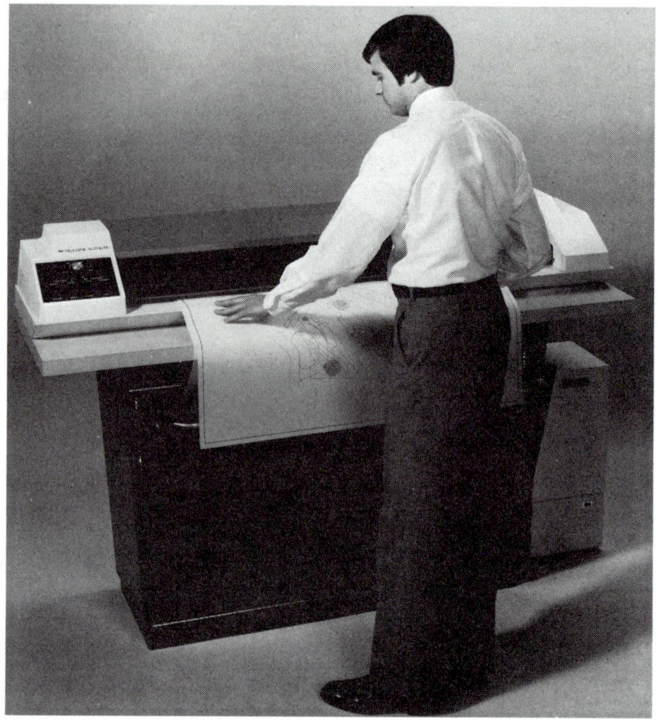

This architect is making a print of a land development site. It is drawn on tracing paper so the light will pass through to the light-sensitive reproduction paper.

paper. When it is exposed to ammonia fumes for development, the lines—which were not exposed because of the opaque pencil lead or ink—turn blue. The exposure and development takes only a few seconds. This is an important reproduction process, because many people in many locations all over the world make parts and produce goods which have to be manufactured to exact specifications. The prints serve as a valuable communication tool, relating such information as shape, size, type of materials, processes to be used, and parts to be assembled. The process of photographically reproducing materials such as these is called *reprographics*.

Transmission. Examples of transmission methods of visual communication surround us every day. Other than messages among individuals or small groups, most communication would fall under the category of *mass communication,* which is the term given to communication directed toward everyone. Although the information may be only local or regional, as opposed to nationwide, it is still for the general population of that community.

Some transmission methods of mass visual communication are television, movies, books, magazines, and billboards. The purposes of mass communication are entertainment, information, and advertising.

The number of books sold, the number of video tapes being used, and the number of people attending movies today show the strength of mass communication for entertainment. At the same time, there is a strong demand for information. More people in the world today are literate than any time in history. Immediate information can serve many functions, including saving lives, forming public opinion and preserving and shaping political structures. The war in Vietnam was the first war to be brought into the living room, via television. For the first time, television provided an opportunity for people to become almost as close to an event as the participants. The availability of information because of mass communication most certainly serves to shape public opinion. Advertising seeks to sell a product, a service, or an idea, and it allows people a freedom of choice in deciding what best answers their needs. Theoretically, this advertising contributes to fair prices because people are able to compare costs of products and services.

"Still image" communications usually have to be placed into a correct position for viewing so transmission can take place. Billboards usually are covered with preprinted graphics which are unrolled from long strips for hanging. Highway signs are attached to posts at the correct traffic locations, and magazines are organized on the newsstands. Without these simple steps, transmission would be difficult.

Where many pieces of information make up a unit, they have to be bound into some kind of order. Examples of this type of communication are books, magazines, and newspapers. As an example, books are divided into two groups: hardbound and softbound. Hardbound books are more expensive because they use more material and require more steps in binding.

Binding a book takes three steps. The first is *forming*. The pages are printed on large sheets which contain from four to thirty-two pages on each side. The pages are not printed in a numerical sequence. A machine is used to fold the sheets several times in a very specific way, down to single page size. After folding, the pages are then in the correct sequence, as they would appear in a book. This is called a *signature* and represents a section of the book. Next, a machine gathers the

Trimming is the last step in the binding process before the cover is attached. When the pages are assembled the edges are very uneven. This machine trims the edges and blows the waste material through the pipe in the back and into a disposal cart.

signatures in correct sequence as they appear in the book. They are then automatically sewn together with cotton or plastic thread. The second step is called *shaping*. The book is then pressed to give it uniform thickness. It is glued, trimmed on the three open sides, and a reinforcing mesh fabric is applied to the backbone. The third is *covering*. The case (cover) is being developed at the same time the book is being shaped. The cover is usually heavy paperboard, covered with cloth, plastic, or coated paper. Casing-in joins the pages and the case by applying adhesive and placing it into another machine where the grooved hinges, or joints, are pressed in. This allows the book to open easily. Softbound books usually are not sewn. The pages are glued, along with the paper cover, and pressed. The entire book is then trimmed on three sides.

Books, magazines, and other printed materials have to be constructed so they function well during transmission, or complete understanding of the message might be lost. It should be noted here that there is a very close working relationship between the communication technology system and the production technology system when mass communication is involved.

The process of transmission involving postal letters illustrates how other technology systems are involved. Letters may be picked up by small trucks. A machine arranges them so they all face one way and automatically cancels the stamp. They are then sorted in another machine which groups them according to destination at a rate of 10,000 letters per hour. Delivery might involve large semitrailers, airplanes, and small handcarts, as well as various kinds of chutes and conveyor belts.

Communication using moving images usually involves electronics in the transmission system; both energy and controls are needed. Examples of moving images would be television and movies.

The transmission of television signals most certainly involves technology. Most television signals are transmitted through the air, up to a distance of about 75 to 150 miles (121 to 241 km). The signal is called an electromagnetic wave; it travels through the air at the speed of light.

There are three methods of transmitting the signals farther than just the local area. Coaxial cable is a heavy cable usually buried underground. Many stations use this method to send programs to their affiliated stations throughout the country. Cables are also used by people in remote areas where the signals cannot be received and by those who wish to subscribe to the service for alternative types of programming. Microwaves are very similar to electromagnetic waves. A system of

This microwave tower is typical of those used in the Bell System's ground communication network. It is used to relay pictures of the world's weather conditions, as sent from satellites circling the earth.

Because signals are transmitted in a straight line, several satellites are required to provide continuous communication.

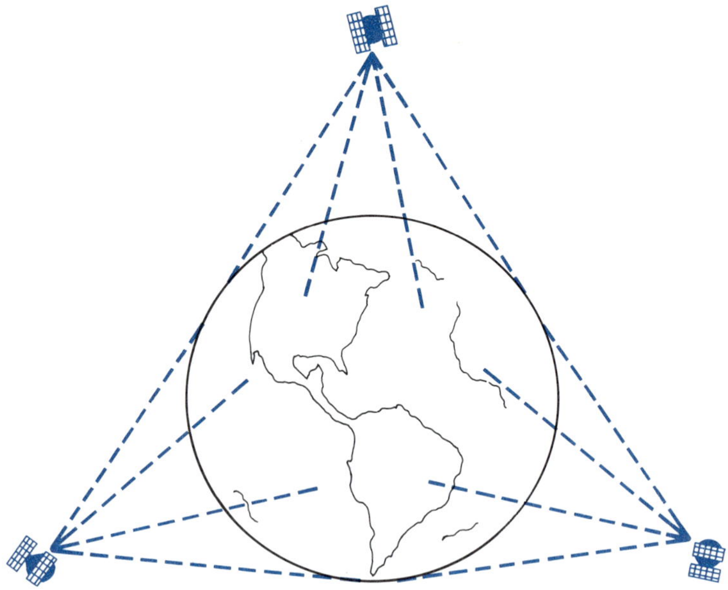

tall relay towers spaced across the country pick up the signals, amplify (increase) them, and retransmit them on to the next relay tower. This allows nationwide coverage. Communication satellites are able to receive and retransmit signals where there are no cables or relay towers. Before we had satellites, we had no worldwide system of direct television communication. The oceans and underdeveloped countries caused a span of too much distance between land-based relay stations.

When a signal is received by the television set, it is split into two signals; one is changed into an audio signal which comes out the speaker as sound, and the other is sent to the video tube to be transformed into a picture. Because the screen generates light much like a fluorescent lamp, there is no need to darken the room for viewing.

The transmission of movies (as well as still images) usually involves projecting the image on a screen to make it larger. There are two basic types of screens. The reflecting screen reflects the light rays from the projector back into the eyes of the viewer. Because the light rays are traveling some distance, it is necessary for the room to be dark in order to see the image. The second type of screen is the rear-view projection screen. This one is translucent (allows light to pass through but cannot

This small rear-view projection system allows these chemists to keep graphic data on file in slide transparencies and review them at any time in a lighted room.

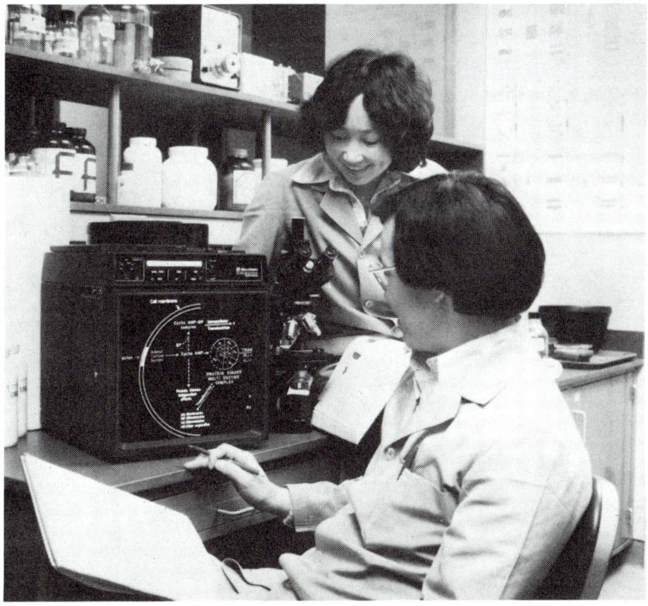

be seen through). The projector is on one side of the screen, while the viewer is on the opposite side. The viewer is actually looking directly into the lens. Because the light waves are shorter and more intense, darkness in the room is not required.

Although it is not a communication method used by everyone, radar certainly is a part of everyone's life. This moving image provides communication in many applications, both civil and military. It is used extensively in air traffic control and on board aircraft to detect storms and changing surface height, such as mountains. Most large ships use radar for navigation, as do harbors for directing traffic in shipping lanes. The United States Weather Bureau also uses radar in weather forecasting.

Radar can be used to calculate both speed and distance. The system works on the same principle as an echo. Radio waves are sent out at the speed of light, 186,282 miles per second (299,792 km/sec). When it strikes an object, it is reflected back like an echo. Because the speed of travel is known and the beam direction can be determined in degrees, the position and approximate speed of the object can be calculated. The impulse of the beam is transmitted back to a video screen, which shows it as a dot or a blip (sudden movement in a line).

Because the echo is constantly being returned from the airplane, its location and movement can be seen on the radar screen as a moving dot.

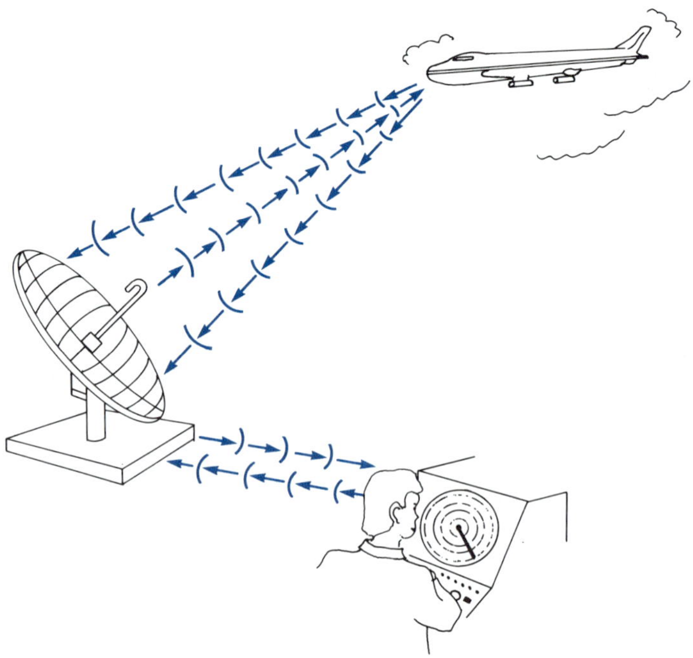

Other kinds of radar can pick up the image in three dimensions by using several beams. This provides information concerning not only where the object is, but what it is.

Photographs can also be generated with radar. Like infrared photography, pictures can be taken at night or in fog, because light waves are not used. The infrared photography and radar, however, may provide different kinds of information, because one generates the image by radiated heat and the other generates it by distance.

Storage and Retrieval. The process of storage and retrieval can be very simple, such as a small file-card system with an index, or very complicated, as with sophisticated computer database systems. The storage and retrieval of visual information is very closely associated with reproduction. For the most part, the original is reduced for storage, because full size storage would require too much room.

Storage refers to the process of keeping the material, while retrieval refers to the method of taking it out of storage. Because the reproduction process is involved, certain kinds of hardware are required.

These photographs are X-ray negatives taken of aircraft parts. The X-rays show small cracks and flaws which can cause the part to fail months after they have been in use. The imperfections cannot be seen under normal visual inspection because they may be inside the part.

Many of the processes involve the use of film in one form or another. This requires a photographic system, as well as a projection system. Several methods of storing information on film are home movies, slides, and filmstrips. These methods usually require a separate developing and projection system. Another common storage and retrieval system is the VCR (video cassette recorder). This method has the image generation and projection capabilities within the same system. Many home systems allow recording from commercial television programs, so they can be retrieved and viewed at a later time.

Industry, business, and other commercial enterprises make extensive use of storage and retrieval systems because of the tremendous amount of information that has to be communicated, as well as stored for many years. As products, procedures, and future plans change, everyone who is involved has to be informed. In today's technological world, these changes take place daily. There are two systems which are used exten-

The micrographic reader-printer allows the microfilm to be viewed on the screen and hard copies made to the size desired.

sively today in commercial applications for storing and retrieving information: microfilm and electronic storage.

Microfilm as a method of storing information has been used for several years. The process of making microfilm is called *microphotography*. It involves reducing an original, such as a drawing or a page in a book, to an extremely small size (an entire book can be stored on a short roll of film).

The microfilm can be stored in a roll, or it may be cut into small frames and mounted on cards so they may be filed. When the film is in this form it is called *microfiche*. Two or three cards, which measure about four inches (102 mm) by six inches (152 mm), can hold the entire contents of a book. This method requires a special machine reader designed for the cards.

Readers are usually desk-top machines that project the microimage onto a viewing screen; the enlarged image can then be read. There are many different types of readers, depending upon whether they accept roll film, cards, jackets, cartridges, or cassettes. The sizes of film usually are 16 mm and 35 mm for roll film and up to 105 mm for other types. Film size is designated in the metric system, with the millimeter size referring to the width of the film.

Another system which is used, where multiple copies are needed, is called the *reader-printer*. When an image is viewed which is to be reproduced, a button is pressed and a hard copy (paper), which is enlarged, is produced in seconds. This process of reducing, expanding, and manipulating copies is called *micrographics*.

Electronic storage utilizes the computer memory or magnetic tapes for storage. This takes up much less space, provides more opportunities for copy manipulation, such as size and editing, and is capable of handling a larger volume of storage. Such a system allows the viewing on a screen of any image within the entire system (which could be several thousand) in only seconds. It can then make a microfilm copy of it and produce an enlarged hard copy. The use of this system, called COM (computer-output-microfilm), is growing rapidly. A consulting firm, for example, might use this system to record information concerning a bid they are submitting to a possible client. The output is the bid, which is written and stored in the system, a microfilm copy is made for the files, and a hard copy is made for the client.

AUDIO COMMUNICATION

Although the first use of technology in audio communication might have been the primitive drum, the world had to wait for the development of electricity before sound messages could be effectively communicated. Electricity is necessary for two reasons in audio communication: (1) Sound waves cannot travel any great distance. They can be amplified (increased) by electricity when they travel through the air or on a wire. (2) Sound waves must be converted at the receiving end if they are to be stored or processed in any way. Electricity can convert them into electrical impulses or radio waves for transmission and processing at the receiving end. As with so many other aspects of our technological society, the Industrial Revolution and the accompanying advances in energy and production gave a tremendous boost to the advancement of communication. The most common examples of audio communication, and the ones we shall touch upon briefly, are the telegraph, telephone, phonograph, and radio.

The technology of sound transmission involves changing various types of waves.

1. A person's voice consists of sound waves, or air movement, created by vibrating vocal chords. The microphone picks up the sound and changes it into electrical waves which cannot be heard.
2. The electrical waves are sent to the control board which changes the waves. This adjusts the sounds, such as volume control and blending.
3. The electrical waves are then changed into stronger electrical waves, called *carrier waves,* at the transmitter. The changed waves are called *radio waves.* Sound waves can travel only a few hundred yards (meters), while radio waves can travel many hundreds of miles (kilometers).
4. The radio waves are then transmitted in all directions from the transmitting tower. Radio waves cannot be seen, heard, or felt.
5. The radio receiver then changes the radio waves back into sound waves, which are immediately heard.

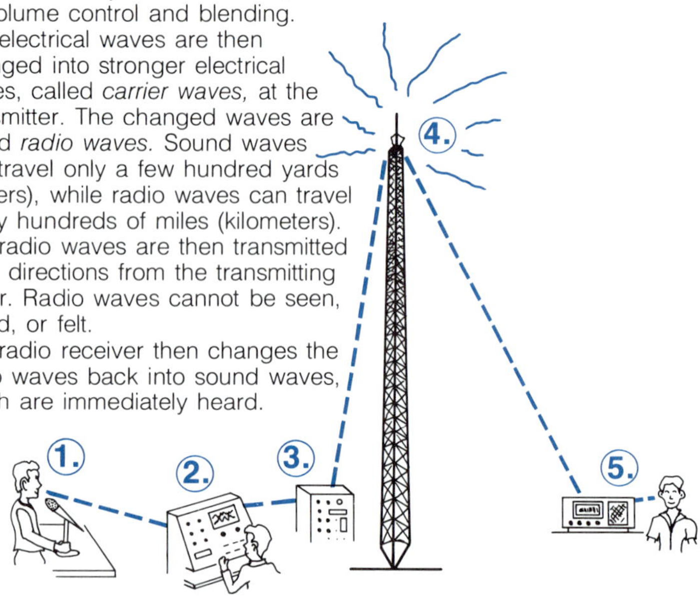

Telegraph

The first effective audio communication was developed in 1837 by three men: William Cooke, Charles Wheatstone, and Samuel Morse. It is interesting that Cooke and Wheatstone developed the telegraph in England and Morse developed the same instrument (with some major differences) in the United States, totally independent of each other. It was Morse, however, who made continued improvements in the system and, of course, developed the Morse Code which became an international language.

In 1844, the first long-distance message was sent from Washington, D.C., to Baltimore, Maryland, by means of a telegraph wire. In a few years many cities were connected by telegraph, and in 1851 (only seven years later), there were more than fifty telegraph companies operating in the United States. In 1856, twelve companies combined into the Western Union

Samuel F.B. Morse transmits the first message—"What Hath God Wrought!"—over a forty-mile (64 km) telegraph line from Washington, D.C., to Baltimore on May 24, 1844.

Telegraph Company, the beginning of a nationwide communication network. In 1866, a transatlantic telegraph cable was laid on the ocean floor, connecting two continents with the first immediate communication system.

Telephone

The next breakthrough in communication came in 1876, when Alexander Graham Bell developed the telephone. A teacher of the deaf from Scotland, Bell came to the United States in 1871. He spent much time experimenting with a telegraph that would send many messages at once. He failed in this attempt, but he accidently discovered the principle that soon led to the telephone. When a metal reed accidently stuck, the vibration sound of its snapping open was heard on the other end of the short wire. This discovery, that vibrations cause variations of electric current, is the basis of our telephones today.

In 1877, the first switchboard started operating, connecting four banks and a factory. Also in 1877, Bell and three partners formed the Bell Telephone Company. In 1880, the American Bell Telephone Company was formed, merging the Bell Company with the telephone interests of the Western Union Telegraph Company. In 1899, the American Telephone and Telegraph Company (who operated long-distance lines) took over America Bell. Thus, A.T.&T. became the parent company of the Bell System. In the short period of time from 1871

The first commercial telephone, which went into use in 1877, had only a single opening for both hearing and speaking, which meant a constant shift of position.

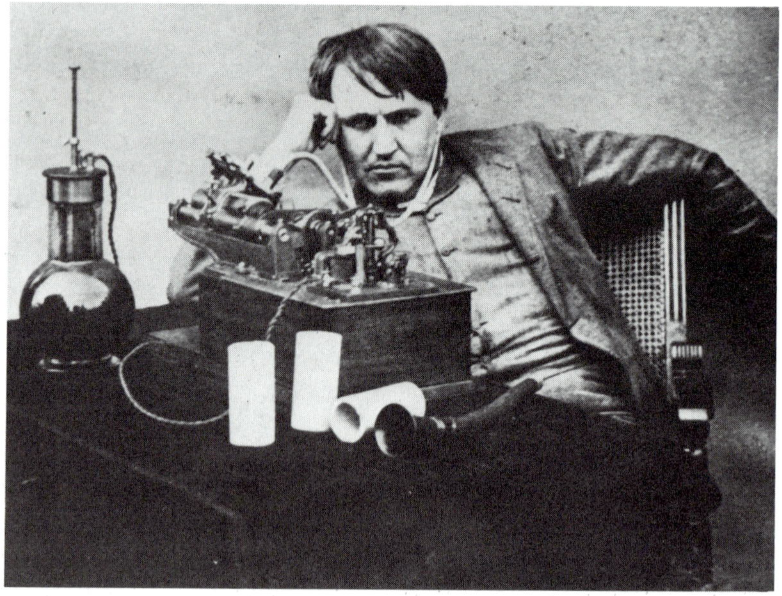

Thomas Edison listens to an advanced model of his phonograph. Notice the recording was done on cylinders, not on the flat disks we use today.

to 1899, the country became equipped with a complete long-distance verbal communication system.

Phonograph

The first practical recording device was invented in 1877 by Thomas A. Edison, an American inventor. The machine used a vibrating diaphragm and rested on a cylinder covered with tinfoil. As words were spoken, the cylinder was rotated. The vibrations of the diaphragm were transferred to the tinfoil, making small dents. When the cylinder was rotated with another needle and diaphragm, the dents caused the process to occur in reverse, causing the original sound vibrations to be heard. Improvements such as flat recording disks (records) and the addition of electricity for amplification in the 1920s came slowly because few people recognized a practical application other than entertainment.

Radio

The first radio was called a "wireless," because it required no wires, like the telephone and telegraph. In 1895, Guglielmo Marconi, an Italian inventor, developed the first practical radio, although it did not transmit voice signals. It really was a wireless telegraph, using the telegraph code. By 1901, he was able to send code signals across the Atlantic Ocean. The new

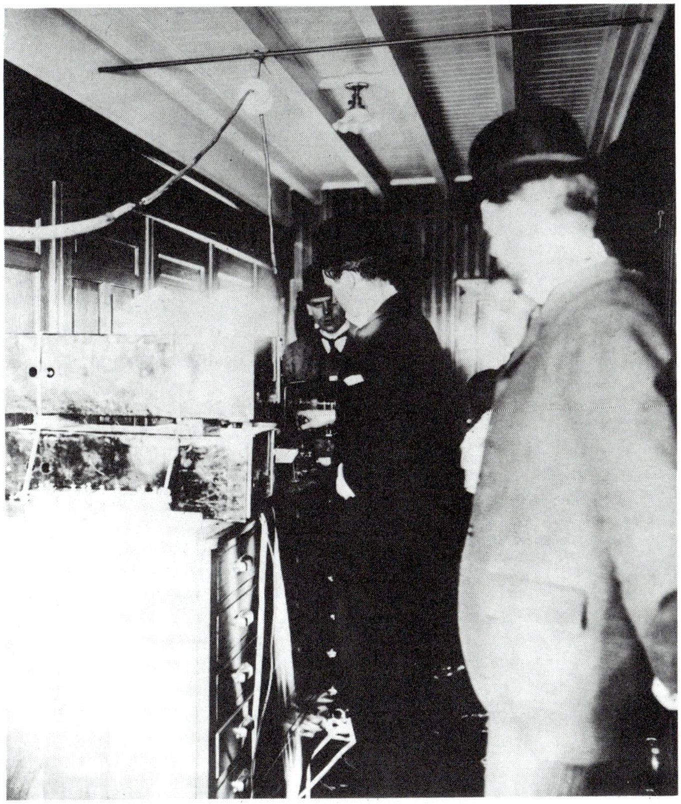

Marconi (in the dark suit) sends the first wireless marine message in America.

system gained use on large ships and in lighthouses at strategic points along the coasts.

The wireless radio used a system which produced electromagnetic waves; these waves travel in all directions. As is the case with many new technological developments, Marconi applied information gained by other people before him, and others applied his ideas to new developments after him.

During the next few years, electrical engineers developed new electronic elements called vacuum tubes, which would pick up and amplify the radio signals. This invention led to the first transmission of human words by radio in 1906. By the 1920s, improvements allowed most people to own a receiver, and there were many commercial radio stations in operation. By the 1930s, two-way radio was a part of standard equipment for aircraft, ships, and emergency vehicles.

These four important inventions: the telegraph, telephone, phonograph, and radio, represent the real beginnings of modern communication technology. The scientific principles involved in each can be found in nearly all of today's communication hardware.

The Hardware and Processes of Audio Communication

Audio communication, like visual communication, can be broken into the three categories of reproduction, transmission, and storage and retrieval. The lines separating the three, however, are much less distinct. Because electrical energy is used in nearly every phase, more than one of these operations may be accomplished using the same process at the same time. For example, it is possible to amplify a speech over a public address system (reproduction and transmission) and at the same time record it for future use (storage).

Reproduction and Transmission. Because the process of reproducing sounds also involves the transmission, these two categories will be discussed together. Today's electronics have revolutionized audio communication. The developments in electronic circuits allow more sensitive sound pickup, wider range of sound frequencies, and greater amplification. The circuits, which make use of electric current, are composed of small electronic components that regulate the current flow. They may slow it down, store it, make it flow in various directions, increase it, or decrease it. By means of this manipulation of electrons, as they flow through the circuit, sound is reproduced.

The reproduction of sound without storage usually involves such methods of communication as public address systems, telephones, and one-way and two-way radios. In some instances today, it is difficult to differentiate between them. It is possible to have a "wireless" telephone in your car, a public address system in your hand (bullhorn), and a two-way radio in your home (intercom which also functions as a one-way AM and FM radio).

A sound wave is a vibration or regular disturbance in the air. The waves travel through the air much like ripples travel in water when a stone is dropped. When the vibrations are received by the ear, they are converted into messages that are sent to the brain for decoding. The frequency of sound refers to the number of vibrations, or ripples, each second. The higher the frequency, the higher the pitch.

The first radios only picked up the middle range of frequencies, which produced a very flat-sounding reproduction.

A ringing telephone creates sound waves like a stone tossed into water creates ripples. Sound waves are of a certain frequency, which we soon learn to recognize as the sound of a telephone.

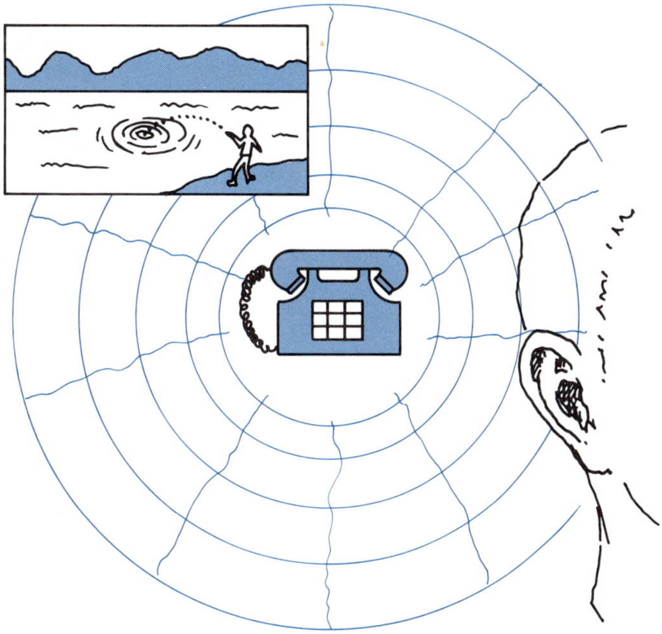

As knowledge of electronics developed, the quality of sound, called *fidelity,* improved because both higher and lower frequencies could be reproduced, making it sound like the original.

Some frequencies can be produced which are too high for the human ear to hear. Ultrasonic waves have such frequencies. These waves are used in more specialized types of communication, one of which is sonar. Like radar, sonar operates according to the principle of echoes, but uses ultrasonic sound waves rather than radio waves. Sonar, which stands for

Sonar provides information on both the depth of water and the exact location of objects under the surface.

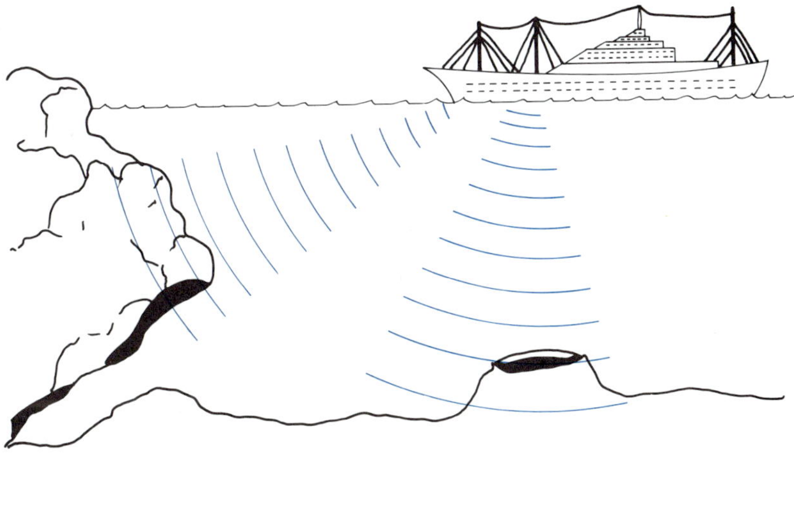

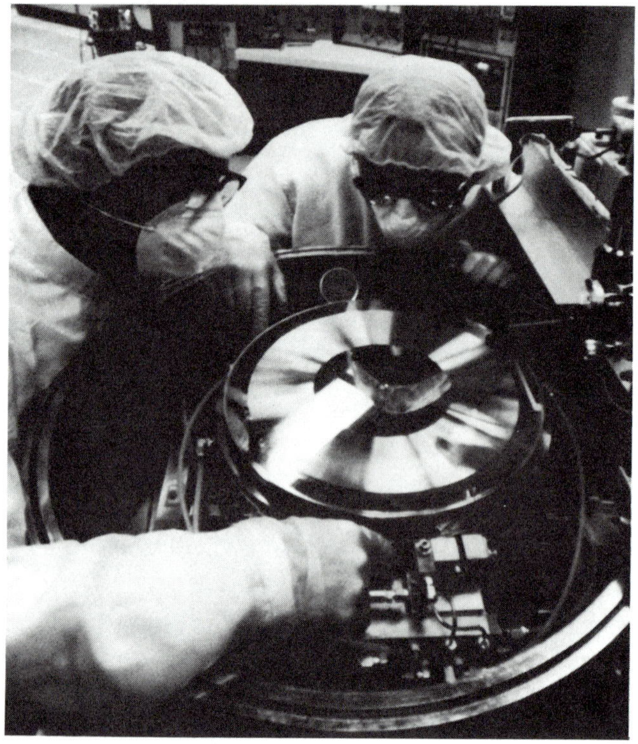

These technicians are examining a master disc from which stereo records are pressed. A closely controlled environment is required to ensure high-quality reproduction.

"sound, navigation, and ranging" is used extensively in water navigation. The signals are generated and received by equipment on board ships or submarines. Because the speed of the waves is known and the time which it takes to bounce back is automatically calculated, distance can be determined. The uses of this method range from military applications to depth finders for fishing.

Storage and Retrieval. The storage and retrieval systems involve the use of recording and playback units. Of the two basic types, the first is disc records. At the recording session a master tape is made. Tapes are used rather than records because they can be edited (rearranged, sounds deleted, and so forth). To make a stereo record, two or more microphones pick up sound from opposite sides. Those sounds picked up on the left side will be reproduced in the left channel and those on the right side will be reproduced in the right channel. Each one is recorded on a separate track of the tape.

After editing, a master disc is cut with a heated cutting stylus. The "cutting lathe" is connected to the tape recorder.

The videotape recorder works much the same way as a sound recorder—by using a recording head and magnetic tape. The recorder is carried over the shoulder while recording. For playback, it is connected to a standard television receiver.

As the sound vibrations reach the stylus, it causes it to cut microscopic indentions into the grooves as the disc rotates (much like Edison's first recorder). The left channel is registered on the inner wall of the groove, while the right channel is cut into the outer wall.

This record becomes the master used to produce thousands of records. A metal copy of the master, called a stamper, is made which stamps the final records. A plastic blank record is placed between the two metal stampers under heat and pressure. The result is a record ready to be packaged and sold.

When the record is played on a stereo phonograph, the process is reversed. The diamond stylus picks up vibrations from each side of the groove. They are sent through the system to be changed back into sound, the left channel coming out the left speaker and the right channel issuing from the right speaker.

The second major storage and retrieval system is magnetic tapes. A popular recording method in the last few years, magnetic tapes both store and retrieve sound on the same machine. Because they are relatively inexpensive, some are made for children.

Magnetic tapes involve the use of a microphone which picks up the sound vibrations and translates them into electrical impulses. From there the electrical impulses are sent to an

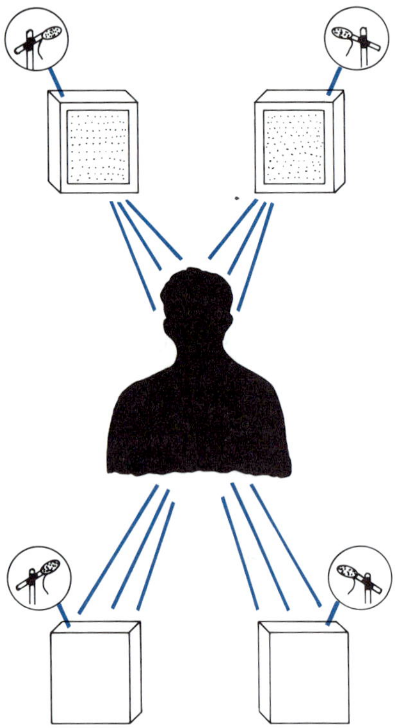

In a quadraphonic system, each recording microphone picks up only the sound near it. When the sound is reproduced by record or tape, only the sound that was recorded is heard on each speaker. This recreates the location of the original sound.

amplifier to be increased in strength. They are then sent to the recording head where they are converted into a fluctuating magnetic field. At this point, a plastic tape coated with iron or chromium oxide passes the recording head. As it passes, the coating is magnetized in varying degrees, according to the fluctuations in the field. The magnetized tape is then a representation of the original sound vibrations.

To hear the recording, the tape is again passed by the recording head which now operates in reverse. This picks up the magnetic field, changes it back to electrical impulses and sends it back to a speaker, which transforms it into sound waves we hear. Speakers vary greatly in size and type, with the more expensive recording units having two or three speakers for each channel. Speakers have a vibrating surface. Large surface areas vibrate much slower, providing lower tones, while small surface areas vibrate with a higher frequency, providing higher tones.

A sound system called four-channel, or quadrophonic sound, is the most true-sounding recording method. It uses four microphones with four separate channels. With four sets of speakers, the listener can get a reproduction very close to the original live sound.

COMMUNICATION PROBLEMS

Progress made thus far in the field of communication in the last half of this century has been phenomenal. Yet progress has brought problems. Because of the rapid development of electronics and space technology, we are now faced with many problems that were inconceivable only a few years ago.

Four forces shape our communication technology. One is technological growth. Some estimates indicate that the cost of electronics may decrease by 20 to 35 percent per year in the near future. Advances in capability with decreasing costs provide for rapid development. This means many new technologies will appear, but the rapid convergence of different technologies competing for the right to function can cause difficulties. For example, the increased popularity of CB (citizens band) radios and similar communication devices has caused some problems with television and radio reception. This has resulted in the breaking of transmitting laws, and has even hindered law enforcement. People have the right to use these functional radios, but many times new systems are not always compatible with other systems and rules.

A second force concerns policies. Because of rapid development of new hardware, it is difficult to estimate the potential impact on society before the new hardware comes into play. Many times growth proceeds without direction. In some instances, no policies are established, in others, policies are extremely outdated. Common carriers (such as telephone companies), for example, are still subject to provisions of the Communication Act drawn up in 1887, and broadcasting still operates under substantially the same provisions as those adopted in the Radio Act of 1927.

The new technologies involving both audio and video recording of commercial broadcasts are causing copyright problems. Home recording systems can produce copies equal in quality to commercial records and tapes. The same is true with programs taped directly from the home television set. The expanding technology has required new copyright laws and new methods of monitoring and controlling reproduction. These efforts are not entirely adequate to the task, however, and the lack of adequate policies prevents the formation of an integrated communication network where all systems function together.

A third force is economic in nature. The time lag between new technologies and their availability on the market is closing. Industries want a return on their investment as soon as possible, as well as being first to gain a large portion of the market. This is understandable, but in some cases little time is available to evaluate the need of the market and the impact a particular application will have on the total communication system.

Another economic problem concerns communication satellites. If a private corporation develops a satellite, is it privately owned, or is it publicly owned, as a utility? If it is publicly owned, should the cost of development be the responsibility of the private corporation or the taxpayers? These questions will need to be addressed as our communication system expands with space technology.

A fourth concern is world interests. There is a limit to the number of satellites that can be put into orbit before they start interfering with each other. Therefore, some countries

One solution to problems involving world interest would be for countries to undertake more joint research and development. This communication satellite is a joint effort of INTELSAT (International Telecommunications Satellite Organization), of which more than ninety countries are members.

will have them and some will not. Who will decide which countries will have them and which ones will be dependent upon other countries? Some countries of the world do not want any outside information, including propaganda and world events, to reach the people of their country. With satellites, the flow of information will be difficult to restrict. Also, underdeveloped countries could develop internal problems when their people realize the quality of life in other countries is vastly different from their own. And because communication satellites are more effective than other methods of transmission, their importance to national security for any country cannot be overemphasized. There is a real concern, therefore, about the vulnerability of satellites to natural disasters, malfunction, or intentional destruction by other countries. A hostile country conceivably cripple its foe by wiping out the other country's communication network.

COMMUNICATION AND THE FUTURE

A term that has evolved in the last few years to describe communications by electronic means is *telecommunications*. This word will be an increasing part of our vocabulary in the years to come. It is estimated that by the year 2000, 75 percent of all communication will involve telecommunications.

Shown here is a small computer designed for interactive learning for young children.

Although it is impossible to accurately predict the future of communications, it will most certainly involve extensive use of telecommunications. The following predictions are actually speculation; however, the required technology is currently available.

Because television involves both audio and visual messages, much of tomorrow's communications will involve a television screen. Cable TV will probably allow two-way communication. Video games may be played with a central computer by cable, while a recording device on the set will be able to pick up still images which can be read (such as the daily stock market returns).

Video discs hold the potential for revolutionizing storage and retrieval capability; they offer many advantages over records and tapes. There are several types, but the most promising is the laser disc system. A laser beam is used to etch the grooves and read back the stored information. Extremely high quality is obtained, because there is no direct contact made with a stylus in either recording or playback. The discs are difficult to damage; they can store still pictures, moving pictures, and sound; and they allow immediate selection of information anywhere on the disc. A single disc can hold up to four billion characters—the equivalent of several hundred books. Although in limited use now, primarily for entertainment, they will be used extensively in education, business, and industry.

There may be unlimited TV broadcasts from which to choose. Communication satellites from other countries will allow their programs to be received in other parts of the world. We also will be able to select channels from nearly any city in our country.

Telephones as we know them may be a thing of the past. Three developments in technology may completely change our system: fiber optics, lasers, and satellites. Fiber optics already are used in place of underground electrical wires which transmit messages by electronic impulses. According to fiber optics technology, a laser which produces a beam of single wavelength light sends light waves through a glass fiber. The glass is clear, with a coating on the outside. As a result, the light is kept inside until it comes out the opposite end, even if the fiber is tied in a knot. This process, already in use, sends messages by light rather than by electricity.

Such technology will allow many more conversations at one time and at a greater speed. It is estimated that a single fiber, which is one-fifth the thickness of a human hair, can replace ten thousand ordinary telephone wires or act as a TV

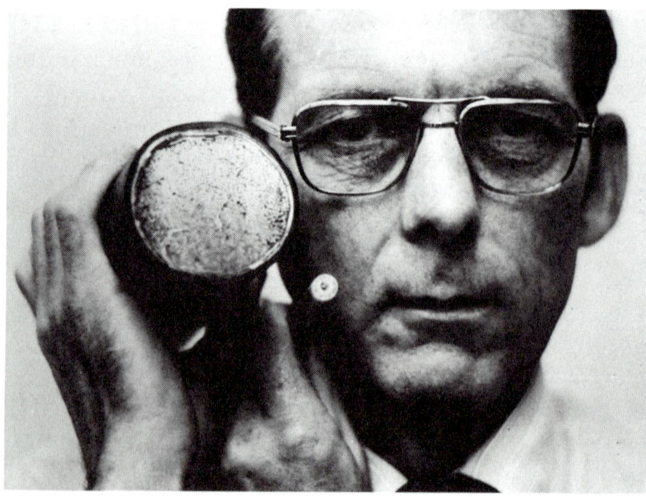

This engineer is holding two telephone cables. The conventional cable on the left will carry the equivalent of 10,800 conversations, while the half inch diameter fiber optics cable on the right will carry nearly 50,000 conversations.

cable which can transfer nearly eight thousand different channels at one time. These are already in use in many of the major cities in the United States. Another advantage to this system is the cost; the raw materials for glass and plastic are much more plentiful than metals needed for electrical wire.

The third technology involves satellites. The use of fiber optics with a satellite system may allow direct dialing to anyplace in the world. An individual may carry a personal communication device, called a *dator*, just as one would carry a purse or billfold. It would have a keyboard, a small video (and maybe a camera), a radio, and a telephone. It will have capabilities of sending, receiving, storing, and processing information. There will be public access to databases. For example, you might want to review something in the Library of Congress. A direct line to the library would allow you to make a selection from a computerized file system, and the information requested is displayed at home on a video screen.

Home computer systems now allow financial calculations and transactions to be done at home. Home banking may one day make the physical building of a bank unnecessary.

This panel dotted with question marks represents a major advance in television display technology. This television screen of the future measures only a quarter of an inch (6.4 mm) thick.

These systems may allow many jobs to be performed at home rather than at an office. For example, a typist might have a word processor at home, with the capability of sending the finished copies directly to the "boss" for a signature within minutes.

One of the major uses of microprocessors (small computers) is in controls. Vehicles of the future may have computer controls that will decrease the chances of traffic accidents and traffic jams. Aircraft most certainly will have more sophisticated computer control to make traveling safer.

There are also definite possibilities of developing the technology to allow deaf people to hear and blind people to see. As our knowledge increases in both technology and physiology, physical disabilities involving the senses will most certainly command more attention.

Finally, it will be possible to have robots to do various jobs. They are already in use in industry today, but they are still relatively expensive for other uses. The robots (which do not really look like a human) can be programmed to perform very specific functions. Experiments are continuing now to perfect a robot which can process verbal commands as well as produce verbal replies.

The progress we make in our communication systems around the world may very well determine the kind of world in which we will live. The hardware of technology is important, but the use people decide to make of the hardware will determine the future.

SUMMARY QUESTIONS

1. Name the three purposes of communication.
2. Name three ways that evolving technology in communication has had an impact on society.
3. Identify how the communication system is also a part of the transportation system.
4. Why was the development of electrical energy as important to communication as the development of the steam engine was to transportation?
5. State a major problem facing the communication system today and identify two hypothetical solutions that might help solve the problem.

6. Identify two new methods of communication which we are likely to see in the future and briefly explain how they may affect our daily lives.

CHAPTER 9

The Production System

In this chapter you will learn:

1. **How the production system is organized.**
2. **How the construction subsystem is a part of many other subsystems.**
3. **Why the processing subsystem is important to the manufacturing subsystem.**
4. **How we get the products we use every day.**
5. **What production problems we will face in the future.**

ANALYZING THE PRODUCTION SYSTEM

The production system is composed of the subsystems of construction, processing, and manufacturing. Each subsystem plays a major role in our technology today and all are closely related in terms of objectives and organizational structure. Three characteristics the subsystems share are the following:

- They all produce some kind of product
- They require continual replenishment of raw and processed materials
- They require a highly developed organizational structure of people, materials, and machines.

The subsystems also differ in a number of ways because of the nature of the products each produces.

The construction subsystem primarily erects structures and prepares surfaces. Its products make up three categories—buildings, nonbuilding structures, and transportation surfaces. A particular project might incorporate more than one category. For example, a dam might be built (nonbuilding structure) with a road on top of it (transportation surface), along with a control booth (building) and a radio transmitting tower (nonbuilding structure). Building construction projects also vary in size, from a house to large apartment buildings or condominiums. Both examples are homes for those living in them, but the construction methods for apartments and condominiums are usually more like those required of other commercial buildings such as factories, hotels, and offices.

The construction system involves the building of structures and surfaces on site, which will either contain something or provide support.

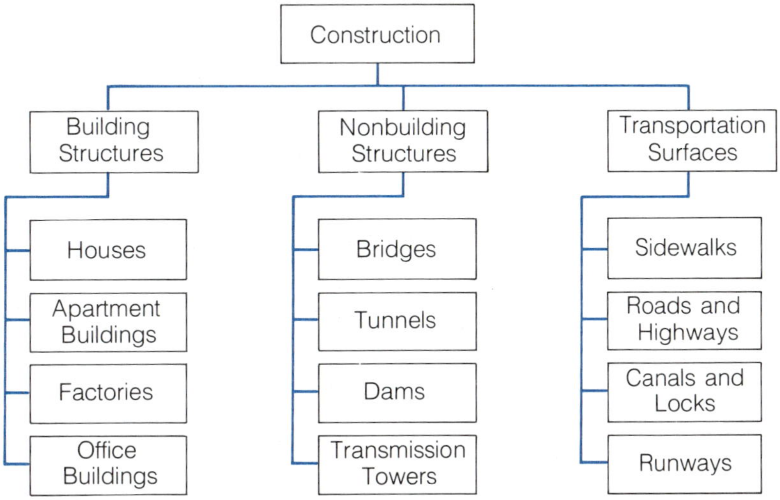

Nonbuilding structures are defined as those which are not intended to enclose or to contain anything. Rather, they are barriers (such as a dam) or supports (such as bridges) and towers.

There are many types of transportation surfaces for both pedestrian and vehicular traffic. An airport runway is a transportation surface for air vehicles. Water vehicles travel more easily if nonbuilding structures, such as canals and locks, are constructed to provide an alternative surface to natural water routes.

At this point you might be wondering why the erection of prefabricated or modular buildings is considered construction when the parts are manufactured in a factory. Or you might be wondering why a tower is considered to be a construction when all the parts are manufactured. The key element in distinguishing construction from manufacturing is the location. Anything permanently built or assembled on the site where it will be used is considered construction. This may be either inside or outside of another structure. It should be noted here that even though large machines may be assembled on a site, they are not considered to fall under the category of construction because they are usually impermanent at this site.

The processing subsystem involves the first steps toward providing a manufactured product. It deals primarily with obtaining the raw materials and converting them into a form which has the necessary characteristics for making them into a product.

Production

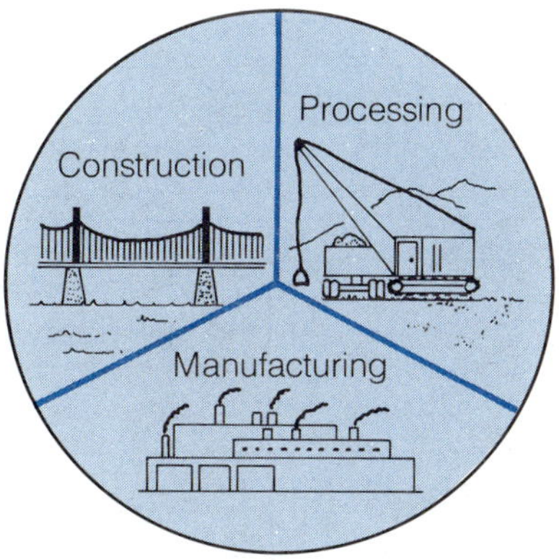

Just as with other subsystems, none of these three production subsystems could function without the other two.

The building of bridges and other types of structures requires extensive use of math and physics, both in the design and in the construction.

The processing system acquires and changes raw materials into a usable form.

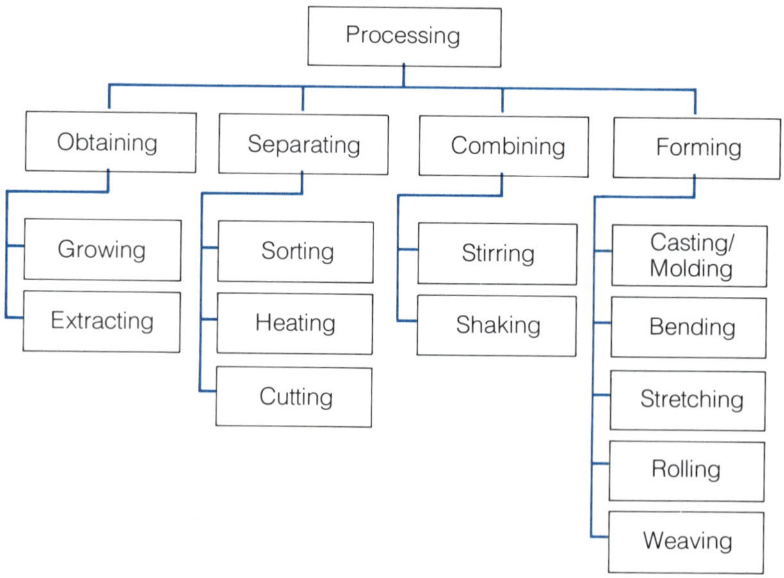

For example, many clothes are made of a material called cotton polyester. These two raw materials have little in common until they become a fabric. The cotton has to be grown and harvested, while the polyester starts as a petroleum product which has to be extracted from the ground. The cotton goes through a number of steps including cleaning, spinning, and bleaching, while the petroleum raw material is converted into a special chemical form. There are many steps involved in developing the chemicals and converting them into a compound which can be spun into thread. When these two materials are woven together into a fabric, the natural material (cotton) and the human-made material (polyester) provide a cloth with characteristics better than either of the materials could produce alone. It should be noted that both materials go through entirely separate operations of obtaining, separating, and combining. When they reach the shaping operation, they are spun into thread and woven into a fabric. At this point the processing stage is complete. The fabric and/or thread is then shipped to the manufacturer who cuts and sews it into a piece of clothing or one of many other cloth products.

Two characteristics distinguish between processing and manufacturing. Processing produces standard materials which are ready to be manufactured, such as fabric, metal rods, dyes, plastic pellets, and various chemicals. Manufacturing fabricates and assembles them into a product. Processing also produces materials which are monolithic (only one at a time) in a continuous process, such as paper, petroleum products, fertilizer, and various other chemicals. Manufacturing usually assembles several of these materials into a single unit.

The manufacturing system involves many methods of combining standard materials into usable products.

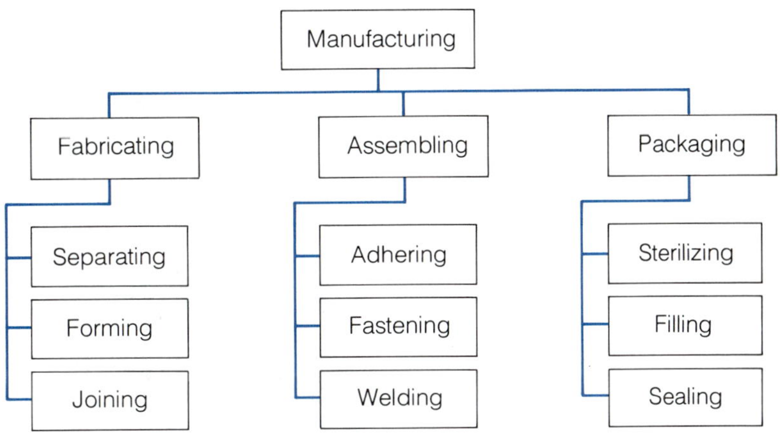

Manufacturing may involve many of the same methods utilized in processing, but the result is a completed product ready for market. The activities involved are fabricating, assembling, and packaging. Fabricating and assembling may, in some cases, be the same operation, depending upon the product. However, the primary difference is that fabricating involves using standard materials to make specific parts or a combination of parts, while assembly involves putting the parts together. The final step in manufacturing is the packaging of products for transportation and sale.

CONSTRUCTION

Much of the knowledge concerning construction of our prehistoric ancestors comes from caves, which, as permanent natural structures, survive to the present, while shelters constructed of reeds and wood, which were only temporary, have disappeared. Early humans were limited in their choice of building materials to those offered by the immediate environment and were also restricted by their lack of tools for heavy cutting and shaping.

The evolution of early construction technology remains somewhat of a mystery. The greatest example of early construction technology is in Egypt. Beginning around 2700 B.C., the massive stone pyramids began to rise. The hauling, cutting,

The Great Pyramid at Giza on the banks of the Nile River is the largest. The base could hold ten football fields and the apex reaches as high as a forty-story building.

The aqueducts served two purposes: they carried water, much like an irrigation canal, and they were wide enough to act as bridges and roadways. Arches were used extensively in construction to hold the weight of the stone.

and fitting of the blocks of stone, which weigh many tons each, almost defy explanation. It would be a very difficult task today, even with our advanced technology of lasers, bulldozers, and huge cranes, to recreate such a structure.

Stone was not available in Babylonia and Assyria, so bricks were made of clay and straw. About 3000 B.C., they were already making arches and decorative walls. A true architectural style evolved in Greece starting about 1800 B.C. This classical architecture was reflected in Roman architecture about the time of Christ and still influences our modern structures, particularly public buildings. High vaulted ceilings, large arches, huge pillars, and massive domes are all hallmarks of this tradition. The Greeks and Romans not only built such buildings as public indoor bathing halls, art galleries, and libraries, but they also built a network of roads and overhead aqueducts. Their construction techniques were so sound that many fine examples still remain today.

Many architectural styles have developed throughout the world. The architecture of some countries has a distinctive style which immediately identifies a particular building as coming from that part of the world. Many of these styles developed as a result of raw materials unique to that locality or in answer to environmental characteristics of that region. Sometimes, however, particular styles simply reflected the decorative tastes of the people.

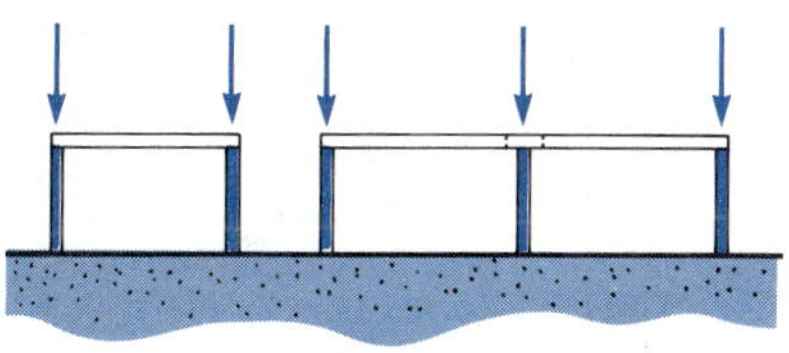

Smaller buildings need no interior load-bearing walls as long as the horizontal member is strong enough. Larger buildings require interior load-bearing walls because horizontal members are too short to reach from outside wall to outside wall.

Although styles varied, most early buildings were constructed in the same way, that is, they used load-bearing walls. In this method, the weight of the building is supported by the walls under each floor. This limited the number of stories a building could have, because the walls would not support the weight of tall buildings.

The first major advance in construction in modern times occurred in 1884, when the first skyscraper—a breathtaking ten stories high—was built in Chicago, using a steel framework, or skeleton, which carried the weight. This method was based on the principle that much more weight can be placed on a vertical member (post) than a horizontal one (wall). Many cities were already becoming highly congested by the turn of the century, and skyscrapers were the answer to demands for space.

The construction of transportation networks was much slower in coming. Most people thought trains were the transportation of the future, so little attention was given to roads.

This photograph shows both early and later tall buildings. Because early skyscrapers had to use heavy, thick walls, all of the weight could not be placed on the outside walls. By designing the buildings in steps, with the smaller sections at the top, the weight was distributed among several interior walls as well. Today's skyscrapers, however, are able to be built without steps because of the use of verticle posts rather than horizontal walls.

Many early roads, made by laying parallel planks of wood, were very rough, but they eliminated the problems of ruts and mud.

By the turn of the century most cities had main streets made of either cobblestone or brick, but country roads were still hard-packed soil and gravel. The advent of the automobile after 1900 served as the impetus for developing construction techniques in road building; the first concrete road was laid in Detroit in 1908. By 1924, only sixteen years later, the United States had over 31,000 miles (49,879 km) of concrete roads. As this expansion took place, it also became necessary to develop the technology providing improved bridges, tunnels, and overpasses.

Construction Today

Most of today's construction uses the same basic materials, regardless of whether it is a building, a microwave tower, or a highway overpass being built. Wood is primarily used in smaller buildings, as support members (walls), and in finishing (paneling, trim, and so forth). The wood may be in the form of lumber, which is used as it comes from the tree, or it may be a processed wood product. An example of a wood product might be particle board, which is made of wood chips. The four feet (122 cm) by eight feet (244 cm) sheets are made by compressing the chips with a glue to hold them together. This product can be made in many thicknesses, will not warp because the chips lay in all directions (unlike lumber which has wood grain running in one direction), is very strong, and provides a use for scrap and leftover wood. Plywood, used for structural work and paneling, is also a wood product. A large lathe peels thin layers of wood off a rotating log. The layers are glued together and trimmed to standard size, which is usually four feet (122

When the pieces of veneer are glued together to make plywood, the direction of the wood grain is alternated, providing additional strength and preventing warping.

cm) by eight feet (244 cm), with varying sizes of thickness. Because it is used in interior and exterior work, it is necessary to have the right type of adhesive. Interior plywood does not require waterproof glue, but exterior plywood does. Plywood can be obtained with a good surface on one side, a good surface on both sides, or with neither side good, depending upon its use. Very rough, low-grade plywood is used in constructing forms for concrete, because it is usually destroyed when it is removed after the concrete hardens.

Concrete is used in nearly all construction work. It is very heavy, strong, and fireproof. It can be cast into most any shape, and it can be used in nearly any environment. Concrete is made of four basic ingredients: cement, water, sand, and aggregate (stone). The key ingredient is cement, which contains lime (a mineral product used in making glass and fertilizer), silica (a mineral used in making glass), alumina (a mineral used in making aluminum and porcelain), iron oxide (a compound of iron and oxygen), and gypsum (a mineral used in making plaster of Paris). It is the cement which bonds the sand and aggregate together. Most material depends upon drying to harden. Concrete, however, hardens by chemical reaction

Concrete has to be cast in forms because it remains fluid for several hours. Even after the forms are removed, the material takes several days to completely harden.

when water is mixed with the cement. This is why footings for bridges, sea walls, and other structures can be cast under water. Concrete is very versatile. By changing the mixture balance or adding other ingredients, it can be made heavier, lighter, or stronger. It can be made to harden more quickly or even to contain small air bubbles, like a sponge, to resist cracking from expansion.

There are two basic kinds of concrete. Reinforced concrete contains steel rods and wire to increase strength; they are cast into the mixture when it is poured. Prestressed concrete contains steel cables. The cables are stretched with large hydraulic jacks while the concrete is poured and hardening takes place. After hardening, the tension is released, allowing the cables to create an inward pressure as the cables attempt to return to their original size. As a result, it makes the entire piece much stronger. This may be done on the site or in a factory, with the pieces being hauled to the site. Many times steel tabs and bolts are cast into the pieces. When the concrete modules are erected, they are then welded or bolted together to make them a permanent part of the structure.

Steel is used extensively in construction for structural members (beams, girders, and so forth), fasteners, and exteriors. Unlike concrete and wood, most steel for construction pur-

This bridge girder is being made of prestressed concrete. The tension on the rods becomes a compression force when the tension is released; the inward pressure makes it much stronger.

poses is usually manufactured to specific sizes before being brought to the construction site. There is little cutting, bending, and fabricating, other than what is required to attach the pieces of steel. Almost all the structural members for a particular bridge or skyscraper are designed and made to exact specifications for that particular building project.

Structural members are usually constructed in various lengths in the form of an *I* or an *H*. This shape gives them strength without the weight of a solid square or rectangular piece. Plates are usually about one-fourth inch (6.4 mm) thick and used for heavy construction in such places as building oil storage tanks and bridge floors.

Steel has a great advantage as a building material because it can be welded. This process melts the joining edges together, making them into a single piece. Many times they are bolted (using fasteners) or riveted (using pins which are flattened on the ends) to hold them in place while welding takes place. As a result, the skeleton of a building may actually be one piece of metal.

Other materials used in construction include glass, ceramic tile, aluminum, plastic, and masonry materials (brick and stone).

All structures have two things in common: a base on which they set and a superstructure which gives them support and shape. A base is usually made of concrete and set underground if the structure is large. If it is a flat slab it is called a

The use of steel structural members, such as these girders, allow wide open areas without supports in the middle. Note the walkway and the pressbox which are suspended from the girders in this athletic facility.

footing. The purpose of a footing is to keep the structure level and to keep it from sinking into the ground as the weight of the structure increases. The structure does not sink, because the weight is distributed over a large area. In some areas, there is no solid soil on which to build until a certain depth is reached. In this case, concrete columns, called piers, are poured so they rest on solid soil or rock. The structure actually sits on these legs which extend deep into the ground. The term superstructure refers to assembled members which make up the skeleton. They may be left open as with bridges, or they may be covered with an exterior skin, as for example, with large office buildings. Columns are vertical members, while beams and girders run horizontally. Girders are usually larger than beams. Smaller, closely placed beams are called joists.

This photo shows a concrete foundation wall with the superstructure built on it. The soil is being backfilled to wall, thus placing the wall several feet (cm) below the ground level when finished.

Smaller structures, such as houses, have a superstructure, but the terminology is somewhat different. Vertical members, two inches by four inches (5 cm × 10 cm) or two inches by six inches (5 cm × 15 cm), are called studs. Horizontal members are called joists, while the roof members are called rafters. A triangular shaped support used in place of a rafter is called a truss. The cross section of a typical wall will show some of the structural members used to make up the skeleton of a house.

Most buildings require several utilities, such as hot and cold water, electricity, natural gas, and telephone. These are

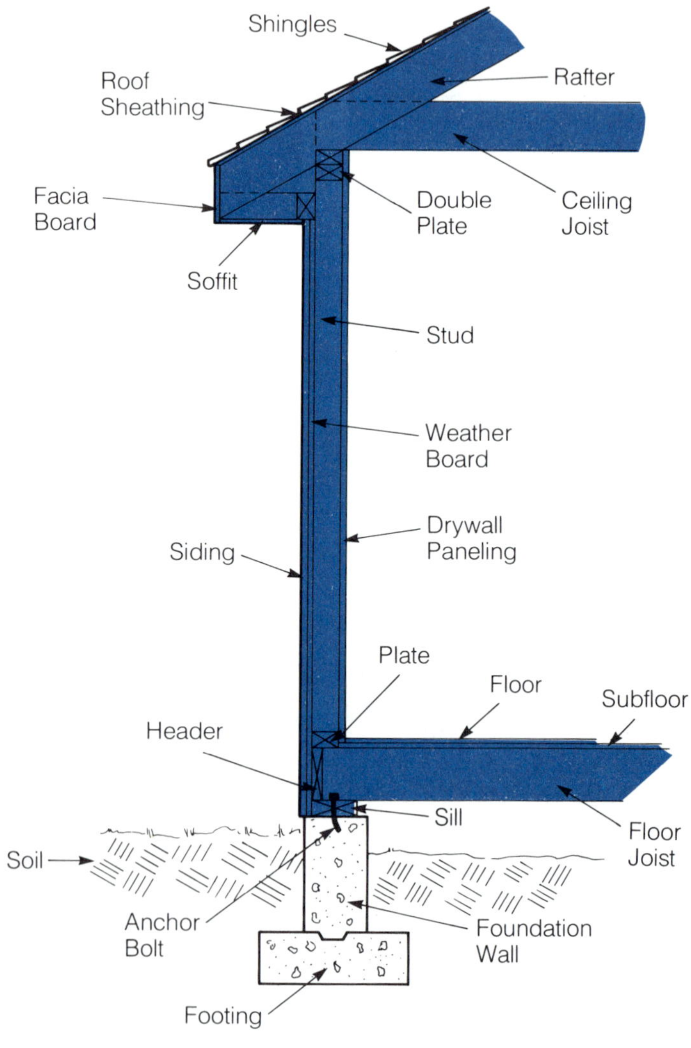

This cross-section shows a one-story building with no basement. A two-story building would be built in the same way, with the ceiling joists of the first floor serving as the floor joists of the second. A basement would require the foundation wall to extend downward, one story into the ground.

installed after the superstructure is completed, but before the walls are enclosed. Other installations which might be required are communication systems, security systems, fire sprinkler systems, special wiring for equipment, and fire alarms. Pipes, wires, and hardware inside the wall are a part of what is called the rough construction (such as rough plumbing). The finished plumbing is what is seen, such as sinks and faucets.

Construction workers usually have a specialty and perform only one type of job. An electrician only does electrical work, a plumber only does plumbing, and a "heavy construction" worker only deals with erecting the superstructure. As a result, there are usually many companies (called subcontractors), each with a specialty, involved in constructing a building.

Transportation surfaces also require a base for support. The preparation of the base involves cutting down high areas and filling low areas to make a relatively level surface. The soil is removed below grade level (surface of ground). This level, called subgrade, forms the roadbed. It is the natural soil which is packed down with heavy equipment. The base is laid on the subgrade, using crushed stone, sand, bitumen (a petroleum product), or cement. This provides support for the next layer and prevents moisture from forming under it. The next layer is called the surface, or wearing course. This is a smooth finished, water-resistant surface. It is usually concrete or a bituminous mixture, called macadam or blacktop. Road projects many times require parallel construction of utilities such as sewers, gas lines, electrical lines, and telephone cables. Ramps, overpasses, and interchanges are also necessary where highways intersect.

Notice the slight slope of the roadway in both directions; this allows water to drain off the surface. Also, the slope is usually increased more toward the inside of curves to help hold vehicles on the road where the force of inertia has to be overcome in turning.

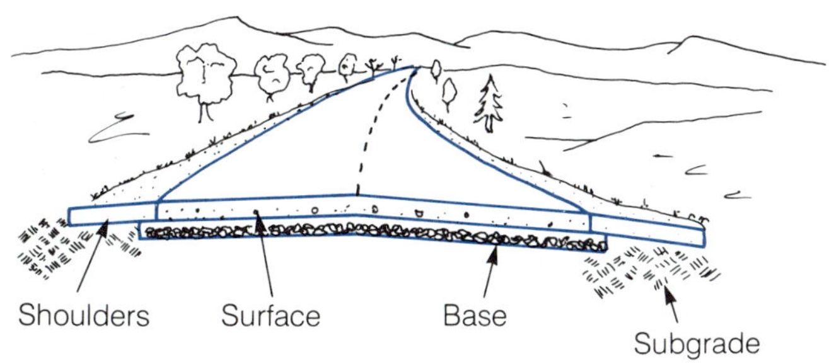

Construction Problems

Current problems in our construction system can be summarized in three broad categories. The first concerns inflexible structures, which in a society of rapid change, may lose their usefulness, become inefficient, and may be too costly for conversion to other uses. It is understandable that as human beings we feel the need to have things around us that are unchanging and permanent. But that impulse is self-defeating if it makes us reluctant to design new structures that respond to changing times. Boarded-up buildings in decaying urban areas testify to our failure to build anew or to adapt old structures.

The second problem concerns traditional thinking, which has slowed the development of less costly and more efficient construction. By nature, the construction system is traditional. Building materials have on the whole remained the same, as have the methods used in creating the structures and surfaces. Years can elapse before a new architectural design is generally accepted. The forward-looking designs of, for example, Frank Lloyd Wright, an American and one of the most famous architects of this century, took years to get some of his innovations accepted as standard practice in construction. Since innovative construction techniques take time before they are accepted, we are just beginning to explore alternative solutions to construction design which will be more efficient and more energy saving. Our problem now is that the traditional methods are becoming too costly, and we can foresee a serious time lag between our future needs and the reality of new innovations.

Third, lack of planning has made it even more difficult to resolve the conflicting viewpoints of various sections of the construction industry. Today, as the population increases and available space becomes scarce, conflicts become more pronounced and the consequences of not settling them more dire. Valuable farm land, for example, is consumed by urban sprawl. Housing is in short supply and very expensive. Transportation systems are outdated or have reached the limits of repair. Steel bridges are rusting away. A new approach emphasizing the interrelatedness of the subsystems is necessary, and a plan has to be developed to insure revitalization as a total system.

This house, designed by Frank Lloyd Wright in 1936, appears contemporary even today. Concrete cantilevered (no support) floors and ceilings, indirect lighting, natural air conditioning, built-in furniture, and the use of natural materials for both interiors and exteriors were among Wright's pioneering ideas in architecture.

Construction and the Future

As dismal as the problems sound, there is hope for the future. Future needs, extending into the twenty-first century, must be identified and examined. Three areas of research and development will have a positive impact on our lives: the development of materials, systems, and structural forms.

New plastics that are lightweight, strong, and even translucent are now available. New urethane foams can be sprayed on a lightweight superstructure for both interior and exterior walls. This has already been pressed into service in disaster areas to provide instant shelters for homeless victims. Strands of glass filaments can be wound around a collapsible mold and covered with plastic resin to provide a lightweight

New building materials make structures more energy efficient and more pleasing visually. This new coated glass reflects the sun's rays, keeping the building cool and light, and repeats the surrounding images, blending the building with the environment.

shell structure. New lightweight steels and new mixtures of concrete hold promise for lighter construction materials. These will carry more weight and will likely have an effect on design of bridges and similar structures. Ferrocement is a very lightweight cement material that can be applied to a steel and wire superstructure, much like a sculptor molds with plaster of Paris. This material, which is promising, has even been used to build boat hulls. As traditional materials are combined with new ones and new materials result from research in chemistry and physics, traditional construction designs will change.

The second area, the development of new systems, refers to the method used in joining components into a completed unit. One of the new systems is modular construction, where manufactured units of complete rooms or complete floors are connected on the site. Once refined, this method should provide a quality building at a lower cost than on-site construction, because of the high volume of completed products and the controlled manufacturing operations.

Other systems which are now available are layered exterior wall sections of fiberglass, which are attached to the superstructure, and collapsible wall sections, which provide flexibility in interior design. A number of energy systems are also under development, including solar and wind. These will change the design of structures, making them more efficient and adaptable to the particular energy system.

This modular house had each room built as a rectangular box at the factory. A crane on site lifted and stacked the boxes. Notice the cantilevered room jutting out to the right with no support.

One example of new structural forms would be the dome-shaped building. These are lightweight and free standing, allowing flexible interior walls and solar heating possibilities. In straight wall construction, severe roof angles will combine both roof and wall into one surface. This requires less construction work and also provides surfaces for solar collectors.

Cellular units of single rooms or multiple rooms, could be in our future. New materials, including steels and concrete, will allow wide expanses by use of larger girders. Superstructures, much like a bridge, could be built to accept these cellular units to make an apartment complex. One possibility

This structure is constructed of prefabricated units supported by a single concrete pedestal. It provides an interior circular stairway, as well as serving as the entrance for all the utilities. The lightweight dome roof is unsupported by interior walls.

Buildings designed for cellular units would contain elevators and other facilities. The crane could be constructed on top as a permanent part of the structure.

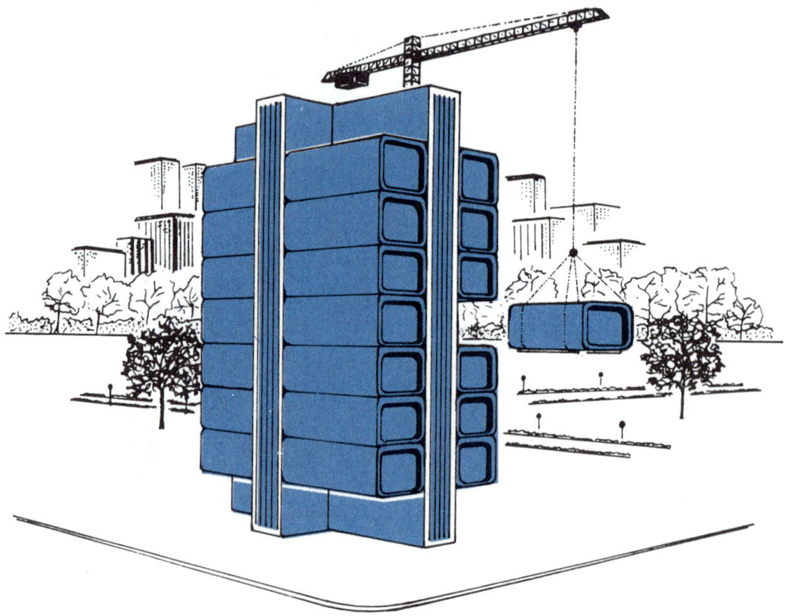

makes this different from modular construction: people may be able to take their own unit with them when they move. The next people moving in could have their unit "plugged into" the structure.

Another possibility, which has already proved to be highly successful on a smaller scale, is the use of container structures. These are huge domes which might cover an entire downtown section, or even an entire city. These could provide a "greenhouse city" with controlled weather conditions.

Integrated, highly efficient transportation systems will also be built, as will underwater structures and space structures. As the world population increases, it will become inevitable that future human existence will have to expand to include both underwater and space habitation. It is very likely

A container structure made of transparent or translucent material would control the exterior environment, allowing the use of solar energy for heating and protection from rain and snow.

that as structures are built for these environments, the designs will have an effect on the structures being built on the earth's surface.

PROCESSING

If processing is defined as that system which incorporates the steps of obtaining, separating, combining, and forming, then the first processing probably was the production of food and its preparation for final cooking. Early hunters and gatherers obtained raw materials from the natural environment for food, and also for clothing and shelter. As early humans domesticated animals and developed farming methods, they slowly began to exert control over their environment. Control over their environment allowed more control over their lives, because the availability of the raw materials was predictable if the correct procedures were followed.

As animal power gave way to mechanical power during the Industrial Revolution, the volume of food production increased. Tractors and other machinery not only allowed more food to be planted and harvested, but they helped make previously unproductive land more productive by building dams, clearing new land, and providing irrigation systems. Not all new ideas and machines worked out, of course. One example of a machine that did not succeed was the electric plow. It was not supposed to be powered by electricity, but to send an electric current into the ground through the cutting shears to make the soil more fertile. This neither put nutrients back into the soil nor made plowing easier.

By the beginning of this century, fishing and forestry were about the only industries that still obtained living raw materials from their natural environments. The forests, however, were being rapidly depleted, and a new endeavor, forest management, now seeks to control that environment. Fishing continues to draw from the least controlled of environments, but it cannot continue as it has in the past, or some varieties of marine life will be pushed into extinction.

Technology plays an extremely important role in the conservation and protection of living natural resources through planned replenishment and harvesting. Tree farming provides a continuous supply of wood for our products and assures us we will not overcut. Overcutting, which has happened in the past, disturbs the surrounding environment, can cause several years of shortage until young trees mature, and can lead to the extinction of some varieties. Unless proper controls are observed, animal and marine life can also be overexploited.

Extraction of nonliving materials became possible as scientific knowledge about minerals and their characteristics developed. Increased demand for certain minerals and other materials encouraged improvements in the methods of extraction. Large offshore drilling rigs, for example, were set up to drill for oil under the ocean when the demand became great enough to warrant the development.

Methods of changing the extracted raw materials into usable forms also evolved in step with scientific knowledge. The processing of raw materials makes extensive use of math, physics, and chemistry. As the processes became increasingly complex, industries saw the need for specific research and development. The application of scientific knowledge in continuous research and development on the same raw materials that

Mechanized tree farming provides wood, at the same time allowing conservation practices that keep the supply constant without depleting the resource.

have been available to humans for thousands of years has led to new cures for diseases, new lightweight materials, and new communication techniques.

Processing Today

The processing subsystem involves obtaining raw materials and changing them into a form which is usable at that stage or is ready for further changes in the manufacturing subsystem. Human progress may be observed by the historical study of processing techniques, because as the techniques become more refined, the potential of the end product is increased. For example, early humans burned coal for heat. With the advent of the steam engine, coal became a source of power with the potential for making new kinds of processing possible. This allowed the development of many other products made from coal.

Today we still use coal for heat and for the creation of power. In addition, we are able to process coal to obtain many valuable products. Examples would be coke, which is used in chemical processing and in making welding gas and steel, and coal tar, which is used in making plastics, drugs, rubber, insecticides, and storage batteries. Additional examples would include the following: processed materials, such as light oil, antiseptics, motor fuel, varnish, and food preservatives; ammonia, which is found in fertilizers, explosives, fire retardants, and refrigerants; and gas, which is used in making sulfuric acid, dyes, cotton products, and fungicides. As with all technological systems, the processing subsystem evolves in response to developments in other systems. The products produced from coal make it possible to process other raw materials that may be made into hundreds of new products.

The processing system begins with the obtaining of raw materials, which are classified as either living or nonliving. Most living raw materials consist of plants and animals, which are grown or collected. Growth occurs in a controlled environment, of which farming is an example. Collection involves acquiring raw materials from the natural environment, as fishermen do. Thus, farmers, fishermen, and foresters are important elements in the processing system. Nonliving materials are extracted from the ground. Solid materials, such as coal, limestone and copper, are mined. Liquids, such as water and petroleum, are removed by drilling.

The transportation system plays an integral part in the total processing system, carrying materials such as coal to and from processing plants.

After the materials are obtained they are separated from accompanying waste material. This is usually done by sorting, heating, and/or cutting, depending upon the material and how it is to be used. After separation, many raw materials are combined, perhaps by stirring or shaking, before the last step of forming is reached. Upon attainment of the final form, the material is either ready to be sold or to be used in the making of other products in the manufacturing system.

Obtaining. Although collecting is a part of the processing system, most raw materials are acquired through growing or extracting.

Growing activities vary greatly, depending upon what raw material is produced. They also vary according to where the growing takes place, whether outside or inside. Outside, plants, animals, and trees are grown on farms and ranches. Environmental conditions, such as weather and growing time influence the work involved.

Farming is a processing industry. On a worldwide basis, it involves more people than any other single industry. Most grain is harvested with a combine. It separates the grain from the stalk, drops the grain in a large hopper, and scatters the rest back on the field.

The growing seasons for crops are usually spring (planting), summer (cultivating and spraying), and fall (harvesting). In regions far north of the equator, such as the northern part of the United States, only one crop is grown each year. In areas closer to the equator, such as the South, two or more crops may be grown each year because of longer growing seasons.

Other environmental conditions, such as disease, insects, and amount of precipitation also have an impact on crops, as well as on animal production. Also, because the availability of animal feed depends upon how well the crops do each year, crops and animals are closely related in terms of production.

The growing season for tree farming is, of course, much longer. Trees take several years to mature, and during this time they must be protected from the dangers of insects, disease, and fire. Protection of the tree crop is extremely important as the forest products industry ranks in the top ten of the United States manufacturing industries.

Inside, a very different kind of growing is done in a controlled environment. In most cases the volume is not as great as, for instance, the thousands of acres of wheat grown each year, but the product may be just as important. Penicillin, for example, is obtained by growing certain kinds of molds which produce the antibiotic. Processors grow the mold in highly controlled conditions in large tanks which may hold up to 15,000 gallons (56,78 l). Many other medicines are also made from plants, which are grown under carefully controlled conditions in greenhouses. The correct amount of light, water, and nutrients ensures a top quality plant for further processing.

This symmetrical colony is an example of the penicillin mold, which is green in color and produces almost all of the world's commercial penicillin.

Extracting is the process of removing nonliving raw materials from the earth through either mining or drilling. Several methods are used in mining, depending upon the depth of the material. Some minerals, such as coal, may be found in veins (layers), while others, such as gypsum, are found in large beds.

Above-surface mining includes dredging, open-pit mining, strip mining, and quarrying. Dredging is warranted when sand and gravel contain heavier minerals. A pond or lake is formed to float a dredge (a bargelike machine), which has a boom (long arm) in front. This boom has a series of buckets in front, which continuously move back to the machine after being filled with sand or gravel. They are dumped into a bin where the material is washed, allowing the heavy minerals to settle in grooves in the bottom of the box where they are saved. The rest of the material is dumped back into the water at the rear of the machine. This continuous process replaces the soil as the machine moves forward. Open-pit mining involves digging a huge hole with large steplike ledges called benches. These circle the tapered hole allowing trains or trucks to bring the material to the top by following them around and up to the top, much like the threads on a screw. The rock is usually broken loose by explosives and loaded onto the trucks or trains with large power shovels (cranes). Strip mining is utilized where the material is close to the surface. This method is used in mining much of our surface coal. The surface, called overburden, is removed by large power shovels, leaving the

This huge machine, called a dragline, is used in strip mining. It weighs 13,500 tons (12,150 t), can reach the length of a football field, and can move 325 tons (293 t) of soil in a single pass.

This open-pit copper mine is more than 1/2 mile (.8 km) deep and 2 1/2 miles (4 km) across. Heavy equipment vehicles travel on the benches.

land with no growth. This has caused major erosion problems in some areas, but most states now require that the land be made productive again after the mining operation. Quarrying is a method of cutting large blocks of material and raising them to the surface with chains and pulleys. This is done in mining such rock as limestone, granite, slate, and other building materials.

Below-surface mining involves digging vertical shafts, horizontal passageways, and large underground rooms to reach the deep lying materials. In "level and shaft" mining, a vertical shaft might run nearly two miles (3.2 km) deep, with horizontal passages running from it at various depths. Because some veins (layers of minerals) run at an angle with the sur-

This underground salt mine at Detroit, Michigan, has rooms 50 feet (15 m) wide and 23 feet (7 m) high, extending more than one mile (1.6 km). The mine is 1200 feet (360 m) below the surface.

A continuous mining machine digs ore or coal loose and sends it to the rear on a conveyer. Because of the danger in unsupported areas, the machine is operated at a distance by remote control.

face, this method allows several passages to intersect the vein. "Room and pillar" mining is used when the vein is horizontal, or parallel with the surface. Large rooms are cut on the same level, with columns being left to hold up the ceiling. In most underground mining, the equipment is powered by electricity. If fuel engines were used there would be the danger of poisonous exhaust fumes or of explosion of the flammable fuel.

Extracting liquid material usually means drilling a hole to the proper depth and capping it, so the flow can be controlled. Some wells may have enough underground pressure to force the liquid out, while others have to be pumped. Petroleum products, such as gas and oil, are extracted in this way, as well as some minerals such as magnesium and salt. Salt, for instance, is obtained by pumping water into the drilled hole where there is a salt deposit. As the salt is dissolved into the water, it is pumped to the surface. It is then processed by evaporating the water, thus leaving the solid salt. Much of our salt today is obtained in this manner.

Separating. Few raw materials are obtained in a pure form. Plants used for food have to have accompanying soil and other foreign matter removed; minerals have to be separated from other minerals; and trees have to be cut into logs and their branches removed. Many methods are used to separate raw materials, depending upon the characteristics and future use of the material.

Three primary methods of separating are sorting, heating, and cutting. The following examples of each do not consti-

Oil wells drill deep into the ocean bed. The helicopter on the landing pad to the left is the transportation link to the mainland for the people who live and work on the rig.

tute a complete list, rather they are a sampling from many types of processing operations.

Sorting involves four methods: washing, screening, floating, and filtering. Washing separates solvent and insolvent materials. The material is dumped into a tank of water to soak and to loosen the solvent material. When it is separated, the insolvent material passes through a high pressure spray to loosen any remaining particles. Vegetables, for example, go through this process to be stripped of any clinging soil particles.

Screening separates materials according to size by using screens with various sizes of holes. As the material is vibrated across the screen, particles which are small enough fall through. As it moves to the next screen with larger holes, larger particles fall through. Material is separated into as many groups as there are sizes of screens. Crushed rock and gravel are separated in this manner.

Floating separates materials according to weight, by taking advantage of the differences in buoyancy. When placed in liquid, some materials sink to the bottom while some float to the top. For example, in making steel, the molten mixture contains impurities called slag. When the mixture is poured from the furnace into a large ladle, the slag rises to the top and overflows into a smaller ladle, leaving a high-quality steel with few impurities. Because some liquids are heavier than others, they also may be separated using this method.

Finally, filtering separates liquids from solids. The filter may be a fabric, or it may be a porous material several inches (mm) thick. It allows the liquid to pass through, either by means of gravity or under pressure. At the same time, it prevents any solids from passing through. One of the steps in processing raw milk is filtering out any foreign matter before it is pasteurized and homogenized.

Separation by sorting is a mechanical process, whereas separation by heating is a chemical process. Heat alters the characteristics of the materials by changing the molecular structures. The three basic processes used in separation by heating are distilling, melting, and evaporating.

Distilling is a process used to separate gas from a liquid or solid. The material is heated to a temperature high enough to evaporate the liquid. The heated liquid becomes a gas and is passed through a condenser which cools it again. This distilled liquid is purer than the original, because it has left behind solids and impurities which would not turn to gas. This method, called simple distillation, is used to obtain pure distilled water.

Butter pours out the tube at the right end of the machine following a process involving several steps of separation.

Fractional distillation is used to make many of our petroleum products. Although the principle is the same, several different liquids are separated from one mixture. Each one has a different temperature at which evaporation takes place, so they are collected in stages, making up fractions of the total, thus, the name fractional distillation. In producing petroleum products, a huge tower acts as the heating chamber. The materials in the raw petroleum which make up the earliest fractions boil off at the lowest temperatures. Their vapors rise the highest, allowing them to be carried off through pipes high in the tower. Separate pipes at different levels carry off the heavier vapors. Naptha and benzine (cleaning agents) are among the earliest fractions, followed by gasoline, kerosene, and various grades of lubricating oils. All of these products can be produced at one time under a continuous process.

Destructive distillation forms a new substance, whereas the other types of distillation only change the characteristics. This method breaks down some materials, causing an irreversible chemical change. For example, when wood is heated without the presence of oxygen, it does not burn. The gas it gives off is used to make wood alcohol, and the decomposed wood which is left is charcoal. Wood alcohol is an industrial solvent, while charcoal is used commercially as a fuel, a purifying agent in filters, and an absorbent of odors.

Melting, the second method of heating, changes solids to liquids. Because not all solids melt at the same temperature, they may be separated by raising the temperature to a point where one material melts before the other. Also, when both become liquids, they may be separated using one of the previously identified methods. After separation, they are returned to two separate solids by cooling.

An example of this process is in the production of aluminum. The raw material taken from the ground is called bauxite. After being mixed with several other ingredients, it is placed in a digester where it is heated under steam pressure. The impurities remain as solids, while the liquid is filtered into a cooling chamber. As it cools, crystals, called alumina, form and settle to the bottom. These crystals are then further processed into aluminum.

The third heating method, evaporating, is closely related to distillation except that the evaporated liquid is not saved. The process is used to obtain solids which are suspended in liquids. Salt is obtained from sea water using this method. It is

This solar salt plant is in the Caribbean. Sea water is pumped into shallow ponds covering thousands of acres, and the sun evaporates the water. The crystallized salt is then scraped up and hauled to storage areas where it is stockpiled.

estimated there are a little more than four ounces (113.4 g) of salt in each gallon (3.8 l) of sea water, making it a rich natural resource. Although there are a number of solar plants in the United States where acres of salt are dried in the sun, the brine (salt water) is usually heated in large pans over an extremely high heat.

Probably the first separating process used by humans was cutting. This method does not change the characteristics of a material, it changes only the size. Shearing, scraping, and

This machine rotates a log against a long knife, shearing off a continuous thin sheet of veneer which is used to make plywood.

grinding are examples of cutting. Shearing involves the use of a cutting edge passing vertically through the material. It could be a single edge or two opposing edges, like scissors. Running wood through a chipper, whose many cutting edges reduce a large log to very small chips, is an example of shearing.

Scraping removes only the surface. A vertical cutting edge is passed over the surface. This parallel motion does not shear, which would make a single clean cut, but separates by pulling the fibers apart. Animal hides, for instance, are scraped, which removes hair and flesh before the hide is eventually made into leather products.

Grinding removes material by making hundreds of very small cuts, which abrade the surface. Grinding might make the surface smooth, or it might remove unwanted material. The grinding instrument, such as sandpaper or grinding wheels, has a rough surface. The texture of the surface being ground will vary according to the desired finish. Narrow grinding wheels are used for cutting material into two pieces. They are used many times in cutting metal rods to length at the end of the processing phase. Wide abrasive belts remove only surface material. They may be used to smooth a wood surface or to remove the skins of certain vegetables.

Combining. After the materials are separated and further processed, in most cases, they have to be recombined, usually by stirring or shaking with other materials to make the final product. Most liquids are combined by slow stirring with a continuously moving paddle in large vats. Rapid stirring can create air bubbles in the mixture and actually retard the mixing process. Most of our liquid foods, such as salad dressings and fruit drinks, are mixed by stirring. The materials may be combined while either cold or hot. Shaking is usually used to combine loose solids. Many times it is also used in conjunction with rotational mixing, where an enclosed drum not only vibrates but rotates at the same time. Examples of loose solids might be animal feed (crushed corn, wheat, or so forth) or breakfast cereals with more than one ingredient.

Forming. The final step in the processing subsystem is shaping the processed material so that it is suitable for sale or future manufacturing. The combining process is usually the final step for liquids, but solids may be made into many shapes, including sheets, rods, tubes, and blocks. There are five basic

A technician prepares to pump milk chocolate paste from this 10,000 pound (4,500 kg) vat where it has been mixed and refined.

Molten steel is being poured into a line of molds as an overhead track moves the heavy crucible down the line.

methods of transforming materials into their final form: casting/molding, bending, stretching, rolling, and weaving.

The process of casting and molding usually involves hot materials. Casting is the term used when a molten material, such as metal, is poured into a cavity and cooled. The result is a shape exactly like the shape of the cavity. In industry, this cavity is called a mold. It may be made of moist sand, or it may be a permanent one made of steel. Sand molds are made by packing the moist sand around a pattern. The pattern is then removed, leaving a cavity of the same shape. Some casting processes are continuous, eliminating the need for patterns. Molding refers to the forming of materials under pressure by hammering or squeezing the hot material into a desired shape. When hot plastic is forced under pressure into a closed cavity, called a die, it is called injection molding. When a material such as steel is hammered into a specific shape, it is called forging.

Bending machines are usually one-of-a-kind machines built for producing a particular shape. There are innumerable shapes that rods and other pieces are bent into for further manufacturing.

Stretching reforms the material, usually metal, by making it thinner. Flat material may be shaped by a machine which forces it into a cavity called a die. The metal stretches as it is shaped into a three-dimensional object. This process is called stamping. Another method of stretching is used in making wire. Continuous metal rod, about one quarter inch (6.4 mm) in diameter, is pulled through a series of dies (small holes), with each one getting successively smaller. The result is a very thin wire, much longer than the original rod.

Rolling is accomplished by passing either hot or cold rectangular steel blocks, called ingots, between large rollers which compress it. With each pass, the space between the rollers becomes narrower and the metal gets thinner and longer. Semifinishing rolling mills roll the ingots into rough rectangular shapes, while finishing mills roll the steel into elongated final shapes, such as steel beams for construction purposes.

Weaving involves the making of fabric. Although we usually think of fabric as clothing and many other household items, there are many other uses. In fact, industrial fabrics make up most of the textiles used today. Filters may be made of glass, plastic, or even metal fabric, while belts, covers, and

The rolling process reduces the size or gives a certain shape to long strips of material. White-hot steel is being rolled into an* H *beam which will be used as a structural support in construction.

aprons for machinery require cotton fabric. Even small aircraft are covered with fabric. Glass textiles are used in making fiberglass boats, as well as covers for other transportation vehicles. Many other textiles are used in manufacturing books, luggage, adhesive tape, and typewriter ribbons. Most processed fabrics are either cut into sheets or rolled into large rolls for shipping to the manufacturing plants. The looms are completely automated. They provide a continuous process, which is programmed for specific colors and patterns.

Processing Problems

The problems facing the processing system are essentially the same problems facing society today: environmental problems, shortage of certain raw materials, and shortage of energy. Each has a definite effect on the other.

When any activity takes place, the environment is disturbed. The removal of raw materials has an effect on ecology, as does transporting, processing, and discarding. The question is how to minimize the effects and plan for restoration.

Most raw materials taken from the earth are nonrenewable. Each year as the world population grows, the demand for these resources increases. Some are running short now, requiring more expensive processes in order to meet the demand. Others are still plentiful, but will not last forever. It is even possible that renewable resources, such as plants and animals,

Technology can be applied to reclaim land which has been disturbed during the removal of natural resources. A hydroseeder sprays a mixture of water, fertilizer, and tree and grass seeds over bare land, reaching areas that could not be reached by foot.

could be in short supply on a worldwide basis in the years ahead.

The energy shortage is a matter of deep concern, because high technology requires much energy. Because most of our energy in the past has come from nonrenewable resources, it is important that new thinking be given to energy consumption. The solution to these problems can be found in directing research toward three general areas.

- Make older processes more efficient. Most all raw materials have to be processed in some way. The cost of a product reflects the cost of its processing from its raw state; efficient processing means less cost for the end product.

The application of new technology enables older processes to become more efficient. A laser measuring system checks the accuracy of a production machine.

- Discover new raw materials and uses. A shortage of raw materials causes a shortage of products. Synthetic materials and processes need to be developed to continue production of today's products and to meet tomorrow's needs.
- Discover new products. Each new product opens up possibilities for others. The laser, for example, is now used in processing many materials. It can produce heat over 10,000°F (5538°C), it can produce holes in high-strength steel, and, yet, it can be used for delicate surgery, such as reattaching a loose retina to the eye.

Processing and the Future

To a great extent, the same three solutions mentioned in the preceding paragraphs indicate the direction of the future. Research on new processes and materials involves work on earth and in space. The lack of gravity in space holds great promise for producing products which cannot be produced on earth. Weightlessness produces unique characteristics in crystal growth, in the mixing of liquids, in separation of liquids and solids, and in solidification of solids. These unique processes will have a great impact on new materials, such as plastics, metals, medicines, and crystals used in electronic components.

An example of a new type of product to be developed is a very high-strength magnet. Metals containing iron can be given magnetic characteristics by the arrangement of the mol-

Research and production in space will become a common element of our technology system.

ecules. If this arrangement can be accomplished in a weightless condition, it is possible that an extremely high-strength magnet could be produced. This would open up many new possibilities here on earth for lightweight electric motors and turbine blades used in the production of electricity.

Although many new methods of creating energy are under consideration, nuclear and solar are receiving the most attention. Although we have had nuclear power for several years, it is still not fully developed as an energy source. There are problems of containment of radioactive materials and waste disposal of spent fuel. It may hold promise as a long-range complete energy source.

Solar energy, because it is unlimited and "clean", probably has the greatest potential for shorter-term energy production. It is estimated that the sun is available on the average of 17 percent of the time in Arizona and only 6 percent of the time throughout the United States. If solar collectors could be moved into space, the sun would be available over 99 percent of the time.

Because of the weightless condition in space, large collectors are possible. One such collector, called a Satellite Solar Power Station (SSPS) is under consideration for the future. It consists of two large rectangular panels, each measuring approximately three and one-half miles (5.6 km) by two and three-quarter miles (4.4 km). The energy would be beamed to earth as microwaves, then converted to electrical energy.

Other space projects in energy will be the development of technology to make generation and transmission more efficient. Plans call for the development of power reflectors in space, which will relay energy from earth to the reflectors and back again to a different location around the world. This would provide a tremendous boost to developing countries, as well as distributing energy evenly throughout our own country.

The problem of ecology, recycling materials, and world food production are all related. There is a growing consciousness of these problems, and research is being conducted to help solve them. For example, recycled crushed glass has been used successfully in making paved roads, and used rubber tires have been successfully converted back to gas and oil (raw materials from which they are produced).

A giant shredder has reduced automobiles to fist-size pieces so the steel can be recycled. A magnetic drum separates the metallic from the nonmetallic material. When scrap is used rather than ore in making new steel, there is a 74 percent savings in energy expended.

Although new hybrids of plants and animals are being sought, the attention for future food production is moving to the oceans. A relatively new term, called aquaculture, means sea farming. The ocean holds a wealth of minerals and nutrients, and much of the sea life, including plants, has not been explored as common food. Plants and fish in underwater farms, under controlled conditions, are being grown now on a limited basis. It has proved to be successful and holds potential for a whole new food source.

The cultivation of water plants is a new technology. The water hyacinth, an aquatic plant which grows in warm climates, thrives on sewage. It absorbs and digests wastewater pollutants, converting the polluted water to relatively clean water. By-product potential for the harvested plants is energy and livestock feed.

The search for new sources of minerals has also entered into ocean exploration. The ocean floor in many areas is covered with black nodules about the size of potatoes. These rocks are very rich in minerals, including manganese, nickel, copper, and cobalt (all valued industrial raw materials), but there are the questions of how to mine them, because most are over three miles (4.8 km) deep, and who owns them, because they are beyond the 200 mile (322 km) limit of the continents. A workable solution to both questions may lie in a joint effort by several countries.

MANUFACTURING

Even though there was a division of labor early in human history—some members of a tribe hunted, others gathered, still others wove—there is little evidence to indicate a division of processing and manufacturing. Even as late as the early 1800s, both were still combined. Those who tanned leather also made the shoes; those who operated ironworks also made the cooking pots.

Until the Industrial Revolution of the nineteenth century, most goods were produced within the homes and shops of craftsmen, who planned, made, and sold the products. At that time, machine labor replaced hand labor and mechanical power replaced human power. Two groups entered the labor force in unprecedented numbers: unskilled workers and female workers. As a result of the new resources of labor, materials/processes, and energy, the factory system rapidly replaced the craft system. It soon became more profitable to perform only one part of the production operation. Manufacturing industries began obtaining processed materials from others, concentrating their own efforts on fabrication and assembly.

Until the 1920s, little consideration was given to the aesthetic design of products. Most were designed by the machine operators, because they knew the capabilities of the machines and materials. The primary concern was function. This situation changed as new machines and metal alloys allowed greater flexibility in design and industries began competing on a large scale for sales. A new job, that of industrial designer, came into being. That person, or group of persons, was responsible for designing a product that would be the best on the

This photograph, taken in 1913, shows workers assembling small parts in a manufacturing industry. Many of the parts were not made in the plant. Rather, they were purchased from other industries which specialized in producing only parts.

market and outsell the competition. To fulfill this purpose, an industrial designer needed knowledge of engineering, economics, and marketing, combined with an awareness of social attitudes—a considerably wider background than a machine operator would have. Today manufacturing industries employ designers of different specializations who identify, design, and develop new products, which enables consumers to choose from many sources and many prices.

As the manufacturing subsystem expanded with the transportation system, more products were shipped from one part of the country to another, and packaging became necessary. The products of a general store of the early 1800s were contained in barrels or piled on the floor or hung from the ceiling. Today, most items are manufactured with the package as a part of the product or as a protective container to be discarded once the product is removed.

Manufacturing Today

The manufacturing subsystem begins with the design phase, which usually involves many people, from engineers and scientists to marketing experts. Ideas are formulated and tested before actual production begins. There are three basic steps. The feasibility study defines the problem, identifies difficulties that might arise with the product or production process, notes the effects of the product on the market, and assesses the im-

Packaging serves several purposes. It allows the product to be stacked and transported easily, it protects the product, and it identifies the contents and special instructions for handling.

pact of the product on society and our ecological systems. Several ideas are analyzed, with the best solutions being identified.

The preliminary design phase settles on the one solution to be followed. This path is evaluated in greater detail in terms of, for example, the way heat, noise, humidity, and vibration may relate to human use. This aspect, called human factors engineering, takes into consideration the biological factors involved when people and machines function together.

The detail-design phase is that portion of the procedure where the engineering design details the final decisions regarding exact size and shape, appropriate materials, the types of machines and processes needed, and the sequence of events for manufacture of the product.

This three-wheeled urban car is near the end of the preliminary design step, having been analyzed in terms of both human factors engineering and environmental impact. It is a full-scale model, 10 feet (3 m) in length, with a canopy that opens like the cockpit of a small jet plane.

The time period involved in these three steps varies greatly according to the product. It could range from only a few weeks to three or four years. Most industries have a research and development facility, which might be a modern machine shop, a laboratory filled with the latest scientific apparatus, or a large outdoor test track. Full-scale or proportional models might be made. A three-dimensional model used extensively during the testing phase is called a prototype. It may or may not look like the final product; its purpose is to demonstrate function. It has its greatest use in engineering design. Computers are of great use during testing. Simulation tests can be run on a computer, which saves a tremendous amount of money because labor and materials have not been used. Hundreds of drawings are made during the design phase. These drawings are reproduced so those involved, which might include factories all over the world, can have copies. These copies, which usually have a white background with a blue line, are the blueprints, or "prints", referred to earlier in this book. Plans for the product are drawn, as well as those for new machines and their placement in the factory in proper step-by-step sequence. Other industries, including the processing industries which supply parts and materials for the new product, also have to have prints.

The next step is obtaining the necessary processed materials and setting up schedules for transporting materials into the factory and storing them. Transportation also has to be worked out within the plant, so the materials flow steadily to the work stations and flow smoothly through the many required operations.

Fabrication. Fabrication refers to the operations necessary to prepare the parts for final assembly. It might incorporate many

Depending upon the product, many thousands of prints might have to be made. As the design changes from year to year, new drawings have to be made and the revisions have to be reproduced to keep all those concerned up to date.

of the same methods employed in the processing system, such as separating, joining, and forming. In addition, it may involve producing subassemblies. These are small groups of parts which go together to make up a single unit; they are usually permanent assemblies, done by using fasteners, welding, or adhesives. The units are then placed in the product during the final assembly process.

Fabrication techniques depend upon the product being manufactured. The list of materials which are utilized is almost endless, because most are used in combination with others, providing new characteristics not possible with each one alone. The four primary industrial materials are woods, metals, plastics, and ceramics.

Because wood was once a living organism, it has characteristics somewhat different from other industrial materials. Its

Fabrication may be done on an assembly line where there are many product units, or it may be done on a single unit. These engineers are assembling a single-unit optical communication device (fiber optics) with prefabricated parts.

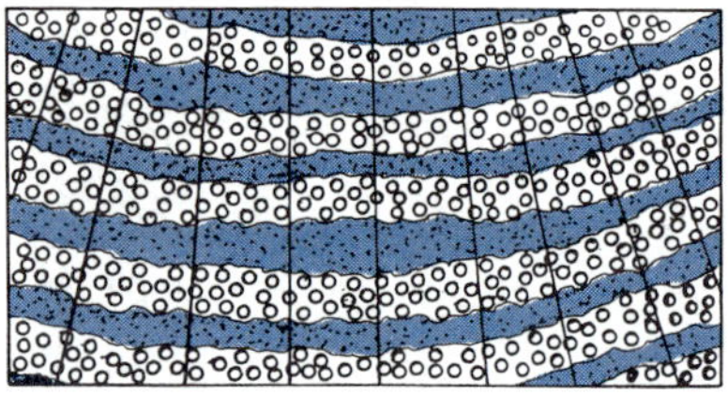

The cell structure of wood can be seen by examining the end grain of a piece of wood that has been made very smooth. A magnifying glass helps, but some woods (such as oak) are "open" enough to show the structure without one.

cellular structure is porous, which makes wood light in weight, flexible, and, at the same time, relatively strong.

Wood is classified commercially into hardwoods and softwoods, depending on the type of tree from which the wood has come. Hardwood trees, such as oak and maple, are deciduous; that is, they lose their leaves each year. Softwood trees, such as fir and pine, are coniferous; they bear needles the year round. Different woods vary greatly in hardness, color, smoothness, smell, and resistance to decay. Usually softwoods are used for construction, and hardwoods are used for furniture making, where a fine finish is required. However, the degree of hardness does not necessarily depend upon whether the tree is deciduous or not. Some hardwoods are soft and some softwoods are hard. The specific variety determines hardness.

The dimensions of lumber are given in the size in which it is rough sawn from the log. Finished lumber is smaller than the size given because it has been planed down to make the surfaces smooth. For example, a 2″ × 4″ refers to the thickness and width of the cross section; the actual size of the finished piece is really 1-½ × 3-½ inches (38 mm × 89 mm). Most lumber comes from the processing system in standard sizes of thickness, width, and length.

Because wood is composed of porous cells, it is able to hold moisture. When it comes from the processing system, it has been dried in a kiln to take out as much moisture as possible to keep the wood from warping because of uneven drying

Lumber is rough-sawn from logs by means of a wide-blade band saw. A band saw uses a continuous blade that operates as a moving belt on two pulleys.

in the air. The final step in manufacturing a wood product is applying a finish to protect the surface and seal it so moisture cannot be absorbed, causing cracking and warping.

A large part of our earth is made of metals, which are present in varying amounts. Each metal has certain characteristics that make it useful for certain applications. One of the most common and most useful metals is iron, which is, however, rarely utilized in its pure form. It is mixed with other ingredients during processing, making it into steel. Most steel is also mixed with other metals to obtain specific characteristics for certain uses.

When metals are combined, the mixture is called an alloy. Metal alloys which contain iron are called ferrous metals and those which do not are called nonferrous metals. Iron oxide (rust) can be seen on ferrous metals; nonferrous metals do not rust. This is why bridges have to be painted to protect them from rusting, while other objects, such as statues and some building materials (usually copper or bronze), do not require painting. Nonferrous metals will tarnish, providing a dull protective finish. Rust does not provide protection, allowing the air and water to continue the erosion process. One well-known exception to the rusting of steel is an alloy called stainless steel, It requires no finish and still resists rusting and tarnishing. In this alloy, chromium and nickel, which are nonferrous metals, provide the antirust characteristics.

Most health care and food processing equipment is made of stainless steel. Milk trucks have a stainless steel tank to prevent any rust from getting into the milk.

Ferrous Alloys			
Name	**Alloys**	**Characteristics**	**Use**
Stainless steel	steel, chromium, nickel	resists heat, wear, rusting, and corrosion	medical equipment, food preparation equipment
Tungsten steel	steel, tungsten, others	very hard, high heat resistance	cutting edges on metal cutting tools
Titanium steel	steel, titanium, others	very light, high strength	aircraft and spacecraft frames
Manganese steel	steel, manganese, others	heavy, tough, and easily worked	heavy equipment, structural steel (beams, girders, and so forth)

Nonferrous Alloys			
Name	**Alloys**	**Characteristics**	**Use**
Bronze	copper, tin	easily formed, inexpensive	statues, coins
Brass	copper, zinc	easily formed, ranges in color from red to yellow	light sockets, decorative items
Yellow gold	gold, copper	bright yellow-orange, expensive	jewelry, dental fillings
White gold	gold, silver, or palladium	silver color, expensive	jewelry
Sterling silver	silver, copper	silver color, expensive, can be highly polished	tableware, decorative items
Pewter	tin, copper, lead	easily worked, can be highly polished	tableware, decorative items

The alloying (combining) of different metals provides characteristics that are not possible to obtain separately.

Metal alloys can have many different characteristics, depending upon the combinations of metals. For example, refractory metals are those metals which have a melting point over 3600°F (1982°C) with some going over 6000°F (3316°C). At the same time, sodium melts at 208°F (97.8°C). The heat-resistant characteristics in the possible combinations of alloys falling between these two extremes are nearly endless. Carboloy, an alloy used in making cutting edges to cut other metals, is nearly as hard as diamond, while solder (an alloy of lead and tin), which is used in joining metals, can be soft enough to mark with a fingernail. Among other important metal characteristics are heat conductivity, electrical conductivity, magnetism, and weight.

A process called heat treating is used in nearly every instance where steel is used. This involves heating the steel to a "cherry red" color and cooling it quickly in a liquid bath (called quenching). The higher the temperature when it is quenched the harder the steel. Heat changes the structure of the molecules, and quenching stops this change at a predetermined point, giving it a specific hardness. A high-carbon steel bar can be made so hard it will break from brittleness, or it can be made soft enough to bend with the hands.

Of all the industrial materials, plastics are probably the most versatile. They are made from many raw materials, ranging from oil and other petroleum products to cotton and cornstalks. Because they are the result of chemical processes, and therefore a synthetic material, many different characteristics can be obtained. They may be made into material hard enough to be used as machine parts or soft enough to be used in making sheer fabrics. Some plastic products are listed below:

- Adhesives
- Aircraft canopies
- Appliance parts and cabinets
- Automobile bodies
- Boats
- Brushes
- Buttons
- Chairs
- Clothes
- Dinnerware
- Electronic components
- Electrical parts
- Food packaging
- Frying pan coatings
- Greases
- Hand and power tools
- Handles
- Imitation leather
- Industrial fabrics
- Insulation
- Knobs
- Lamp shades
- Life jackets
- Light fixtures
- Luggage
- Machine parts
- Oils
- Optical lenses
- Packaging
- Paints
- Paper coatings
- Phonograph records
- Plumbing pipes and fixtures
- Printing plates
- Protective coatings
- Refrigerators
- Rope
- Shoes
- Squeeze bottles
- Swimming pools
- Table and counter tops
- Telephones
- Tires
- Toys
- Washing machines
- Wheels

There are two basic classifications of plastics: thermosetting and thermoplastic. Thermosetting materials can be

These gears are made of nylon, a thermoplastic material. They are very tough and, unlike metal gears, require no lubrication. They are commonly used in tools and household appliances.

melted only once. After they are melted during processing they only get harder as more heat is applied. Examples of products where these are used would be paints, electrical components, dinnerware, and kitchen countertops.

Thermoplastic materials can be melted as many times as desired. The melting temperature depends upon the type of plastic, but they all return to a solid state once they are cooled. These plastics can be found in many automobile parts, windows, fabrics, product packaging, and toys.

Because all plastics are at one time in a liquid state, or can be reheated to attain that condition, most products are molded into their final shape from a liquid. Very little cutting and machining is required on plastic materials.

Manufacturers usually get their plastic materials from the processors in either large sheets, small pellets about the diameter of pencil lead, or rolls (fabric, films, and threads). The manufacturer may use the sheets in fabrication by bending, cutting, and drilling the material. The pellets are fed into machines which melt and form them into a mold, making such products as toys and soap bottles. The plastic films may be used to seal out moisture and protect surface finishes, while threads and fabrics are woven and sewn like any natural cloth.

Glass is a very versatile ceramic material. It not only can be heated and formed or cast, but it can be blown in various shapes.

Ceramics are a material made from clays and rock which fall into a large class of minerals known as silicates. These were formed millions of years ago when the earth's crust cooled and hardened during its transformation into a solid mass. Silicates can be found in nearly every part of the world.

Although ceramics are usually thought of as being used in making pottery and dinnerware, most are used in industrial applications. In addition to the obvious industrial uses of cement and brick in construction, ceramics are used in many other forms.

Glass is a ceramic material. In addition to our normal types of glass, some are highly heat resistant, allowing vision into machines and furnaces while providing protection for the workers. Tempered glass can withstand hard blows without breaking and is used where breakage is possible. Safety glass, such as we have in our automobiles, consists of two layers of glass with a clear sheet of plastic between. This will not pre-

This industrial grinder is using a large diameter grinding wheel to put a round, smooth surface on a smaller cylinder. Most grinding is done to produce smooth surfaces or remove only a small amount of material. The liquid spray is a coolant that carries away the heat generated by the cutting.

vent breaking, but will prevent the glass shattering into flying pieces. Additional industrial uses of glasslike ceramics are containers for holding acids and other corrosive materials, laboratory test equipment, and surface finishes that resist extreme heat.

Most insulation materials are made of ceramic materials, including fiberglass (fibers), asbestos (hard sheets), rockwool (blankets or rolls) and vermiculite (lightweight granules). These are used in preventing temperature extremes, either hot or cold. Most processes produce extreme temperatures, and insulators protect the workers, the machines, and the product.

Ceramic lubricants are used where pressure and heat are too high for normal petroleum lubricants such as oil and grease. Probably the best known ceramic lubricant is graphite, a very fine black powder. Silicone oils have a silicon base rather than a carbon base and will withstand temperatures up to 600°F (316°C). Molten glass is also used as a lubricant in extruding (squeezing through a die) steel shapes.

Because of their extreme hardness and ability to withstand heat, ceramics make excellent abrasive materials. Abrasive paper (sandpaper) and abrasive belts consist of small sharp pieces imbedded in a coating to hold them secure. Grinding wheels are molded with the abrasives and a hard bonding material. The abrasives used most are silicon carbide and aluminum oxide. Flint and garnet paper may be purchased in most stores for craft and home use, but their use in industry is limited because they are too soft. Diamonds are the hardest natural material known. Industrial diamonds (synthetic diamonds) are used in making very smooth cuts in hard materials.

Fabrication techniques involve many types of machines and tools. Some are standard and ordered from manufacturers much as we would order an item from a mail-order catalog. Others are custom-made, being designed and built for very specific operations. Some perform only one operation, while others are built to be extremely versatile.

Modern manufacturing processes require the use of three types of machines: simple machines, complex machines, and machine tools. As noted in the first chapter, the term simple machines may be used to describe most hand tools, where the user provides both the energy and control. A complex machine was defined as a combination of two or more simple machines, designed to make more efficient use of mechanical advantage. Complex machines may be powered by the user or

This complex machine is called an **impact wrench.** *It is powered by air pressure, but the user has complete control of it.*

may use another source, such as electricity. The user, however, retains complete control.

The third type of machine, which has not been discussed before, is called a machine tool. This piece of equipment usually combines several complex machines and operates on its own power without continuous human control. There are more than 400 kinds of machine tools, ranging from a small bench-mounted unit to a machine that may be over three stories high. Some machine tools are designed to perform one specific operation, while some are able to perform several different operations, in sequence, on the same part. These are called work centers. They hold the part being machined, rotate it at the correct angle, choose the correct tool, mount it, and proceed with the tooling operation. Upon completion, the part is automatically repositioned, another tool selected, mounted, and the second operation is completed. Several such sequential operations can be performed before any human input is required. There are several basic processes accomplished with

A drill press drilling a hole in a steel casting is classified as a machine tool. It is powered by electricity, but the feed (pressure of the drill) is automatically done by pneumatics (air pressure).

A lathe rotates the part to be turned. The operator is running the small cutting tool up to the part, where the machine can be set on automatic feed, making a continuous smooth cut.

machine tools. Drilling and boring refer to making holes with a rotating tool, while the work (the piece being machined) is held stationary. Drilling involves making a hole, while boring involves enlarging one which is already drilled or putting a smooth finish on the sides. These machines are called drill presses and boring machines.

Turning produces round and cylindrical parts. The work is rotated, while the cutting edge moves back and forth parallel to the part. Screw threads can also be cut with this machine, which is called a lathe.

Milling may be used to make a surface flat or cut channels and various shaped slots. The cutter rotates in one position, while the work is passed under it. This machine is called a milling machine.

In grinding, an abrasive wheel is rotated and moved over a stationary piece of work. This operation produces a very smooth finish. The machine is called a surface grinder.

Metal forming encompasses many different forming processes and machines. Examples of a few might be machines which punch small parts from sheet metal, squeeze hot metal into molds, and hammer parts into the desired shape. Some of these machines are called punch presses, drop hammers, and shears, depending upon the process involved.

Presses are used for many purposes. They may be small or stand more than two stories high. Large presses may produce well over 2500 tons of pressure.

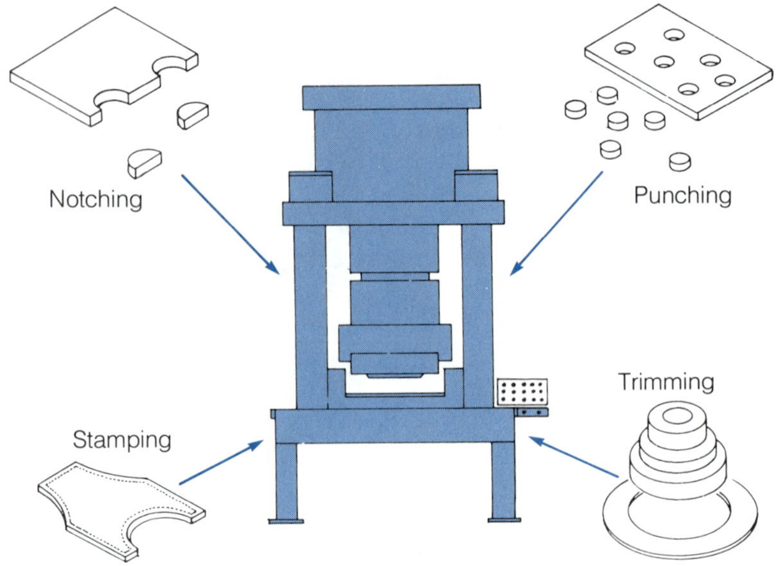

A number of new machine tools use sources of energy other than mechanical. For example, electrical discharge machining uses electricity to remove metal, ultrasonic drilling uses sound waves which vibrate at ultrasonic frequency, and chemical milling uses chemicals to etch through the metal. Because these processes can be controlled very closely, they are used where great accuracy is required or where very small parts have to be produced.

This laser is cutting very intricate designs into the extremely thin metal. Notice the sharp corners which can be cut. Because the heat is concentrated in only a small spot, it will not melt, warp, or distort the thin material.

Another process involving high technology is the laser, which is used for drilling, cutting, and welding. It works on the principle of focusing parallel light waves into an extremely small area. This is accomplished by using a lens, which bends the ray inward to one spot. The intensity of the light can be made so great that it not only can melt metal, it can completely vaporize it. Because the focusing can be done very accurately, the process is used for very precise work.

There are four general categories of machine control: manual, semiautomatic, automatic, and cybernetic. Manual control is direct from the user to the tool. Examples would be a hammer or a bottle opener, where arm movement controls the pressure exerted.

Semiautomatic control uses simple machines. An electric can opener may use a small wheel to rotate the can as it is being opened, while a lever may be used to start and stop a lawn mower. Semiautomatic means that once the machine is turned on, it runs by itself until it is shut off by the operator.

Automatic machines are those which are started in operation and no direct control is needed thereafter. They are able to perform tasks repeatedly and have the capability of turning themselves on and off. Examples might be a clock with a timer or an air pump which turns itself off when a certain level of pressure is reached. Automatic machines are used extensively in manufacturing, where the same operation is required continuously.

Cybernetics refers to two-way communication between user and machine or machine and machine. It is a computer-controlled process that can be programmed to provide feedback to indicate any changes that should be made. Human machine communication gives the user information concerning the situation at any predetermined time. Machine-machine communication usually involves one machine which controls another, or even several other machines, and can provide backup machines in case one malfunctions. A simple example of cybernetics at home is the heating thermostat and the forced-air furnace. When the thermostat registers below a predetermined temperature, a signal is sent to the furnace, which warms up a few minutes before the blower starts, so cold air will not be blown through the hot air ducts. When the blower comes on, heat is then blown into the room. When the thermostat registers an adequate amount of heat, it sends another sig-

The computer numerical control (CNC) operation shown here demonstrates the pattern of movement that has been programmed into the system. Operations of this type always require a safety shield for protection from flying metal chips; the guard has been removed here to allow the photograph to be taken.

nal to the furnace, and the heat is shut off, but the blower continues to run a few minutes until the furnace is cooled down to a safe temperature. This two-way communication between machines may be repeated as often as necessary to keep the house warm.

The trend in manufacturing today is away from the single large computer of earlier years. Instead, many smaller computers are being linked together to provide a complete system of controls. This provides more flexibility and instant backup, if needed.

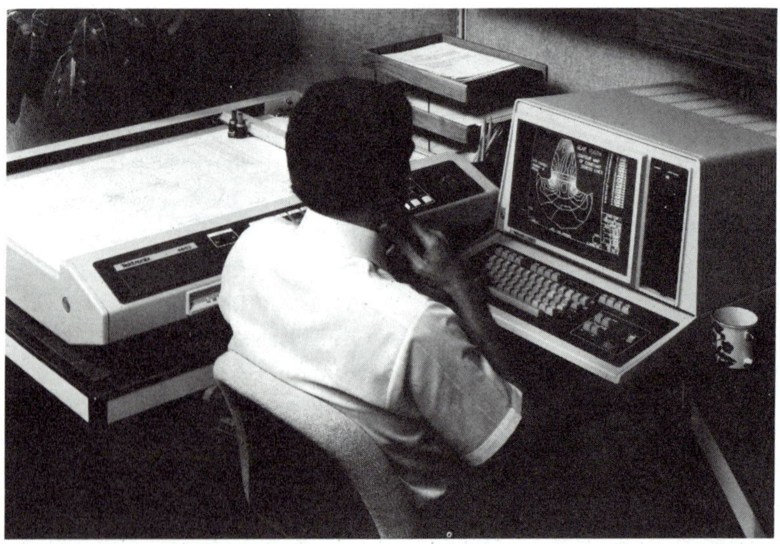

Computer-aided design is a good example of cybernetics, because it uses what is called interactive graphics. *That is, the operator is able to communicate with the machine by means of a computer language which gives a visual readout. The plotter on the left automatically draws what is seen on the screen.*

The application of computers to manufacturing is essential. Nearly every phase of the system is involved in one way or another. There are several terms used which identify their use:

- Numerical control (NC) has been used for several years to control machine operations. Commands are usually given by means of a punched tape (much like the paper rolls in a player piano).
- Computer numerical control (CNC) is the same basic process as NC, only there is the capability of feedback from the machine to the computer, so changes can automatically be made as the work is being done. Communication is accomplished electronically.
- Computer-aided design (CAD). This system utilizes the computer in designing products and machines.
- Computer-aided manufacturing (CAM). The organizational structure of fabrication and assembly is determined and controlled using this system. It integrates the various components, such as materials, machines, and work flow, into the system to obtain a smooth-flowing operation.

Assembly. The process of assembly brings together the materials and standard parts from the processing industries, as well as the fabricated parts and subassemblies from previous manufacturing operations. These might very well come from many locations throughout the world. Much planning is required, because all the parts have to be collected before assembly can start. Also, enough parts have to be stockpiled so that there is a continuous flow of materials into the assembly system. Transportation within the factory also has to be well organized. Materials and parts have to be fed into the line without interfering with the assembly process. Finally, the time it takes for each operation has to be identified. If this is not incorporated into the planning, "bottlenecks" occur where slower operations cannot be completed fast enough, while faster operations get farther ahead. Computers aid in this monumental task of planning, by allowing data to be analyzed and simulation programs to be run. A simulation program provides the results of a theoretical assembly system, without actually going through the real thing. In this way, problems can be identified and changes made.

Adhering, fastening, and welding are three means of assembling the parts of a product. Almost all adhesives used to-

Modern assembly plants use many types of power tools. Rather than having stationary tools using belts and mechanical energy, it is often an advantage to have flexible tools. These are powered by hydraulics (liquids under pressure) or pneumatics (air under pressure). These tools can be identified by hoses rather than by belts.

day in industry are made through chemical processing. Most could be called a plastic adhesive, because they result from plastics processing. Some common types of structural adhesives which may have familiar names are epoxy, vinyl, acrylic, and urethane. Most adhesives cure by heat or by chemical reaction, rather than by air drying. As a result, many have to be mixed with a catalyst, or hardener, before they are applied. After hardening, they can withstand several thousand pounds per square inch (kPa) of pressure or stress. Because adhesives are a chemical product, many different characteristics are possible.

As with adhesives, there are many kinds of fasteners designed for specific uses. Threaded fasteners usually are screws, or bolts with nuts. Unthreaded fasteners are usually pins, nails, and wire fasteners. A key advantage of fasteners is that they allow disassembly for service and repair of the product.

Welding involves the application of high heat to a joint, making the edges melt together to form a single piece. The two most common materials that are welded are metals and plastics. When metal is welded, a third metal is usually added by means of a welding rod to provide enough molten material to fill the entire joint. When welding is completed, there is, technically, no joint.

Heat may be applied in three ways. Gas welding uses a mixture of flammable gas and oxygen, which creates an ex-

A laser welds a part which will be placed into an assembled car. The laser is on the right, and the heat is at the tip of the point near the center of the picture.

tremely high heat. Arc welding employs a high-amperage electrical source. As an electrode (welding rod) is placed near the part to be welded, an electric arc jumps to the part, because the part is connected into the electrical circuit. The arc is hot enough to melt the rod and the part at the same time. A second type of electrical welding, called spot welding, is used in attaching thin pieces of metal. Two electrodes, end-to-end, are brought together after the sheet metal is placed between them. The electric current welds them together in that exact spot. The third method, laser welding, is relatively new and is not used as extensively as the other methods. To date, it has been utilized mostly in welding small-scale components, where great accuracy is required and where heat surrounding the weld area cannot be high.

Another method which is being used more today is called weld-bonding, which combines adhesives and spot welding. The combination causes a stronger joint than the two can produce separately.

The application of machines depends upon the system of fabrication and assembly used. Single-unit production, which usually involves only a few people using any number of tools and producing one product, is rarely seen in modern manufacturing, with the exception of some of the smaller engineering companies which produce models and prototypes. Multiple-unit production involves several people using several machines and produces many units of the same product. The term typically given to multiple-unit production is mass production. Any number of different systems may be used in the mass production process, with more than one being used on the same product. The use and movement of workers, materials, and machines at any given time depends upon the product. In general, mass production may be classified into two categories: custom production and line production.

Custom production involves producing a number of products, with each being designed and built to a customer's own specifications. For example, a machine tool builder might be producing several large machines to be used in factories. However, each factory wants the machine to make its own product, which means every machine is different. When finished, the machines may look alike, but the operations they perform and the parts they produce will be different. In most cases, the custom produced product involves workers moving to the product as the work requires. Various systems, such as electrical, mechanical, and hydraulic, have to be designed and built to function together. The workers are usually highly trained specialists, and as production takes place, each performs a special part of the assembly operation.

Line production involves producing many products that are basically identical. Unlike custom production, which may be accomplished in many different ways, line production is a specific method where the product moves past the worker. Each worker performs a specific operation on every unit that passes. As a result, most workers are not required to be highly trained. The sequences of steps are identified so there are few delays in waiting for a certain operation to occur. As a result, line production is a rapid, high volume method of producing products. Line production can have elements of custom production, however, in that the characteristics of the product may be flexible. For example, cars coming off the assembly line can vary in color, as well as in many other ways.

Few industries have been able to combine the manufacturing of custom products and standard products. One exception, however, is the automotive industry. Cars are built to the specification of each customer, such as type of engine, color,

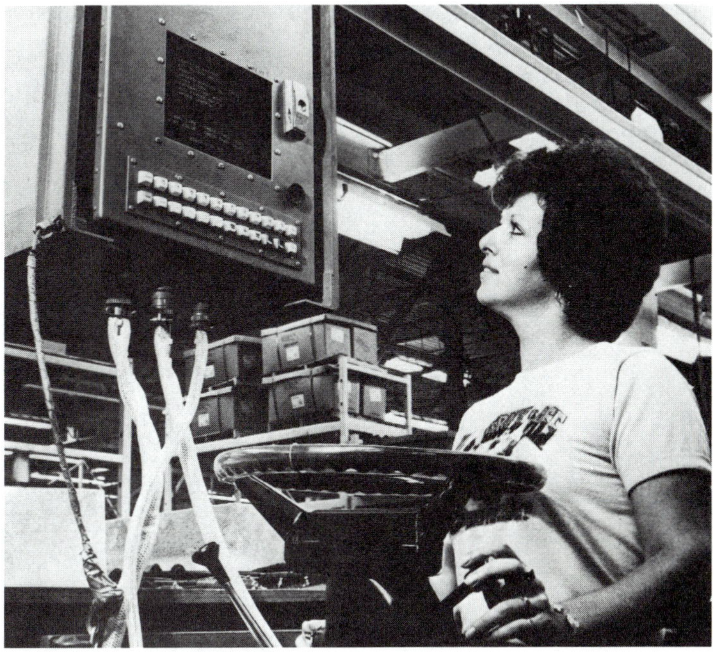

Not all assembly-line workers assemble parts. This inspector is operating a computerized system which checks forty-five different circuits (depending upon the cars' options) in the steering column. The test results are read on a TV monitor.

and accessories, yet they are run through the same production line as the standard cars. The reason for this success is a large, well-organized system of control, which makes extensive use of computer applications. The system incorporates many computer subsystems, ranging from marketing analysis, to parts inventory, to machine control.

This shipping package for grapes is made of molded expanded polystyrene. It is inexpensive, lightweight, moisture-proof, and shock-absorbent.

Packaging. The last step in manufacturing usually involves some form of packaging. If it is a large unit, such as an engine, it has to be crated, using wood and other support materials. If it is a small item, such as a pencil sharpener, it has to be placed in a box printed with the trademark, advertising, and other information. Packaging serves three purposes: (1) it provides protection for the product during transportation; (2) it serves as a container for the product during the time it is being used; and (3) it provides information concerning the product, such as name, description, and use.

Packaging utilizes nearly all the typical industrial materials, such as glass, metal, paper, and plastics. Glass and metal are used primarily for bottles and cans which contain food or corrosive materials. Paper and plastic are versatile, paper is

Workers keep pace with modern technology by becoming skilled in the use of computer hardware. Retraining and upgrading skills may help eliminate job dissatisfaction.

usually in the form of paper board or corrugated board for larger boxes. This material is limited in that it is hard to form into shapes other than regular geometric shapes. Its advantage is that it is relatively inexpensive and can be used for larger objects. Plastics are the more versatile, because the material characteristics allow unlimited shapes. They even can be molded to resemble the actual product. Because the material is durable, the package can become the case in which to store the product.

It is very likely that because of the specialized nature of packaging, some are manufactured in a different factory and shipped to the product manufacturer ready to be used in the assembly process.

It should be noted here that all products follow at least some of the steps of manufacturing just discussed. It is important to understand there is a very close working relationship between the various production systems before, during, and after the manufacturing steps, even to a point of overlap. There also is the continued interface of all the technical systems: transportation, communication, and production.

Manufacturing Problems

Although there are engineering problems to contend with, the major problems encountered in manufacturing are primarily social and environmental. Four general problems can be identified. One is the attitudes of workers. This problem surfaces in poor workmanship, high absenteeism, and low output. This, of course, has a direct effect on manufacturing productivity. As productivity goes down, the price of the product goes up to cover the increased production costs. Among some of the reasons given for these and similar problems are job dissatisfaction, a feeling of being "locked" into a job, boredom, and low pay. Whether these problems are real or a symptom of a larger problem is pure speculation. With more emphasis on productivity, however, some industries have no choice but to utilize more high-technology fabrication and assembly techniques.

A second problem is consumer attitude, which has developed in only the last few years and reflects a lack of understanding of technology. Many think that with our level of technology, we can easily accomplish anything we want. Industries are expected to produce products that will function indefinitely under all conditions and be completely safe, whether they are used correctly or not. Proof of this attitude is the increasing number of product liability court suits. Some

Robots will be used in the future for monotonous or dangerous jobs. They require no protection from toxic chemicals such as these.

industries are reaching the point where they can no longer afford the legal costs of defending themselves.

Third, highly trained workers are needed. As technology advances, there is an increased need for technically trained workers. This means that workers will be learning on the job, in both formal and informal situations. Workers who do not have the ability or the inclination to continue their learning will likely not be working in production. This presents a critical situation for both the worker and the industry. Indications are that in the future, we will experience a labor shortage at the same time we have a high unemployment rate.

The fourth problem is environmental. In recent years there has been considerable discussion concerning our ecology problems, and as a result of applying new kinds of technology, many of these problems have been alleviated. However, a new problem has surfaced recently concerning the interior environment. Concern is now being focused on the long-term health problems of the workers, as a result of working with a variety of industrial materials over a period of many years. Research is being conducted to determine the scope of the problem.

The health and safety—both physical and mental—of workers is becoming more important to management because they have a direct relationship to worker satisfaction.

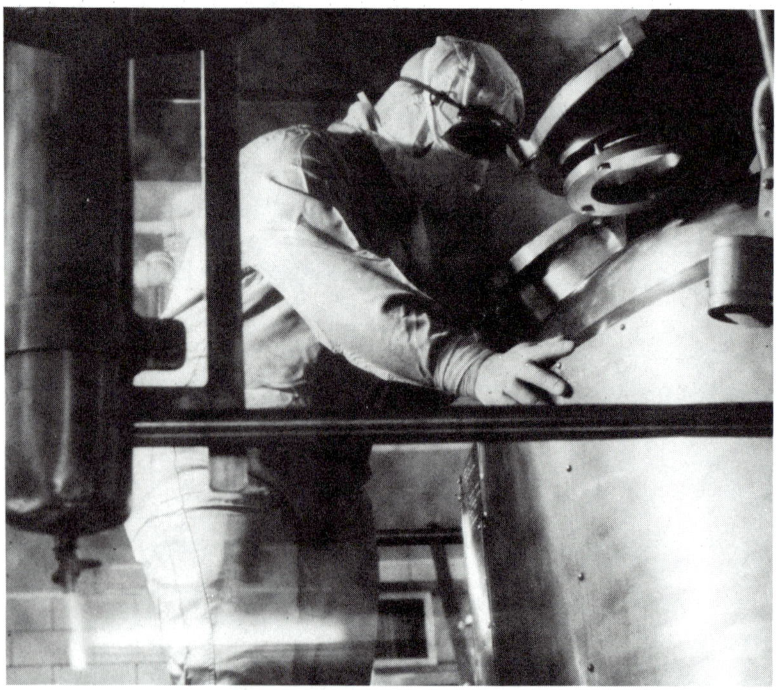

Manufacturing and the Future

Manufacturing is currently in a rapid state of change, and the future will certainly bring more changes. The major emphasis will be on increasing productivity, which will be accomplished by changes in both the organization of the system and in the hardware.

Several organizational changes will be implemented. For example, a four-day workweek of ten hours a day is a possibility. Also, there will be various alternative methods of material and product flow considered. One such plan, which has been used in a few other countries, moves the workers with the product, rather than past them as in the present line-assembly method. This allows a single worker enough time to accomplish more than one operation, while the line continues to move. These and other plans may help to make production more efficient and contribute as well to an improved worker attitude.

The changes in hardware will come with unbelievable speed. The following selected list of predicted changes came about as a result of a study conducted by the Society of Manufacturing Engineers and The University of Michigan. The list is by no means complete, but these examples will serve to demonstrate the direction in which we are moving.

Manufacturing Systems and Material Removal. The following conditions are predicted:

Welding robots represent the assembly line of the future. They are automatically welding this car body together, applying a total of nearly 3,000 welds to each car. A completed car comes off the assembly line at a rate of about one a minute.

- Machining cutting speeds (other than grinding) will be increased.
- Sensors will be developed which will be able to accurately measure the roughness of a surface at the same time it is being cut.
- New physical and chemical machining processes will be developed which will cut nonmetallic materials faster than conventional methods.
- Computers will accurately predict the manufacturing cost, based on a description of the part being produced.
- New abrasive materials will be used.
- The percentage of automatic part loading and unloading (placing the part in a machine and taking it out) devices on stand-alone machine tools will increase.
- The use of computer systems connected to subcontractors (other factories) will reduce the use of blueprints.
- The laser will be used extensively for accuracy control during machining.
- The noise level of machine tools during production will be reduced by at least 50 percent.
- Numerical control machines will be connected to a computer controlling several machines.
- Manufacturing systems will utilize automatic guided vehicles (AGs) between work stations which will be computer controlled.
- The accuracy of computer controlled machine tools will be increased.
- Computer networks will be able to locate every individual part in a manufacturing facility.
- New machines will have automatic gauging capabilities of machined parts. As a result, incorrect parts will be reduced.

Assembly. The following conditions are predicted for the assembly process.

- The number of subassemblies in a final product will be reduced because of new materials and processes being used.
- The use of structural adhesives will replace fasteners in many products, such as appliances and automobiles.
- The assembly systems will make extensive use of computer controlled automatic part inspection.

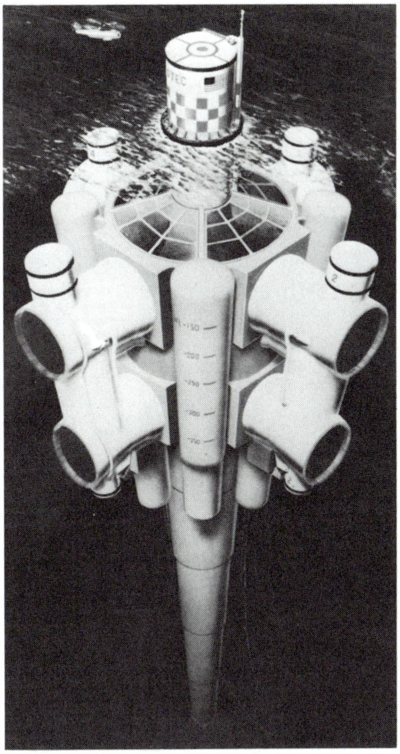

A new energy source called **ocean thermal power** *is currently being investigated. A floating platform uses the temperature differential between the surface water and deep ocean water to vaporize and condense ammonia. The vapor turns huge turbines which generate electricity. Each platform could produce enough electricity to satisfy the needs of a city with one hundred thousand residents.*

- Computer sensing will allow a robot to pick up and arrange parts which are scrambled in a container.
- Parts storage and retrieval will be integrated with the assembly process through continuous computer control.
- Machine vision will allow machines to adjust to conditions, based on visual inspection of parts.
- Products and parts will be designed to be handled automatically during assembly.
- Assembly systems will make use of robot technology. Sensory techniques will enable them to approximate human capabilities.
- Direct labor involved in the assembly of small components will be replaced by automation.
- New assembly systems will be designed by computer-aided design techniques.

The technology of the "factory of the future" is already available. Integration of all elements of the company will be accomplished with common databases. Jobs will change, and workers will have to accept the reality of the situation. Predictions to 1995 indicate that 50 percent of the direct labor in automobile final assembly will be replaced by programmable automation. It can readily be seen that the future will certainly bring new products and new job requirements which will have a direct effect on our lives.

PRODUCTION AND THE FUTURE

The discussion thus far shows that the future of the world depends upon the production system. As the population increases, the demand for food and other necessary products increases. If these demands are to be met, the limiting factors of natural resources and energy will have to be overcome. New applications of existing low-use resources will have to be developed, and new and more efficient methods of utilizing high-use resources will have to be identified. It is evident that new energy sources will have to be developed and existing sources will have to be made more efficient to meet the demands of medium and high technology. The solution to the energy problem for the near future involves the combined application of several efficient sources, in addition to more stringent conservation methods. If these two problems can be resolved, the production system will have the potential of satisfying critical needs on a worldwide scale, providing a better life for all human beings.

SUMMARY QUESTIONS

1. Identify three ways in which the three major subsystems are related.
2. What part does the construction subsystem play in the transportation subsystem?
3. What do farming and mining have in common?
4. State the differences between a simple machine, a complex machine, and a machine tool.
5. Why are many of the manufacturing problems related to workers rather than hardware?
6. What will be the impact of computer applications on manufacturing processes of the future?

Classroom teaching may include many different approaches, but one common thread should be active student involvement.

PART III

Technology: The Materials, Tools, and Procedures for Instruction

CHAPTER 10

Drawing and Measuring in the Classroom

In this chapter you will learn:

1. The names of common drawing instruments.
2. Some of the basic techniques used in making drawings for purposes of communicating and planning.
3. The names of common measuring tools.
4. Some of the basic uses of measuring tools.

DRAWING

Purposes of Drawing

Drawing may be called the language of technology. Everything that is processed or manufactured has been in a drawing somewhere. Drawings involve using a system of symbols which are a part of a graphic code. Ideas and representations of actual objects, or future objects, can be communicated by means of this graphic code. Drawings serve four purposes.

Engineering drawing requires special skills and knowledge of the drawing instruments and graphic symbols.

An artist's rendering shows a conception of the future as the Space Shuttle places an earth resources satellite in orbit.

First, they may be used in working out design problems; changes can be easily made until the design is suitable. Second, they communicate the shape and size of objects, as seen from any angle. Third, they are able to show things that cannot be easily seen, for example, the inside of an automobile engine after assembly or an arrangement of atoms under an electron microscope. Fourth, drawings show not only past and present objects, but images of the future as well. An architect, for example, uses drawings to show the client exactly how the future building will look.

Drawing Tools

Drawing Board. The drawing board is usually a soft wood, table-top-size board which may have metal edges. A piece of heavy paper should be taped to the board while drawing to protect the board and to provide a good drawing surface if thin drawing paper is being used.

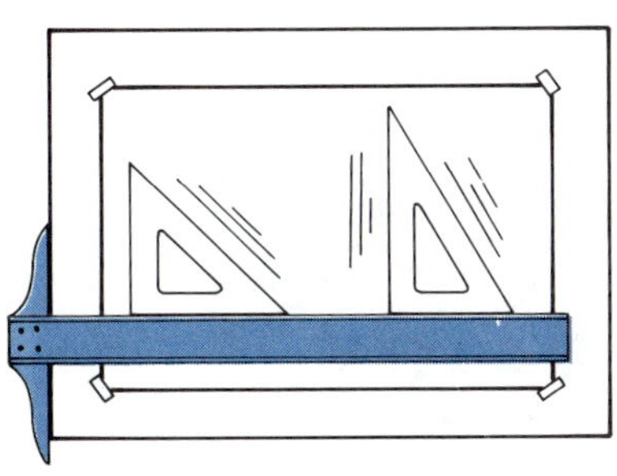

The triangles are placed against the T-square and are used as straightedges for drawing vertical and inclined lines. The T-square is used to draw horizontal lines.

T-Square. The T-square is used to draw horizontal lines and provide an edge on which the triangles move. The head is always placed on the left side of the board for right-handed people and on the right side for left-handed people. It should be kept tight against the board edge at all times. Some boards have a parallel rule which is permanently attached. When this is the case, a T-square is not needed.

Triangles. Triangles are used to construct vertical and angled lines. These lines are identified in degrees from the horizontal lines made with the T-square. Two triangles normally are used: one has 30° and 60° angles, while the other has only 45° angles. They should be used while they are resting on the blade of the T-square. The triangles are usually made of clear plastic and are available in many sizes. Eight inch (203.2 mm) triangles are usually used with a board and T-square.

The triangles may be used on either side, depending upon the required angle and direction of the line. The angle is identified by the number of degrees it varies from the horizontal. Angled lines other than those on the triangles usually require a protractor.

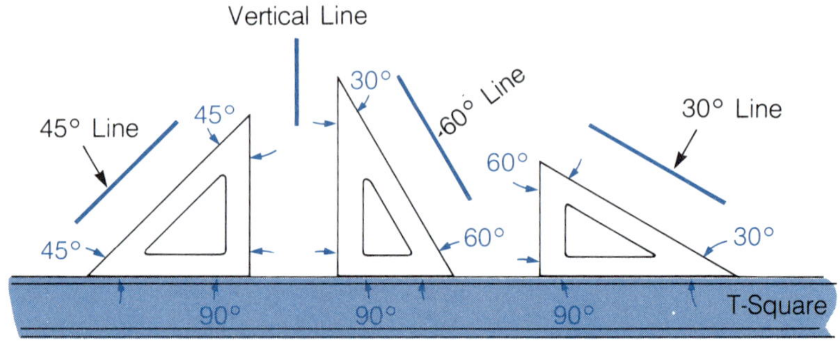

Compass. A compass is used to draw circles and arcs. When adjusting the radius, the distance between the needle and the lead point should be taken from a scale. The lead should be

A slide compass has a direct reading of the radius. It is rotated between the thumb and forefinger.

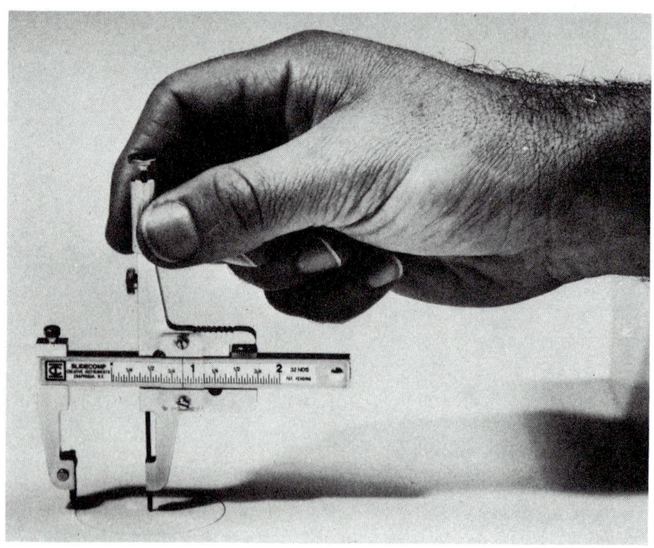

sandpapered to a sharp chisel point. This may have to be done after each circle is drawn, because the lead wears down very rapidly. To draw with the compass, place the knurled grip between the thumb and forefinger and rotate the lead point around the needle point.

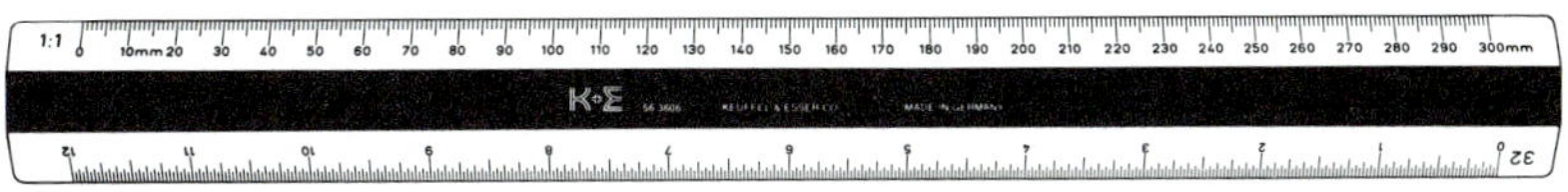

Measuring instruments using both the English system and the metric system should be available to the students. This flat scale has both systems.

Scale. There are a number of different kinds of scales, allowing the drawing to be scaled up or down in size, but reading the scales can become rather complicated for elementary students. Probably the best one to use in the elementary classroom is the flat plastic scale, which has inches designated on one edge and millimeters and centimeters on the other. It is easy to manipulate, and it can be read very easily.

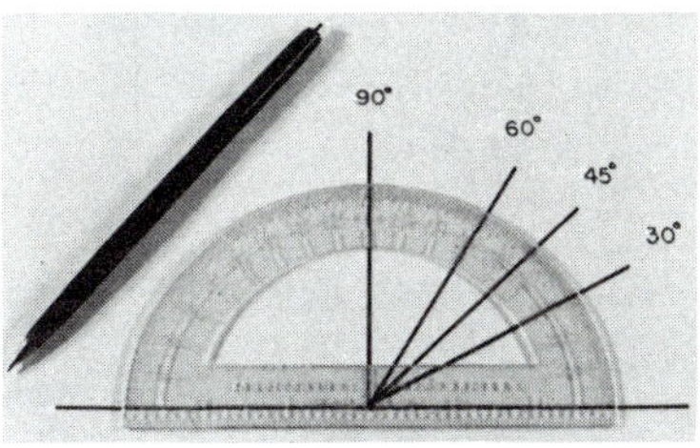

Many things can only be measured in terms of angles. A protractor is a valuable instrument for teaching both drawing and concepts dealing with measurement.

Protractor. The protractor is used to obtain angles which can not be identified with the triangles. A large, clear plastic instrument with large numbers and divisions is best for elementary students.

Templates. Templates are flat, clear-plastic drawing instruments, which have various sizes of holes. They may be any number of geometric shapes, including round, elliptical, square, and rectangular. These are especially well suited for elementary student use, because they require less small muscle coordination than the compass and triangles.

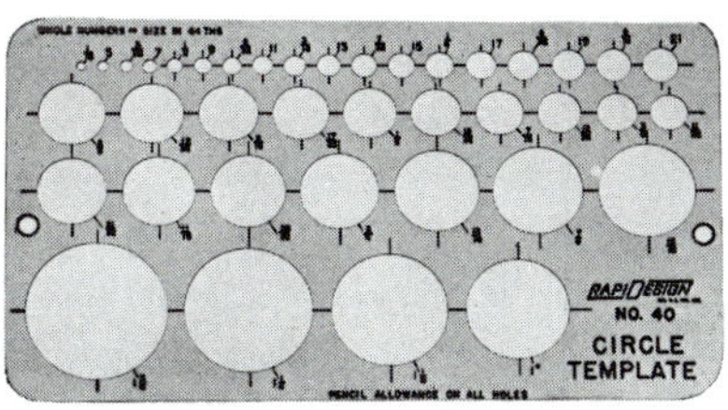

Templates allow elementary students to draw with more accuracy.

Pencils and paper. Although nearly any pencil will work for drawing, there are various grades of lead ranging from soft to hard. A medium pencil is best for general drawing. It should be kept sharp by rotating it on a small piece of sandpaper. The pencil should also be rotated between the thumb and fingers as lines are being drawn. This prevents a flat edge from being worn on the lead, which produces an uneven, fuzzy line.

Although nearly any kind of paper is appropriate, graph paper works well for elementary students, because both vertical and horizontal lines are already drawn. It also is easily used with a standard ruler when drawing instruments are not available.

Shape Description

The views necessary to represent an object depend upon the object itself. Flat objects, such as roads, need only one view, as seen, for instance, on road maps. Because some objects are very complicated to draw as they would actually be seen, a system of drawing three different views at one time is used.

This system, called *orthographic projection,* is part of the code, or standard, used in industry today. *Ortho* means straight, *graphic* refers to drawing, and *projection* means one view may be drawn from another. For example, if the front view of an object is drawn, the side view may be drawn beside it by projecting light lines from the front. Since the height of the front and the side are the same, there is no need to remeasure the side view to construct it.

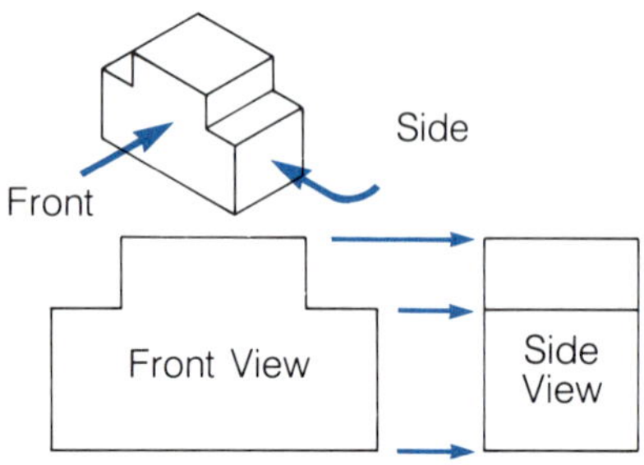

The side view is the same height as the front view, so it is not measured; it is simply projected across.

Imagine a triangular-shaped block inside a glass box. As the block is viewed through the front glass panel, a specific shape is seen. If the block is viewed through the side panel, however, a different shape is seen. The same can also be said for the top panel.

In order to see these views at one time, the panels are opened outward, until they are parallel with the front panel. As a result, the top view is above the front and the right-side

The glass panels on which the various views are projected are called **picture planes.** ***They act much like a photograph of the object taken from a vantage point straight out from each surface.***

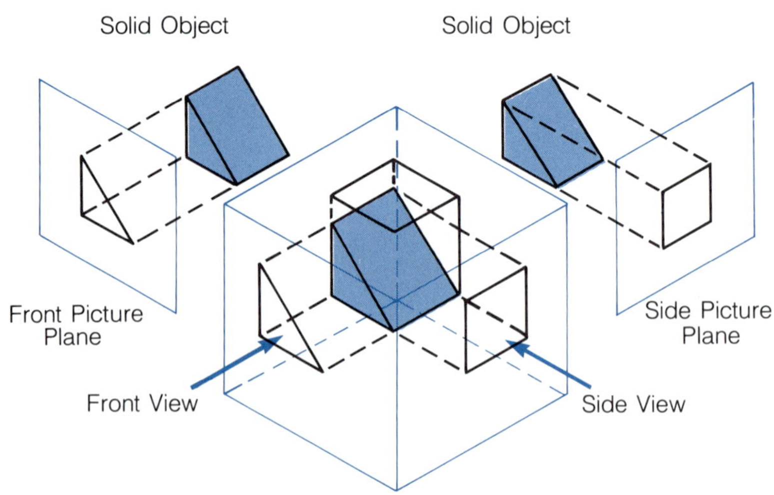

(A) A three-dimensional object can be unfolded so all sides may be seen in one view. (B) The unfolding process is much like unfolding a box and observing the pictures on all the panels at one time.

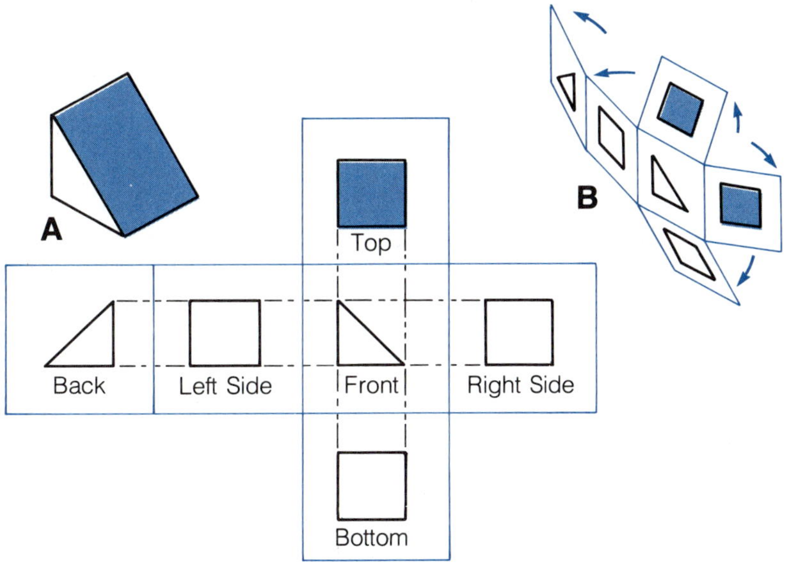

view is to the right of the front. This is the arrangement of views which is always used. The left-side view is to the left and the bottom view is always under the front view. However, only three views are usually used: top, front, and right side.

The front view is usually drawn first, because it generally shows the most distinguishing characteristics of the ob-

The three main dimensions are height, width, and depth; note the depth dimension in the top view is the same as the depth dimension in the side view.

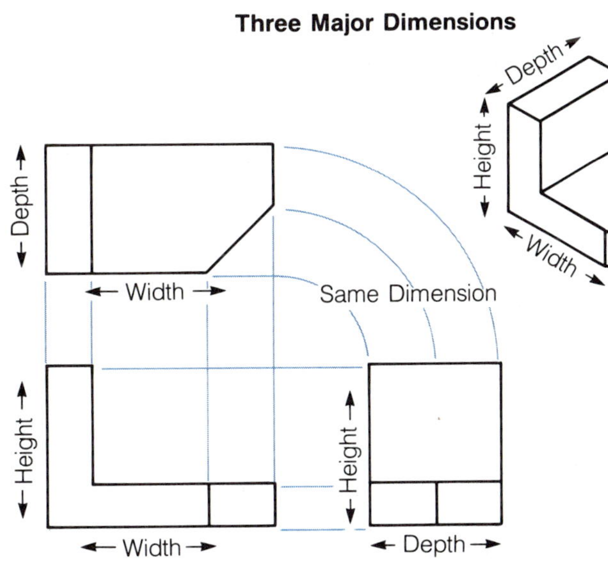

ject. The other two views are then projected from the front. It should be noted that the front and side have the same height, the front and top have the same width, and the top and side have the same depth.

Size Description

The size description determines whether two or more pieces will correctly fit together. To depict complicated drawings accurately, specific symbols and systems are used in dimensioning. For example, certain lines are heavier, while some are lighter. Hidden lines, which cannot be seen in a given view, are dashed, and extension lines are drawn outside the object, so the dimensions are not on top of each other. The following rules indicate the most important drawing symbols.

1. Construction lines should be very light. A hard pencil should be used.
2. Object lines (visible lines) should be heavy. A soft lead pencil should be used.
3. Hidden lines should be medium, with 1/8 inch (3.2 mm) dashes and 3/32 inch (.8 mm) spaces. A medium lead pencil should be used.
4. Center lines are thin and sharp. They are used to show the axis of a symmetrical part or feature. The short line is 1/8 inch (3.2 mm) with a 1/16 inch (1.6 mm) space on either side. A hard lead pencil should be used.
5. Extension lines are medium, and they should leave a 1/16 inch (1.6 mm) space between the object and line. A medium lead pencil should be used.
6. Dimension lines are medium, with the arrowheads approximately three times as long as the width. A medium lead pencil should be used.
7. Extension lines may cross other lines, but dimension lines may not. Center lines may be used as extension lines in order to dimension the axis.
8. Dimensions are placed outside of each individual view and between the three views, when possible.
9. All dimensions should be placed so they read from the bottom of the drawing.
10. Smaller dimensions are placed closer to the drawing, while larger dimensions are placed further out. This prevents dimension lines from crossing other lines.

This illustration should be studied in detail to understand which lines are shown and which are not, as well as how dark the lines should be drawn.

Line Use

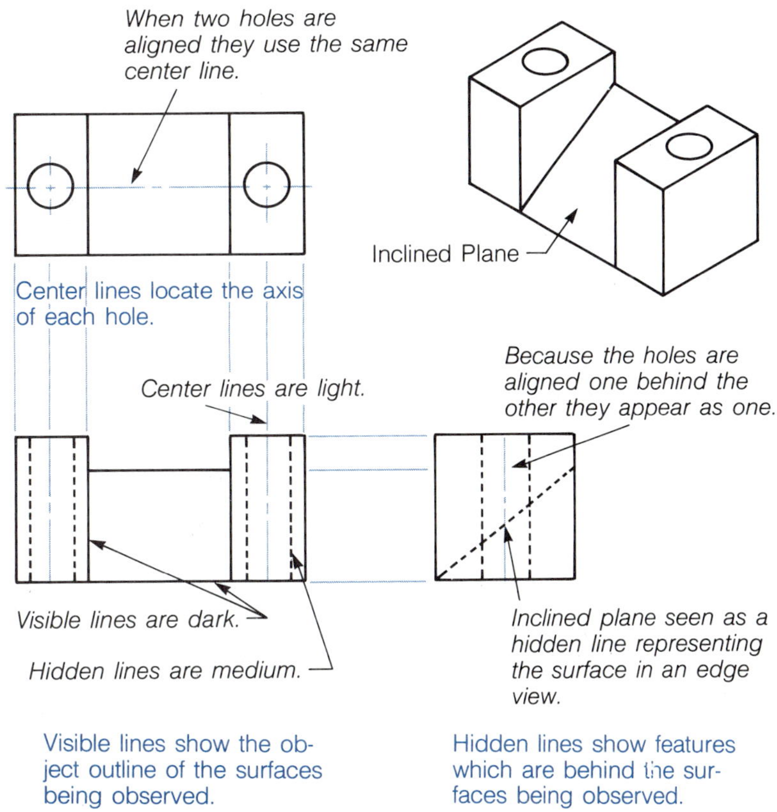

11. Circles are dimensioned in diameter. This is usually done with a leader (a line and arrowhead which points to the axis of the circle, but stops at the circle).
12. Arcs are dimensioned in radius, while angles are dimensioned in degrees.

The extent to which the elementary students might use all these rules will depend upon the grade levels and the objectives of the teacher. It is very advantageous for the teacher to be able to use the tools and drawing symbols effectively. They are a tremendous help, not only in planning activities, but in making signs, bulletin boards, teaching aids, and charts and graphs.

Column A gives the correct placement of lines and may be compared with the incorrect drawings to the right.

Placement of Extension Lines

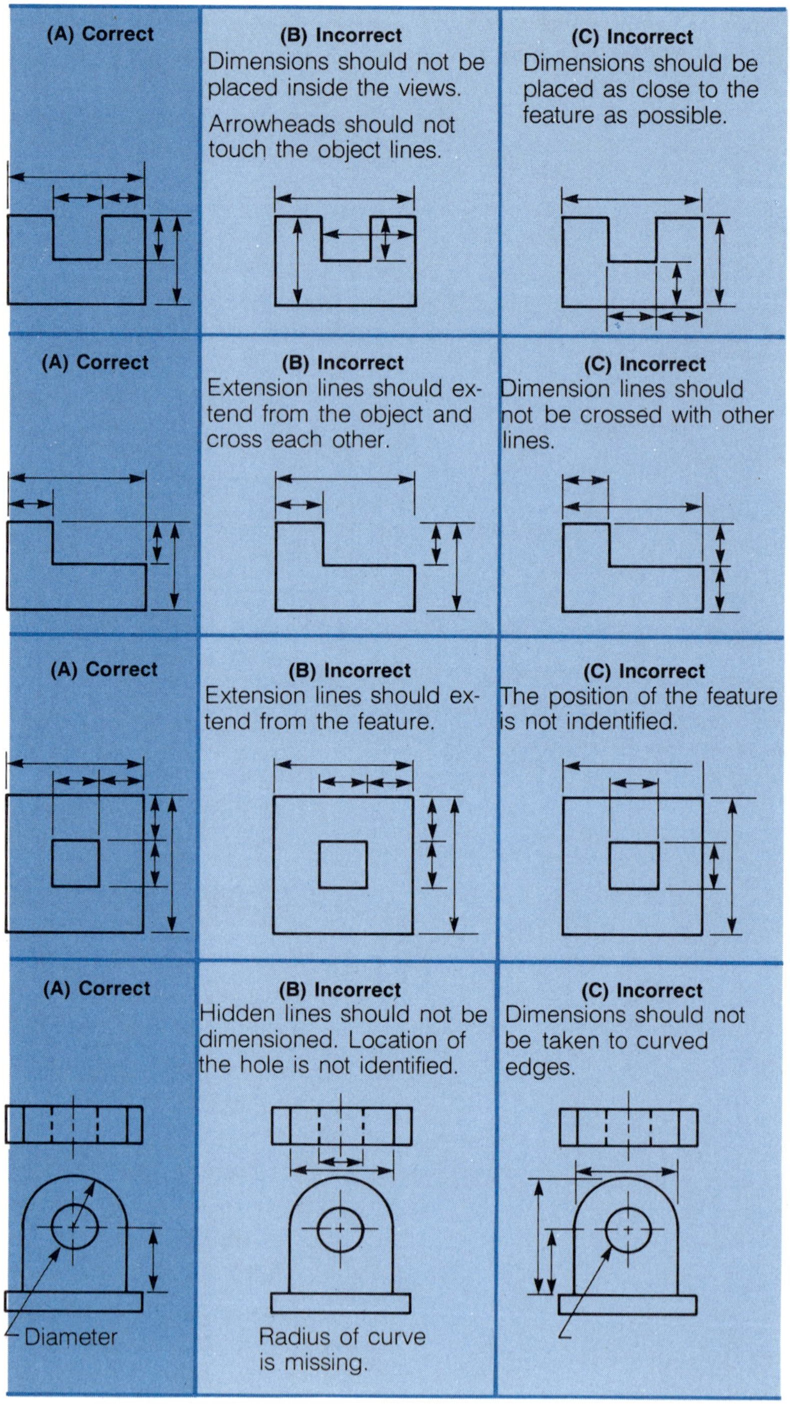

This illustration shows different methods of placing dimensions on the drawing when the space where they should go varies in size.

Dimension Placement

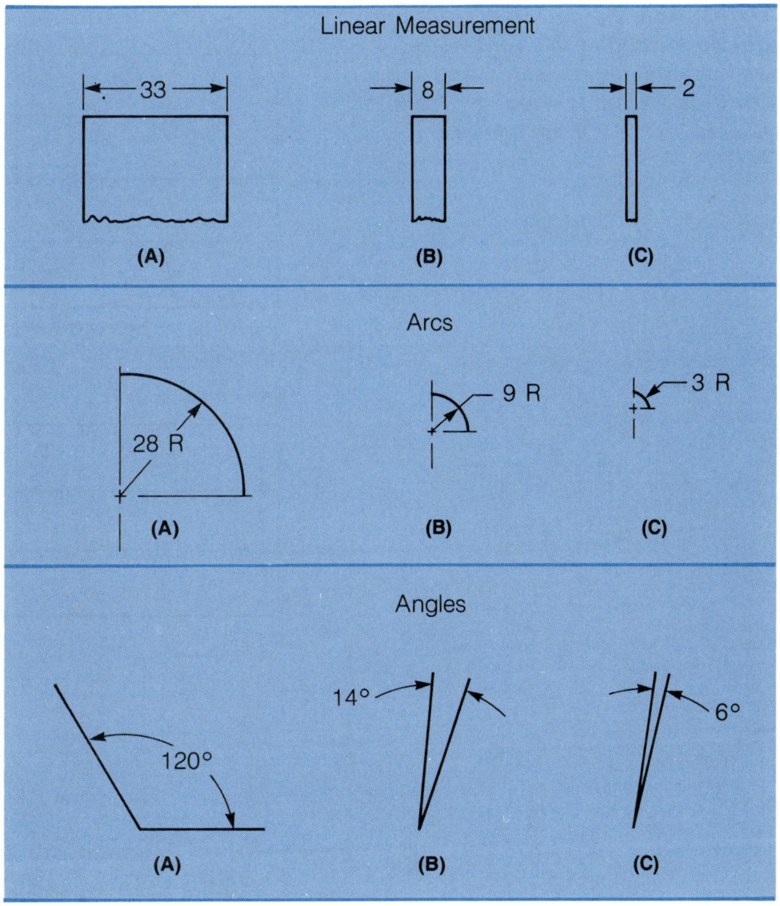

MEASURING

Measuring Instruments

The following measuring instruments are generally used in measuring objects rather than being used in drawing. It is very common, however, for them to be used in taking measurements which will be used in a drawing. Most of the tools may be obtained in either the English system or in the metric system, while some may have both systems.

Each feature is dimensioned in the view that best describes the shape. Note that dimensions are not duplicated in different views; one overall height, width, and depth is all that is needed. These dimensions are in millimeters.

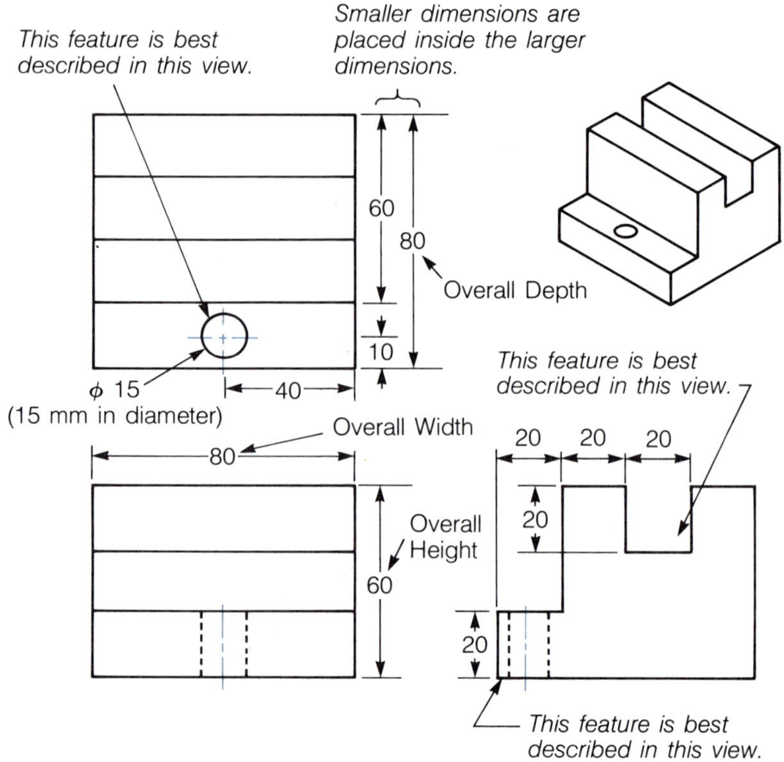

Bench Rule. A bench rule tool may be made of wood or steel, and is usually 12 inches (304.8 mm) in length, although they do come in standard sizes by 6 inch (152.4 mm) increments. The 12 inch (304.8 mm) steel rule works well in the elementary classroom, because it is easy to handle and not easily damaged.

Push-Pull Rule. This steel tape ruler provides measurement capabilities of 6 feet (1.8 m) up to 100 feet (30.5 m). This rule can be damaged because of its flexibility, so it should be used only when longer measurements are required.

Try Square. Try squares come in several sizes, but most are 6 inches (152.4 mm) to 10 inches (254 mm). They have a handle which acts as a guide, allowing lines to be drawn 90° to the edge of a piece of material.

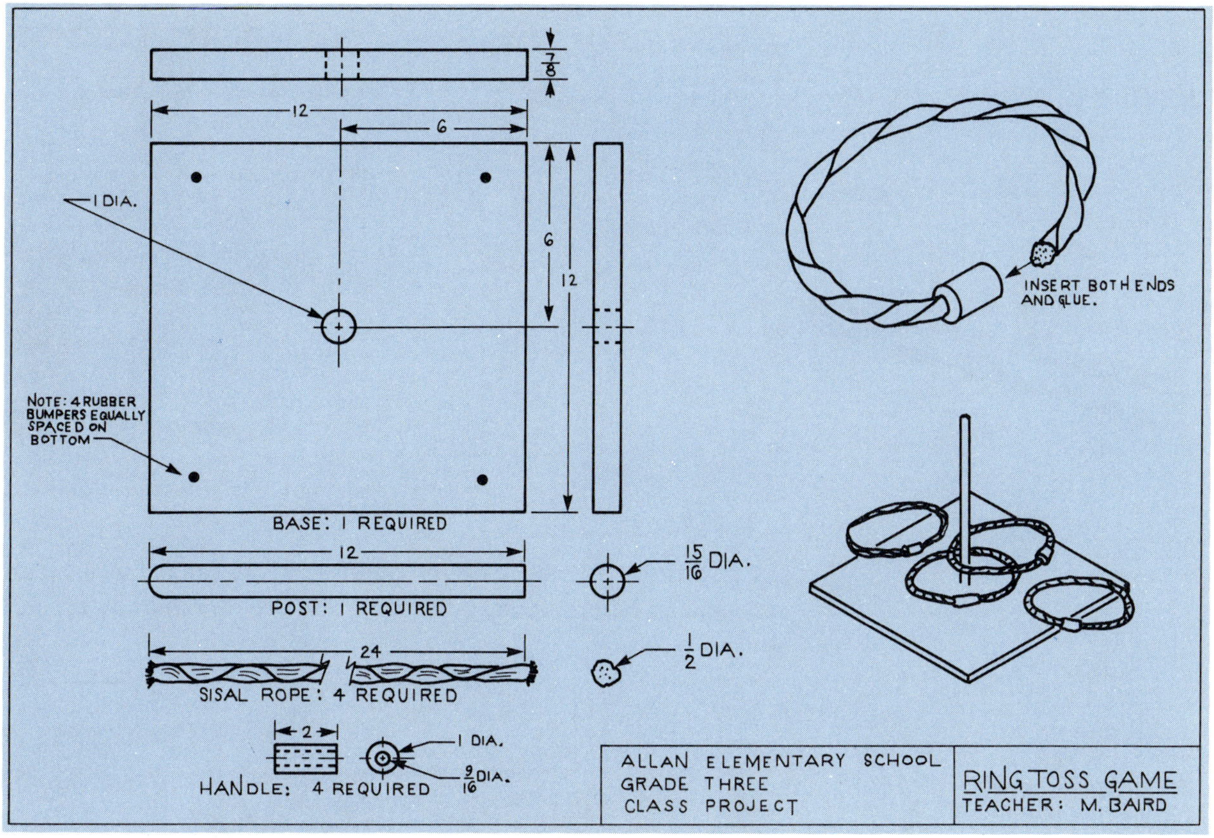

A complete drawing has all the information necessary to make the object, including the dimensions, names of the parts, materials, and any special assembly techniques. These dimensions are in inches.

Combination Rule. A combination rule is a steel rule with attachments. They slip on and are held securely by a long narrow slot in the rule. They allow measurements to be made and lines to be drawn at angles to the rule edge. The rule also has an attachment to find the center of a cylindrical object.

Level. A level, as its name indicates, shows whether a surface is level. It does not produce numerical measurements. Levels come in a wide variety of sizes. An easy length for elementary students to handle is 24 inches (609.6 mm).

Framing Square. Used primarily in the construction system, a framing square is a heavy steel square, measuring 18 inches (457.2 mm) on one blade and 24 inches (609.6 mm) on the other.

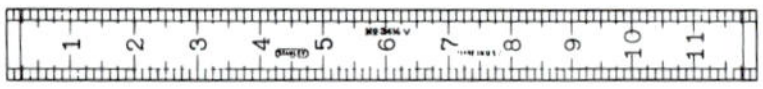

Bench rule

Push-pull rule

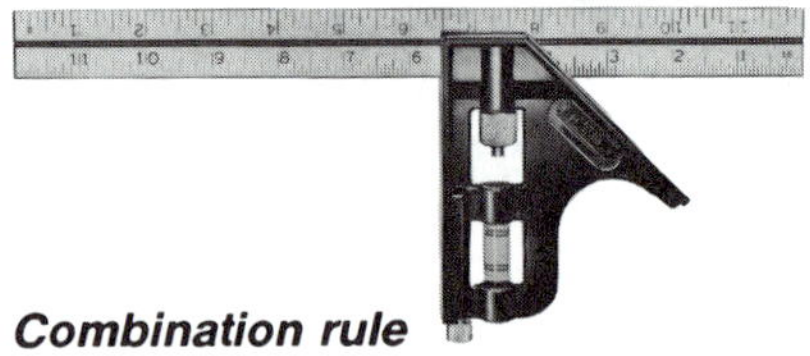

Combination rule

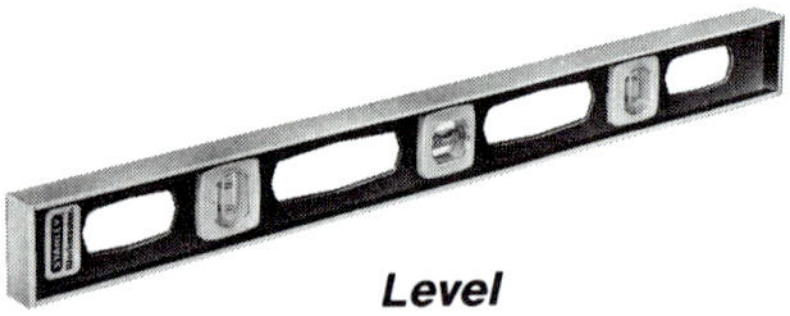

Level

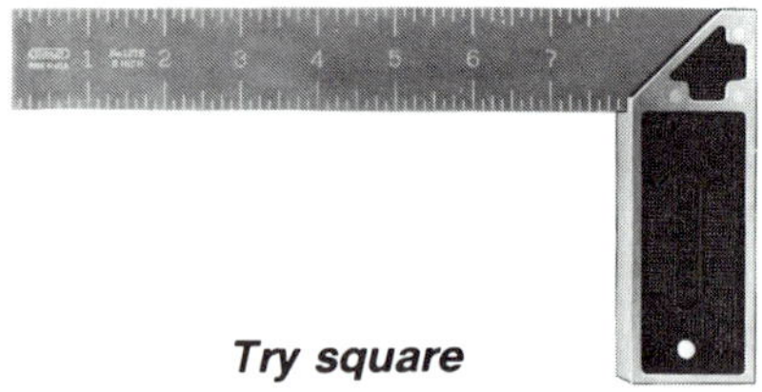

Try square

In addition to measuring 90° angles, it also may be used in measuring other angles, although there are no moving parts on the square.

The framing square works well for making longer right angle lines.

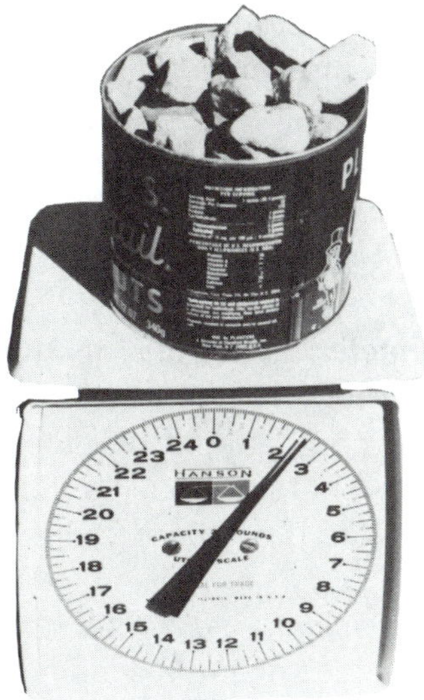

Ordinary kitchen scales may be used in many ways in the classroom. Although not as accurate as balance scales, they are easier for younger elementary students to handle.

Balance Scales. A very good measurement instrument for the elementary classroom is balance scales. Students can determine actual weight, volume/weight comparisons, and specific gravity. There are also many other types of scales which could be used, which would be somewhat less expensive.

Thermometer. There are many types of inexpensive thermometers. These are necessary in the classroom for determining insulation and conductivity characteristics of materials,

The students here are comparing the temperature outside and the temperature of the water in a bowl inside the "student made" solar water heater.

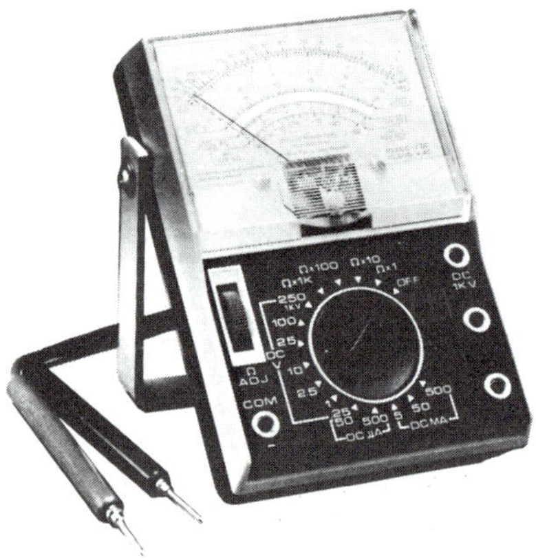

Experiments with electricity should be confined to low voltage, direct current (batteries) because of safety considerations. This meter will measure only direct current.

checking predetermined heat, and generally learning how to read temperature scales.

Measuring Cups. Measuring cups are not only useful in the kitchen but also in the classroom. Several activities will no doubt require the measuring of liquids and loose solids.

Ammeter. An ammeter measures the flow of electrical current through a conductor. It may be used in the classroom to make comparisons of materials and check simple circuits. The instrument requires close supervision, because it can be easily damaged if not used correctly.

Microscope. Although it is not considered a measurement instrument, the microscope is an essential tool for the students to use in making observations and comparisons of materials. For example, weaving may mean very little to a student until a handkerchief, or similar fabric, is observed under a low-power microscope. These also should be used under a certain amount of supervision.

Magnifying Glass. Like a microscope, a magnifying glass is important as an instrument for observation and comparison. Several smaller magnifying glasses, called hand lenses, are considerably less expensive and provide flexibility in classroom activities.

The microscope will enable students to observe and study elements of their environment they have never seen before.

SUMMARY QUESTIONS

1. Identify four purposes of drawings.
2. Explain what is meant by the statement, "Drawing is the language of technology."
3. State why templates might be better for elementary students to use, rather than compasses and triangles.
4. How are drawing symbols similar to letters and words which are in a written sentence?
5. Why is it important for elementary students to be able to draw and use measurement instruments?
6. List two instances when a microscope might enhance the study of technology.

CHAPTER 11

Learning about Raw Materials in the Classroom

In this chapter you will learn:

1. **How steel, plywood, and plastics are processed.**
2. **How to incorporate the study of the acquisition and processing of raw materials into classroom work.**

OBTAINING RAW MATERIALS

Although most teachers may mention that raw materials are used to make products, few devise activities that show the methods of acquiring those materials. Obtaining raw materials in the technology system is usually a large, outdoor operation. Ways to simulate those operations in the classroom are not immediately apparent. By simplifying processes and substituting materials, however, teachers can illustrate certain basic methods. They can, in addition, use pictures or models of the processes and display samples of the actual raw material to help tie the simulation and the reality together. Most raw materials are obtained by either growing or extracting, and classroom activities would likewise fall in one of these two categories.

This observation box has two glass sides. Students can observe the entire plant during the growth stages if the seed is planted next to the glass.

Growing

An observation box with narrow glass sides allows an excellent opportunity to study the growth of plants. The students plant the seeds in the soil next to the glass and as the plants grow, the roots can be observed. Beans and radishes are recommended because they grow rapidly.

Another type of plant that can be grown in the classroom is bacteria. Mix unflavored gelatin according to the directions on the package and add sugar. Pour the mixture into the lid of a can to set. When this host material becomes solid, the students may touch it, leaving several fingerprints. Place the lid in a small container (such as a shoe box), and leave it in a warm dark location for about five days. The bacteria growth can then be seen under a microscope. The dark, warm, moist environment is conducive to bacteria growth, while the sugar which was added to the gelatin provides food.

The teacher can use this opportunity to draw distinctions between useful and harmful bacteria. Most students know bacteria can make us ill, but few realize that bacteria is grown and used in the processing of such products as, for example, buttermilk, cheese, sauerkraut, and vinegar. The bacteria causes a chemical change which is called fermentation. It is also used in the treatment of sewage to make the water pure again, and is a part of the process utilized in producing a chemical called methane gas. This chemical compound is used in making acetylene gas for welding, methyl alcohol, which is antifreeze, and commercial hydrogen, which is utilized in converting vegetable oils into semisolid shortenings used in cooking.

Fungi (mold) can also be grown in much the same way. Lightly moisten a slice of bread and let it lay in the open air for a few minutes to pick up airborne spores. These are single-cell organisms produced by fungi. When they land on damp food, they begin to grow tiny threads which become roots and live off the nutrients in the food. Enclose the bread in a plastic food bag and replace it in a warm, dark location for about a week. Mold should start to appear in two or three days. The students

A worker mixes a large vat of cottage cheese, made by adding a harmless bacteria to milk, which, with the help of other additives, thickens to a semisolid mass.

A technician examines a wafer which contains hundreds of integrated circuits of "computer chips." They are so small that even a speck of dust can cause one of the chips to fail.

may observe the growth by inspecting it each day after the mold begins to appear. The mold may be black, blue, or cotton-soft white, depending upon the type of spores in the air. Molds have many commercial uses, including their role in producing penicillin.

Extraction

Crystals, which can be grown in the classroom, can serve as an aid in the study of minerals and extraction. Borax, salt, even silicon chips used in microprocessors, are all made of crystals. To make borax crystals, you will need a container for heating water, a hot plate, small glass jars, paper clips, pencils, string, pure borax, and water. Powdered borax can be found where detergents and soaps are purchased, but it should not contain any detergents or soaps. The teacher only should heat the water and pour it into the jars. The teacher or students then stir as much borax into the jars of hot water as will dissolve (hot water holds more borax than cold). Several small jars may be used with different types of food coloring for variety.

The paper clip dangles in the solution from a string tied to the pencil, which is placed across the mouth of the jar. As the solution cools, the borax cannot remain suspended in the water, so it forms crystals on the paper clip and string. Cooling will take several hours but may be hastened by placing the jars in a refrigerator or setting them outside if it is cold. When the crystals form, remove them carefully and lay them out to dry. After they dry, students can handle them and observe them under a low-power microscope, magnifying glass, or hand lens. While living organisms grow from seeds or eggs, nonliving substances are shaped by means of crystallization. These crystals grow by adding layers of atoms on top of other layers. Because they are arranged in rows, they have flat surfaces and sharp edges.

Borax, which is closely related to salt, is a mineral compound found in the southwestern part of the United States. Usually extracted by strip mining, borax is used in soaps, water softeners, medicines, ceramic glazes, glass, and food preservatives. Less than 5 percent of salt itself is used to season food. Most is used in commercial processing, for example, in the making of preservatives to prevent bacteria growth and in the making of soda ash, to help produce glass and soap. Salt is also used in producing caustic soda (lye), which goes into paper and rayon, and in producing chlorine, which makes fabrics white and purifies water. An extremely important use of crystals today is in the making of microprocessors (small computers). The small silicon chip of crystals, about one-quarter of an inch (6.4 mm) square, contains many electronic elements, such as transistors and resistors. Several chips may hold tens of thousands of these elements. The crystals are grown in layers under closely controlled conditions.

For further study of extracted materials, students could bring in samples of rocks. A display table for the rocks would help in making visual comparisons, as well as provide a central location for resource materials. After classifying the collection into groups of igneous (once molten), sedimentary (once part of living material), or metamorphic (once of a different composition) rock, each sample may be studied in terms of its use. For example, granite is an igneous rock which contains quartz, feldspar, and mica. Quartz is used in making electronic components and in grinding stone, glass, and mortar. Feldspar is used in making pottery, porcelain, and china, and mica is used in producing electronic equipment, insulation, and rubber tires. Thus, study would emphasize the locations of the various minerals, the methods of extraction from the earth, and the uses of the material.

Part of understanding the uses of minerals is knowing about the characteristics of the raw materials as they are reflected in the products in which they are used. Below we discuss three materials—steel, plywood, and the plastic, polystyrene—and the processes that transform them from a raw state to a product.

CHARACTERISTICS OF STEEL

Materials may be classified as living (organic) or nonliving (inorganic) resources. Metals are considered inorganic, because they did not develop from living organisms.

Open-pit mining requires heavy equipment. Trucks transport the ore to either a shipping or rail center for further transportation, because it is too heavy to be transported via roads.

Iron and steel (which is made from iron) are probably the most useful and cheapest metals known today. Nearly everything we come in contact with has some connection with iron. Iron is a required nutrient for plants, animals, and humans, and it is also present in many medicines.

Iron ore, one of the earth's most abundant materials, makes up about 5 percent of the land surface. It is either open-pit or shaft mined where there are large deposits. Making iron and steel can be compared to baking a cake, in that both involve ingredients combined according to what may be called a recipe.

The Ingredients

Iron Ore. The ore may range in color from red to black, depending upon the kind. Some kinds have to be separated from other materials before they are shipped to the iron and steel mills. This separation process may be very complicated and include magnets and heat or it might be as simple as washing the ore with a stream of water.

Coke. Coke is obtained by heating coal in the absence of oxygen. The process releases gases from the coal, which does not burn because oxygen is not present. The coke oven by-products are used in making many other products, such as explosives, synthetic rubber, medicines, and dyes. Part of the gas is also used as fuel to heat the ovens. Coke serves two purposes in making iron: as it burns with the air forced into the furnace, it changes the iron oxide (rust) in the ore to pure iron; and the tremendous heat which is generated melts the ore and all the impurities.

Limestone. Limestone helps to remove impurities by causing them to melt at lower temperatures. The impurities, called slag, are lighter than the molten ore, so they float to the top, where they are drained off. Slag is used in making cement and insulation materials.

Air. Iron making requires much air because oxygen is necessary. The oxygen combines with carbon in the coke, producing carbon monoxide. This combines with the oxygen in the ore to form carbon dioxide, which separates the pure iron from the iron oxide as it melts.

Water. Water keeps the furnaces from overheating. The iron and steel industry in the United States uses enough water in one day to supply the daily needs of five cities the size of Chicago.

The Recipe

The recipe for the ingredients needed to produce one ton of iron calls for 4 tons of air, 1/4 ton of limestone, 3/4 ton of coke and 1 3/4 tons of iron ore. The required heat reaches nearly 3000°F (1649°C).

There are a number of different types of furnaces, but the blast furnace is the largest, sometimes being taller than a fifteen-story building. Every four or five hours, molten iron is tapped (drawn off) through a hole close to the bottom. The white-hot iron runs through a trough to the ladle cars. These are railroad tank cars shaped like a bottle. They take the mol-

An oxygen furnace is tipped over to pour molten steel. Its size can be estimated by comparing it with the man on the catwalk to the left of the picture.

ten metal to a mixer, which keeps the metal in a liquid state until it can be used in the steel-making furnaces.

The recipe used in making steel depends upon the kind of steel being made. There are many combinations, or alloys, each calling for specific ingredients and amounts.

Metals such as chromium, nickel, manganese, and vanadium might be used, along with scrap steel and iron, molten iron, and other ingredients. The mixture is very closely controlled. If you baked a cake with proportional accuracy in measuring, the recipe would call for one-thousandth of an ounce (.00003 kg) rather than teaspoons and measuring cups.

When the liquid steel is drawn off, it is poured into molds to cool. The resulting large ingots of steel may weigh as much as 2¾ tons each. When these are ready for use, they are

The rolling mill can reduce a large piece of metal several inches (cm) thick to very thin material. Since no metal is removed, the thin material is much longer, requiring it to be coiled into a roll.

placed in soaking pits, which return them to a white-hot temperature, but not molten. They are then sent to the rolling mill, where they are compressed by huge rollers into a variety of shapes, including slabs, sheets, strips, bars, structural beams, and railroad rails. At this point they are ready to go to the manufacturer.

Students can analyze pieces of steel and other metals using microscopes, magnifying glasses, and scales to observe characteristics such as weight, color, hardness (by scratching with a nail), and the presence of iron oxide (rust).

CHARACTERISTICS OF PLYWOOD

Plywood and other wood products are made from renewable, living raw materials. Because plywood is made from a living organism, it has entirely different characteristics from steel. Types of plywood are classified as either softwood (used in construction and structural assembly of smaller products) or hardwood (used in paneling and veneering). It is also classified as interior and exterior, according to whether waterproof adhesive is used.

Uses of Plywood

Because wood is made of cells aligned primarily in one direction, solid wood tends to warp as it absorbs moisture. It also is relatively weak if the force is not in the direction of the cells. Because layers (plies) are alternated so the grain direction (arrangement of cells) of one layer runs at 90 degrees to the grain direction of adjacent layers, plywood has many advantages over solid wood:

- It warps and twists less
- It shrinks and swells less
- It cracks less around the edges
- It permits an expensive look at less cost because hardwood veneers (thin sheets) may be on the outside and cheaper material on the inside
- It can bend into curved shapes

Plywood is used in nearly all large wood products, such as houses, boats, furniture, and packaging.

Plywood comes in standard thicknesses by eighths of an inch (3.2 mm). The thickness of each ply and the number of plies depend upon the thickness of the standard piece.

Making Plywood

Plywood is made in three steps: (1) the logs are obtained, (2) the veneers are cut, and (3) the plies are laid up.

Obtaining the Logs. After the tree is cut, the trunk, which may be 200 feet (61 m) or more in length, is cut into smaller logs of 8 to 40 feet (2.4 to 12 m), so they can be transported to the sawmill. They are dragged to a central location by tractors and cables. There they are loaded on trucks, railroad cars, or barges. At the mill, they are dumped into a large pond of water, called the log pond, where they are kept until being pulled into the mill for sawing. The pond protects them from fungi, insects, staining, and fire.

Cutting the Veneers. The logs used to make plywood are selected for quality and straightness. They are run through a debarker, which strips off all bark, using either sharp knives or streams of water under extremely high pressure. Veneer may be obtained by sawing, which is done to brittle woods that might crack and splinter; by slicing, which moves the log against a stationary knife and is used for the fine-figured woods in furniture and wall paneling; or by rotary cutting, in which a log mounted in a huge lathe is rotated against a long

Workers inspect the veneer as it comes from the lathe. The grain pattern can be seen at regular intervals, representing one rotation of the log.

knife. A continuous ribbon of thin wood is peeled off the log. About 90 percent of all veneer is rotary cut.

Laying Up. The veneer has to be dried, trimmed, and matched for wood grain characteristics. It is then coated with the proper adhesive and laid up (laid in place) with the other plies, alternating the grain direction. The plywood is then placed in a large press with many other sheets under heat and pressure. After the adhesive is cured, the sheets are removed, dried, trimmed, and sanded. At this point they are ready to go to the manufacturer or to be sold as is.

Most plywood sheets are 4 by 8 feet (1.2 × 2.4 mm) and range from 1/4 of an inch (6.4 mm) to 3/4 of an inch (19.1 mm). Most sheets have three, five, or seven plies; some may also have a solid wood core, with one sheet of veneer on each side. Other dimensions are produced but are not used as frequently as the "standard" sheets.

CHARACTERISTICS OF POLYSTYRENE

Polystyrene is a plastic, and, therefore, a synthetic material obtained through chemical processes. The steps to produce polystyrene are much different from those to produce either steel or plywood.

Polystyrene is a plastic made from benzene and ethylene, which are produced from petroleum by fractional distillation. Because of the close link between petroleum, chemicals, and plastics, many large corporations produce all three.

Obtaining Raw Materials

The raw materials are obtained through extracting petroleum by drilling. The gas-and-oil mixture is processed by fractional distillation. Two products which result are a liquid called benzene (used as a solvent and cleaner) and a gas called ethylene (used in making antifreeze and other industrial products). These two are used in making polystyrene.

All matter is made of molecules, which are invisible particles. The arrangement of these molecules (how they are connected together) determines the characteristics of the matter. Synthetic materials can be imbued with certain characteristics by causing a specific arrangement of the molecules. Each molecule contains atoms, which are linked together with other molecules in a chain. Each link is a molecule called a monomer (*mono* means "one"). The entire chain is called a polymer (*poly* means "more than one"). Producing synthetic resins involves building these chains, or polymers.

To produce polystyrene, the ethylene (gas) is bubbled through the benzene (liquid), producing a liquid called ethylbenzene. This is heated to a high temperature and combined with specific metal oxides (compounds) which remove some of the atoms. Molecules of styrene are formed by the remaining atoms. Other chemicals are added, and the styrene is then placed under heat and pressure. This process, called polymerization, causes a chemical reaction, which links the styrene monomers together into chainlike molecules of polystyrene.

The polystyrene is then ground into small pieces ready to be sent to a manufacturer. Polystyrene is clear, colorless, and hard. It can be made stronger by adding fiberglass, softer by adding other polymers, and colored by adding pigments. Examples of products which are made of polystyrene would be food containers, appliance parts, plastic lenses, bottles, and electrical insulators. In addition, it is used in packaging, model airplane and car kits, and picnic jugs and coolers.

We have provided details on the production of steel, plywood, and polystyrene to enable teachers to draw from an informed background when deciding which concepts to emphasize in classes on raw materials and production. Other products may also constitute topics for class reports and individual study, and the level of detail may be made commensurate with the age and interests of the class.

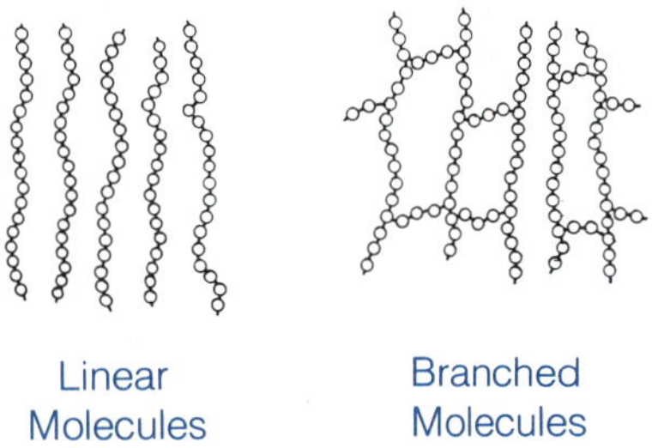

Plastics with linear molecules are usually more flexible, while those with branched molecules are more dense and rigid.

These small pellets of polystyrene are shipped in bulk to the manufacturers to be made into plastic products. Each pellet is less than the thickness of a pencil lead.

The versatility of polystyrene can be seen in this part for a chair. It can be colored to match wood, allows great detail, is water and chemical resistant, and is very rigid and tough.

CORRELATING A CLASSROOM ACTIVITY WITH THE STUDY OF RAW MATERIALS

Many activities can be conducted in the classroom that tie directly into the study of raw materials. Some may be considered involved, while others quite simple. The following activity illustrates how a seemingly typical study of foods may incorporate the study of raw materials processing. Making ice cream is certain to interest students, and they learn from the process itself.

Making Ice Cream

A 5 quart (4.7 l) hand-crank ice cream freezer is necessary for making ice cream, as are supplies and ingredients. The students mix the ingredients, supply the energy for freezing, and have the opportunity to sample the final product. This activity lends itself not only to the study of eating habits, meal planning, food preparation, and nutrition, but also to the study of raw materials and how they are processed. A discussion could involve the simple machines used in making the ice cream. These might include can openers, mixing spoons, and a hand-operated egg beater. In addition, the freezer mechanism should be examined so the gears can be seen. A discussion may be held concerning how the crank, which rotates in a vertical pattern, uses the gears to turn the ice cream container in a horizontal pattern.

Required Materials. Every material used in making the ice cream has to be processed from a raw material. Because most of them will be eaten, they have to be cleaned (eliminating dirt and unusable parts) and cooked.

The materials used in making ice cream come from animals, plants, and minerals. The following list presents important facts about where we get our foods.

- Cream. The cream has to be separated from whole milk and also heated to kill harmful bacteria, a process called pasteurization. Half-and-half contains half cream and half milk. This product comes from animals.
- Eggs. Eggs need little processing, other than cleaning and storing in a cool place.

- Vanilla. A climbing vanilla vine, a member of the orchard family grown in tropical areas, is the source of vanilla. Vanilla goes through many processes, including chopping and cooking. The resulting extract provides flavoring for chocolate, ice cream, candy, and pastry.
- Sugar. Although sugar may be obtained from maple trees, milk, and honey, most is made from sugar cane and sugar beets. Processes include chopping, pressing, and cooking.
- Table salt. A seasoning to bring out flavors, salt has to be cleaned and packaged before it is safe to eat. This mineral is extracted from either the ground or from salt water.
- Industrial salt. This does not have to be table salt, because it is not eaten. The type of salt used on roads and sidewalks or in water softeners works well and is less expensive. If industrial salt is used it should not be eaten.
- Ice. Any kind of ice—ice cubes, icicles, snow, or commercial ice—may be used, because it too is not eaten; it is to provide the freezing temperature. Ice is solid water. Water may be found as a gas, liquid, or a solid.

There also are a number of facts concerning the freezing which are important: the ice cream is a soft ice cream rather than a solid block, because the temperature is not low enough to freeze it solid; the crank turns the paddle inside the container, which breaks up hardened chunks of cream and allows the mixture to freeze evenly; and the ice is packed around the

Because of the ice and salt water, making ice cream is an excellent outdoor activity. If it is done indoors, the freezer may be set in a plastic dishpan to catch the salt water as it melts and runs out the drain hole.

The ice cream recipe for this kind of freezer may be the same as that used for refrigerator ice cream. Although the ice cream will not freeze solid, it will freeze enough so that it can be eaten with a spoon.

ice cream container, with salt being mixed in as the larger container is filled. When salt is mixed with ice it produces a brine (salt water) which has a lower freezing point than that of water. As a result, as the ice melts, the temperature of the water is lowered to below freezing. This causes the ice cream to freeze. As the ice melts and runs out the overflow, more ice and salt are added to keep the temperature low.

For full participation, the students measure and mix the ingredients and provide the power behind the crank. Freezing will require about forty-five minutes.

If a hand-crank freezer is not available, an electric freezer could be used. If neither are available, individual makeshift freezers can easily be made in the classroom. A gallon (3.8 l) milk container is cut down to about four inches (102 mm) in height. A metal food can is set in the middle with crushed ice and salt packed around it. After the can is filled with the ice cream mixture, a standard-size plastic lid is placed on it. The lid should have a small slot cut in the top which is just large enough for a wooden tongue depressor to be slipped in. The depressor is periodically moved back and forth to keep the mixture smooth while freezing takes place. The mixture will have to set in the brine for approximately two hours, with more ice and salt being added as needed. Because the solution will overflow the sides as more ice and salt are added, it should be done in a sink or large container.

SUMMARY QUESTIONS

1. Name three classroom activities that would teach students about growing and extracting.
2. Identify three elements which making steel and baking a cake have in common.
3. State three advantages plywood has over solid wood.
4. Why could an increase in the price of crude oil cause an increase in the price of a picnic jug?
5. Why is salt necessary in freezing ice cream in a container?
6. Why is it possible to produce plastics with so many different characteristics and why is plywood relatively limited in its range of characteristics?

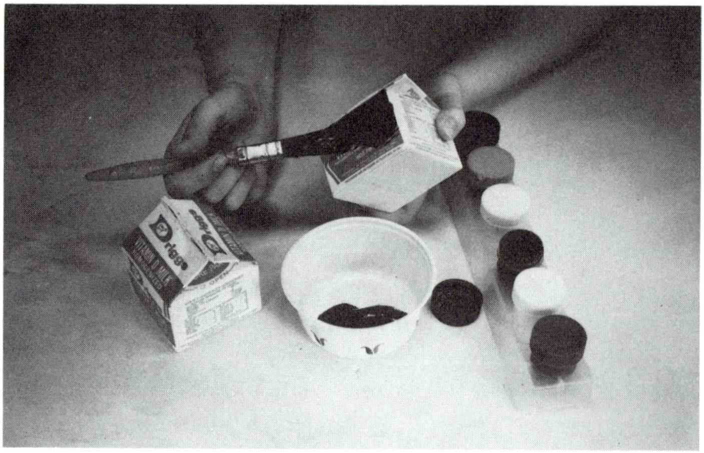

ALUM

CHAPTER 12

Processing Materials in the Classroom

In this chapter you will learn:

1. **How the study of processing may be incorporated into classroom activities.**
2. **What equipment, tools, and supplies might be utilized in the classroom to help teach about processing.**
3. **The names and uses of several common tools.**

Processing includes changing raw materials into a usable form. Separating, mixing, and forming can be demonstrated and experienced in the classroom with a minimum of equipment and supplies.

SEPARATING

Material separation activities should be geared toward developing concepts concerning how separation is accomplished. In addition, concepts should be formulated concerning the characteristics of materials and how methods used in separation are determined. After they are separated, the next steps of recombining and forming usually occur. These processes may also be a part of the classroom activities, either separately or in combination with the separation activities.

Sorting

Sorting may involve many different activities. The following are only a few examples of what could be planned for a classroom experience.

Washing. Washing is probably one of the easiest activities because of the wide choice of materials. Among the topics to discuss might be the reason why water under pressure (spray) loosens dirt and other foreign matter faster than soaking. This touches on the way molecules hold matter together and how energy can be used to accomplish work. A second topic could focus on the role of chemical action (energy) in cleaning and

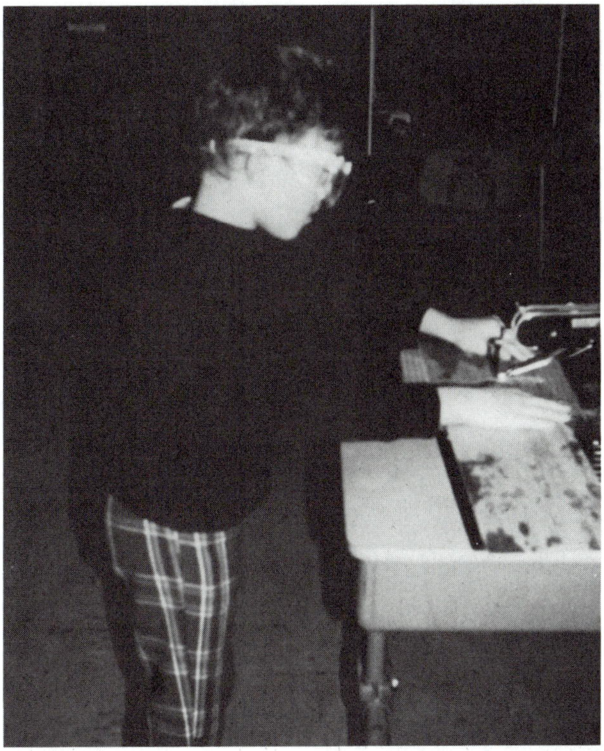

A student cuts a piece of wood on a small vibrating jig saw. These saws are safe for use in the elementary grades. As with many other activities, however, proper instruction and **direct teacher supervision** *should be planned into the lesson or unit.*

washing. Most students know what soap and detergents do, but few realize what is happening when cleaning takes place.

Technically, cleaning products are detergents, with soap being a detergent. Soap is more of a natural substance, being made of either animal fats or vegetable oils. Detergents are usually considered synthetic, sometimes being made of petroleum by-products, as well as of many natural materials. Both work, however, in much the same way. They cause the surface to become wet faster, allowing water to enter small cracks and between the strands in fabrics. Surface tension holds water molecules together. Soap and detergent lowers the surface tension, allowing the water to "breakup" and spread into small areas. Soap and detergent molecules also attach themselves to small dirt particles. As the surface being cleaned is rubbed or agitated, as in the motion of a washing machine, the water molecules are forced out, pulling the particles with them. These particles are suspended in the water until they are flushed away.

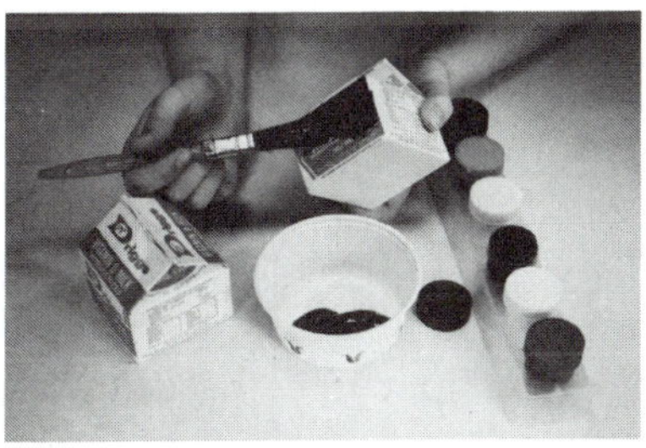

A student is able to paint a wax-coated milk carton with tempera paint because a small amount of liquid soap has been added to the paint, reducing the surface tension and allowing the paint to cover the surface rather than form beads.

Two students filter water through fine screen, which leaves a layer of pulp—a filtering process used in recycling paper.

Screening and Filtering. Used in one form or another in nearly all processing of raw materials, screening and filtering may be done in the classroom with various sizes of window screen, cheese cloth, or sieve, depending upon the material being separated. Screening and filtering devices may be made by attaching fabric or screen to wooden frames with staples. Screening separates particles and small pieces according to size. Several sizes of particles may be obtained by using different screens with varying sizes of grids. The primary difference between screening and filtering is the size of holes in the screening device. Filtering usually involves separating solids and liquids, using a material which prevents even very small particles from passing through.

This student has built a screening device composed of progressively smaller stones, gravel, and sand. When the water containing mud and other foreign matter is poured into the unit, it screens out progressively smaller particles until only water comes out at the bottom.

Floating. Floating, along with the opposite action of settling, can be demonstrated by mixing small particles of different weight materials into a liquid. As more water is added to the container, the lighter materials are separated by overflowing the edge. The liquid may then be drawn off or poured out, leaving the heavier particles which have settled to the bottom. The same process may be used to separate two liquids, such as cooking oil and water. Food coloring may be used to make the visual difference between the two liquids more dramatic.

Flocculation can easily be demonstrated by students using two very common household chemicals: lime and alum.

It is possible to have minute particles in a liquid, which do not rise to the top or sink to the bottom. They are said to be suspended in the liquid. Many times it takes a chemical additive to pull the particles from a suspended state. One such additive can be easily demonstrated in the classroom.

Mix a clay soil with water in a clear glass quart (l) jar. This makes the water murky and difficult to see through because of the suspended particles. Next, stir in one tablespoon of equal amounts of powdered lime (used as a garden fertilizer) and alum (used in cooking). Within a few seconds the two chemicals combine and expand, becoming heavy enough to settle. As the chemicals drop, the suspended particles attach themselves and are carried to the bottom. The result is clear water. Lime is an alkaline made from limestone. It is used as a fertilizer, in the manufacture of glass and steel, and as part of bleaching powders in some detergent products. Alum is a salt made from bauxite (the same raw material from which aluminum is made). It is used in making medicine, cosmetics, baking powders, and plaster of Paris. The process of combining these chemicals into a heavy mass is called *flocculation*. Several different chemicals are used as flocculants in large waste-water treatment facilities. They act in much the same way as the chemicals in the classroom demonstration.

Heating

Heating is a common method of changing the characteristics of materials. It is a very versatile method because the amount of heat can be regulated to accomplish the desired change.

Distilling. Separation of liquids by heat can be demonstrated by building a simple distillation unit. A teakettle containing tea or coffee is heated on a hot plate. A plastic hose is attached to the spout with heating duct tape and run through a pan of ice water placed at a slightly lower level. The end of the hose is placed in a glass container slightly below the ice water. The process simply involves evaporation and condensation. The heat changes the water to steam, which is cooled in the condenser (ice water). Upon cooling, it condenses back into water, filling the lower container. As a result, the water is separated from the tea or coffee and has no odor, taste, or color of the original mixture.

A worker uses a portable electric arc welding unit to repair a heavy construction vehicle in the field. The electricity generated by the power unit in the truck creates enough heat to melt the extremely thick steel.

Melting. A coping saw with an old blade illustrates the process of melting. The blade is carefully heated over an alcohol burner. It is then used to cut pieces of expanded plastic, such as that found in craft stores, or the type of plastic used in building insulation panels. The material is easily cut without teeth on the blade, because it is cut by melting. The surface can be observed and compared with a surface resulting when a piece is broken. Technology uses melting to cut thick pieces of steel in the same way, using a very hot flame or an electric arc.

Evaporating. This separation process takes a little longer to accomplish if only natural evaporation is used. Dissolve table salt into a small container of water, and pour it into a shallow dish or table saucer. Place it by a window where the sun will evaporate the water. After evaporation, the same amount of salt can be taken from the dish. This procedure has many uses in processing materials which are in a solution. However, heat is generally used to hasten the process by boiling. This rapidly converts the liquid to a gas, leaving only the solid material. Many of our salt products are obtained in this manner.

Evaporation is a step in the process of making pure maple syrup. The sugar water is drained from maple trees and boiled until it becomes thick. If it continues to boil, it becomes maple sugar.

Cutting. This method of separation is a large category because of the number of different tools. We do not attempt a complete list, rather we concentrate on common tools which are easy for elementary students to use.

Shearing. Shearing, in its broadest sense, refers to a cutting action which separates material with two cutting edges (scissors) or one (a single saw tooth). Shearing does not necessarily mean a very clean cut; it may be rather rough, as the result of a tearing action. Shearing involves the use of sawing, shaping, and drilling tools.

Several different kinds of saws could be used in the elementary classroom. A crosscut saw is sometimes called a hand saw. It is an all-purpose saw for wood and other soft materials. Although designed for straight cuts across the wood grain, it can be used to cut with the grain. A panel saw is a shorter, fine-tooth crosscut saw. It works well for elementary students because of its size. The back of the blade is reinforced on the back saw. It has a thin, short blade with a square end. It is used for fine cutting of wood and soft materials. This is prob-

The choice of which shearing tool to use depends upon the material to be cut.

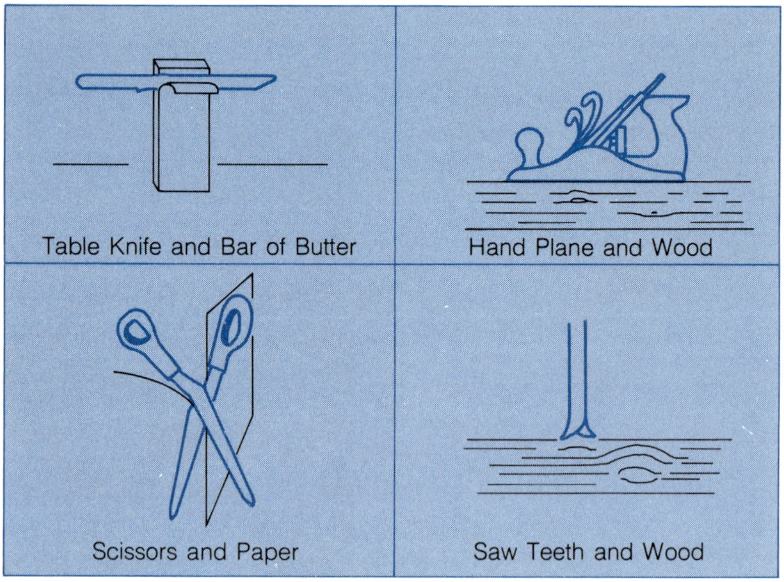

ably the easiest saw for elementary students to use. It also makes only straight cuts. The rip saw is a woodcutting saw used in cutting with the grain. It does not work well in cutting across the grain or in cutting most other materials. The shape of the teeth are different from the crosscut teeth. Curves may be cut with a coping saw. It has a very thin blade which is designed for wood, soft materials, plastic and very thin metal. The blade can be turned at an angle to the saw, allowing cuts to be made parallel to an edge. A hack saw has a thin blade designed for cutting metal, but it may be used on nearly any material. Large curves may be cut with this tool.

A few basic principles should be observed when using saws. In starting the teeth into the material, the thumb of the free hand should be placed alongside the blade to keep it steady. Only backward strokes are used until a saw kerf (cut) is started. For younger students, a small block of wood clamped next to the sawing line works well as a guide in place of the thumb. The saw angle for crosscutting should be about 45 degrees, while the angle for ripping should be about 60 degrees. The back saw, coping saw, and hack saw may be used at right angles to the material being cut. A smooth full stroke should be used, using most of the teeth on the blade. The motion should come from the shoulder and elbow, not from the wrist.

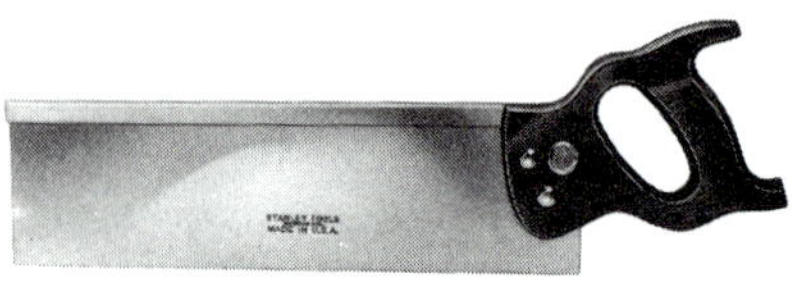

Back saw

Panel saw

***The material should always be clamped tight, whether by a vise, clamp, or other method. The height should also be low enough to be comfortable for the student.* Safety glasses are always required.**

Shaping tools are used to change the basic shape after cutting to the correct size. The plane is a woodcutting tool with a cutting wedge that can be adjusted for the depth of cut. There are two basic types. The smooth plane is used to cut with the wood grain and the block plane cuts across the end grain. Although the block plane is smaller in size, it is not recommended for most elementary students in attempting to plane end grain because of the manual dexterity required. Planes are used in making flat surfaces and rounded edges.

The surform plane is an easy shaping tool for elementary students to use. It may be used on most soft materials, including the end grain of wood. It has a blade with perforated

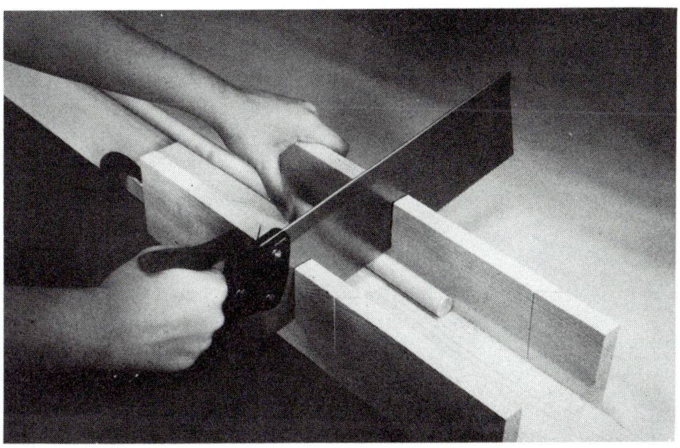

In sawing, cut only on the forward stroke, releasing the pressure on the back stroke. The forward stroke should be firm and smooth; it should not force the teeth into the wood.

Surform plane

openings, which provides many small cutting edges. This is an excellent tool for rounding edges and is somewhat safer than the other types of planes. Teachers should emphasize to the students that they do not rub their fingers across cutting edges, that they always lay the plane on its side to prevent damage to the blade, and that they secure the work in a vise, or other holding mechanism, so both hands are on the tool, providing better control and more safety.

In using the Surform plane or any other cutting or scraping tool, the strokes should always be forward, with both hands behind the cutting edge.

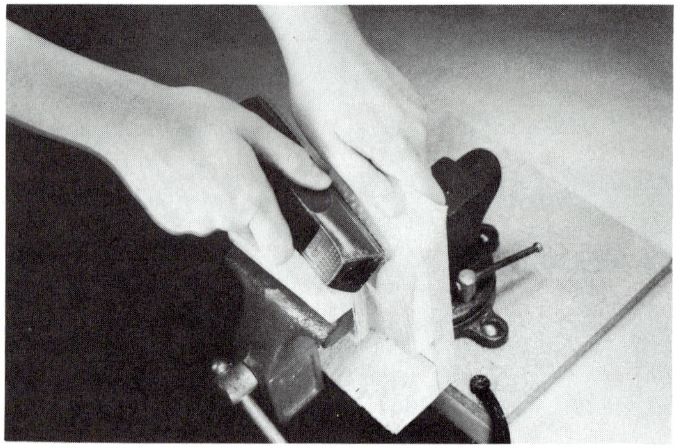

Drilling tools are used in making holes too small to use a saw. A brace and bit is used for wood only. The brace is the handle and the bit is the cutter. The bits come in various sizes. As pressure is applied, the threaded end of the bit helps pull the cutting edge into the wood.

A hand drill may be used on nearly any material except hard metal. It has a small gear which is turned by a revolving handle. The cutter is called a twist drill. Hand drills come in many sizes, but most will accept only up to 1/4 inch (6.4 mm) diameter twist drills.

The push drill looks much like a large screwdriver. The small drill rotates as the handle is pumped up and down. It is used in making pilot holes for guidance of fasteners, such as nails and screws, which makes it easier for them to penetrate and prevents splitting of the material.

Procedures for drilling holes are as follows. Always keep the cutting tool steady and at right angles to the surface of

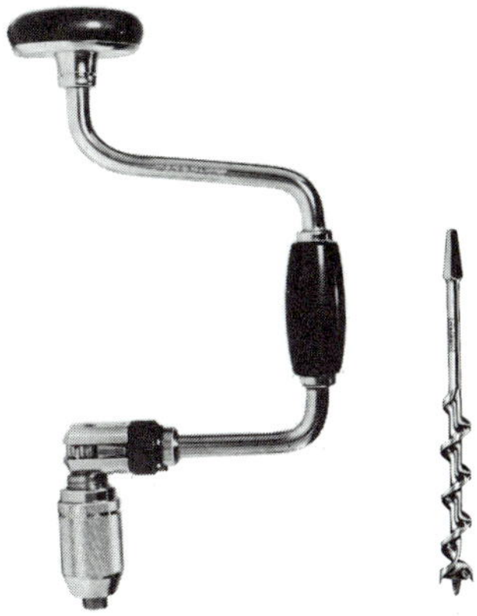

Brace and bit

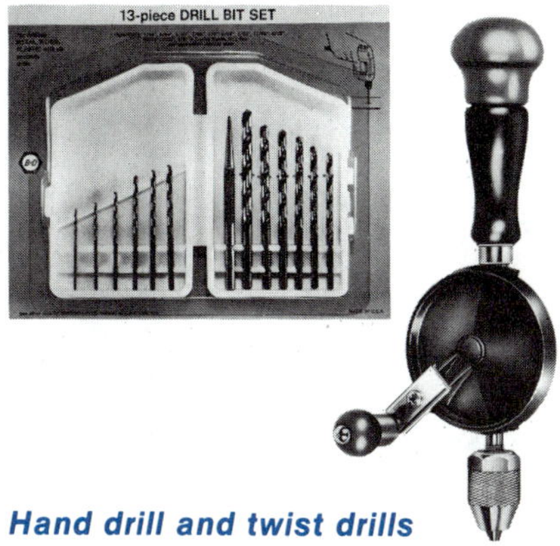

Hand drill and twist drills

Push drill

the material being drilled. Side movement will cause a larger hole to be cut. When using a brace and bit, stop when the screw comes through the opposite side. Reverse the direction of rotation once or twice to bring the screw back into the material, and continue the original direction of rotation as it is drawn out. Then start drilling from the opposite side. This will prevent splitting of the wood fibers. If a scrap piece of wood is used to back the piece being drilled this procedure is not necessary, because the hole continues into the second piece as if there is only one piece of material. When using a hand drill and twist drill, a small hole should be started with a nail or punch. The twist drill has no screw like the bit to help in

Elementary students can drill horizontal holes much more easily than they can vertical holes. This V board has a hole in it to allow horizontal drilling. The piece being drilled is clamped to the board so both hands are free.

getting it started in the exact location. When pulling the drill out, continue the original rotation direction while withdrawing.

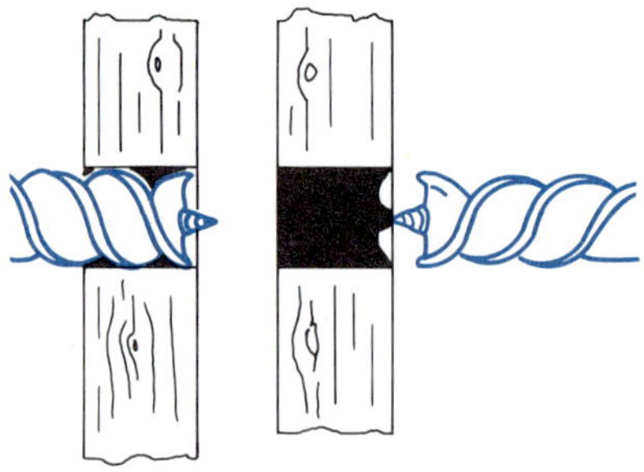

A.

B.

(A) The direction of cut is reversed when the tip first comes through the opposite side. (B) Backing with a second piece of scrap wood does not require reversing the direction.

Starting a hole with a nail or punch before drilling keeps the point of the drill from moving off center when it first turns.

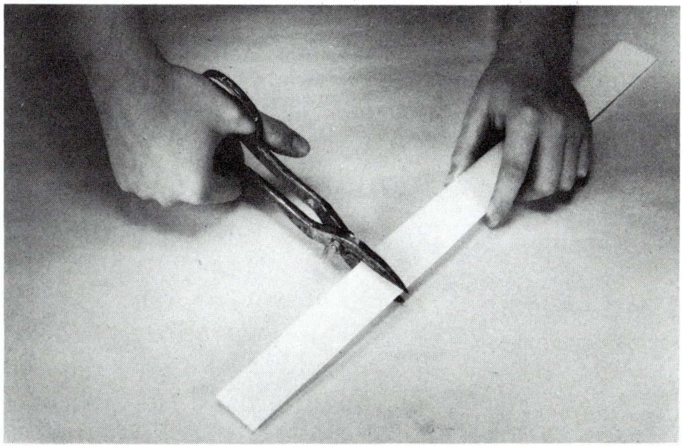

Tin snips can be used for cutting many types of material. The smaller the snips, the easier they are for elementary students to use.

A very versatile tool which is used quite often in the classroom is the tin snips. They are considered shears, much like scissors, but they have several advantages over large scissors. They are not dangerous, because they do not have sharp points. A sharp pair of snips, however can cut paper, as well as cardboard. They also are able to cut sheet metal, wire, and many kinds of plastics.

Scraping. Scraping tools are designed primarily for smoothing material surfaces. Files come in several sizes and grades of roughness. The choice of which one to use depends upon how much material is going to be removed. They also come in various shapes, such as flat, half round, round, and square. There are different files for soft materials and hard materials. How-

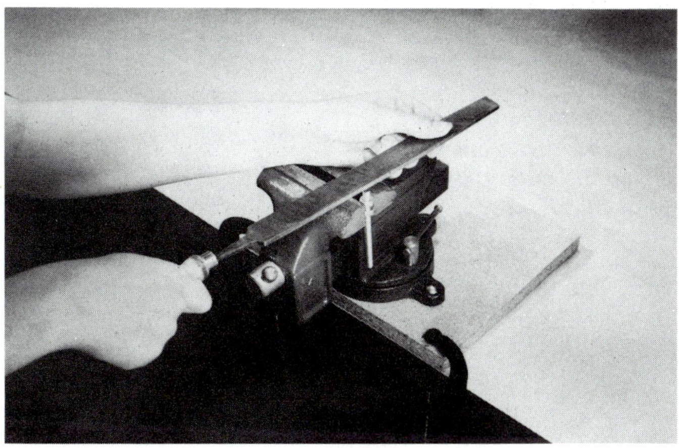

Files are helpful in smoothing and shaping many materials. The correct method of holding the file is shown, and the cutting should be done only on the forward stroke. **Files should always have a handle.**

ever, the best kind for the elementary classroom is the metal file, because it can be used on any material. Files should not be used without handles.

Probably the best-known abrasive is abrasive paper (sandpaper). It may also be obtained as abrasive cloth, and abrasive metal, which uses small metal points rather than minerals glued to the surface. The paper and cloth are consumable, while the metal sanding surface is not. All come in various grades of coarseness.

Some aspects to keep in mind when smoothing surfaces are as follows. All filing, scraping, or sanding should be done only in the direction of the wood grain when working with wood. If the material has no grain, it makes little difference which direction is used, as long as the direction is consistent. Coarse abrasives should be used, followed by medium, fine, and extra fine. Each succeeding finer grade takes out the scratches left by the last one, resulting in a very smooth surface. Abrasive papers and cloth should have the dust knocked out of them periodically to keep them from loading (caked particles stuck in the rough surface). This can be done by lightly hitting them against the work table or a scrap of wood.

Although most power tools would be too dangerous in the elementary classroom, there are three which work well under proper instruction and supervision. The first is the electric drill and stand, which is, in many ways, safer and easier for the students than attempting to drill larger holes by hand. The work is clamped down and all the student is required to do is

A student shapes a piece of expanded plastic insulation with abrasive paper, which may also be used for smoothing. Flat surfaces can be obtained by wrapping the paper around a small wood block so the cutting pressure is evenly distributed.

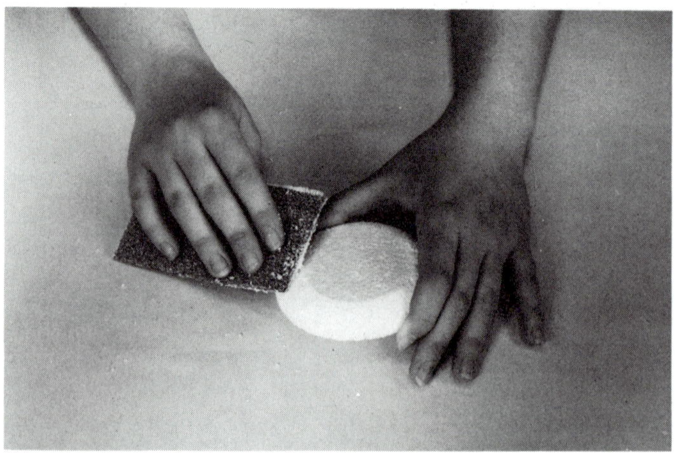

With the electric drill switched off, a student lowers the drill tip to the wood to line it up with the correct spot to be drilled. The wood will then be clamped tightly to the drill table before drilling is started. **Direct teacher supervision is required.**

pull the handle down. This tool allows all students to make larger and deeper holes in a shorter time. Close teacher supervision is, of course, necessary. Second, the jig saw has a stand and cuts with a rapidly moving reciprocating action. It has a small guard to keep fingers away from the blade. It cuts curves and straight lines very easily. This should also be used only with strict teacher supervision. Some saws are equipped with an additional plastic guard around the blade which prevents fingers from coming near it. Third, the vibrating sander can be

The regular jig saw (sometimes called a scroll saw) is larger and will cut heavier material than the vibrating jig saw. Notice the plastic guard around the blade to ensure safe use. **Direct teacher supervision is required.**

This plastic guard is raised and lowered according to the thickness of the wood. It allows the wood to be pushed through, but fingers cannot get close to the blade. It can be easily made with acrylic plastic and standard hardware.

used without direct teacher supervision. The vibration movement is very slight and poses no danger. Proper instruction, however, is necessary.

Power tools are faster, more accurate, and a higher level of technology, but they are not necessarily suitable for all students. Physical and mental maturity rather than specific age should be the determining factor. It should be noted, however, that power tools may have an even greater advantage with some handicapped students, who may experience success they would not otherwise have if they had to rely on manual dexterity.

The vibrating sander should be used with very light pressure. It can be moved in any direction, but sanding in the direction of the grain is preferable.

Grinding. Grinding is closely related to abrasive papers, only in this case the abrasive is part of a wheel. The power tools used for grinding are not appropriate for the classroom. The concept of grinding, however, may be taught by means of abrasive papers. Many times both the paper and the grinding wheel employ the same abrasive material, and the tearing action is the same, with the primary difference being the speed at which it takes place.

MIXING

Mixing is a rather simple process with most compatible materials. The materials may be in either dry or liquid form.

Stirring and Shaking

Most activities involving stirring and shaking for purposes of mixing materials can usually be done by hand. A plastic dishpan is convenient to set the stirring container in, eliminating extensive cleanup.

When mixing is required, a kitchen blender, often used in food units and science experiments, is a convenient power tool for the classroom. An electric mixer also can be used.

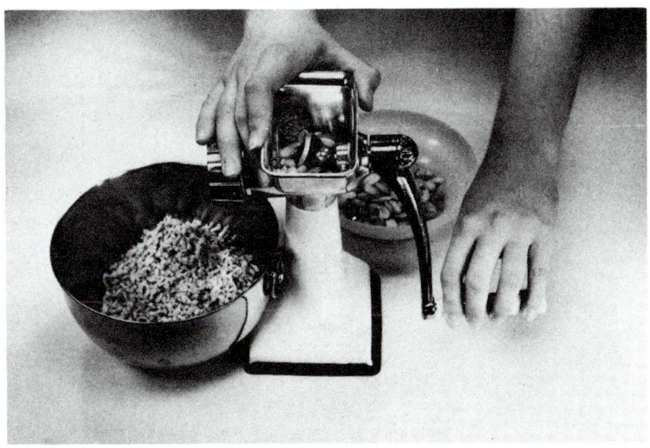

Another household tool useful in the classroom is the food grinder. A student grinds peanuts which will be mixed with cooking oil to make real peanut butter (an activity which the whole class is sure to enjoy).

FORMING

These activities would involve the shaping of materials in ways other than removing material, as in cutting.

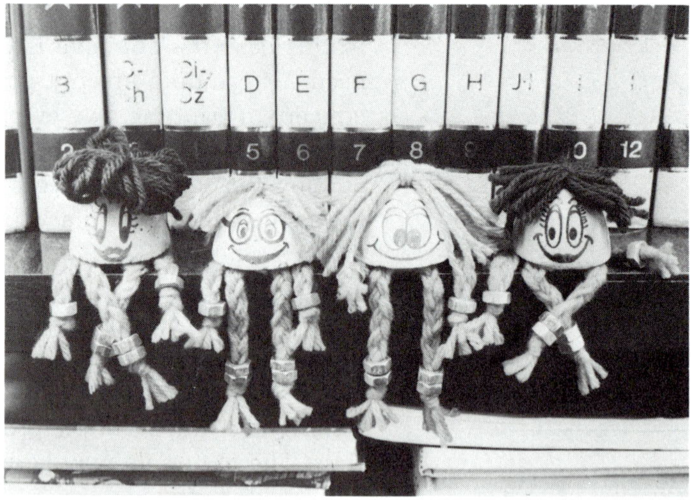

Book gnomes may be cast of either plaster of Paris or fine-grained concrete. They are painted and given arms and legs of braided yarn, with small machine screw nuts for hands and feet.

Casting and Molding

Casting activities can be conducted in the classroom with plaster of Paris. Commercial molds are available at local craft stores, or nearly any type of ordinary container will work. Illustrated are two book gnomes cast in a plastic egg carton. Concepts concerning mixing of materials, hardening time, and finishing (painting) may be taught.

Molding can be demonstrated by using some of the commercial clays for making small ceramic articles. Some can be hardened by simply placing them in a kitchen oven. A molding compound which can be made easily in the classroom and

A student molds salt dough into various shapes.

does not require heat, is salt dough; it does, however, require several days to dry if the object is very thick. It lends itself well to making objects 1/4 of an inch (6.4 mm) in thickness, or less. If thicker objects are desired, it may be molded around other materials, such as tightly wadded paper with an outside layer of aluminum foil, or pieces of preformed expanded plastic. The recipe for salt dough is as follows:

Salt Dough Recipe

1. Salt 2. All-purpose flour 3. Water

Mix one part water, one part salt and two parts flour into a doughy mixture. Knead the dough until it is smooth. A small amount of flour on the hands and table top will prevent sticking.

The material may then be molded by hand, rolled into strands, or rolled flat with a rolling pin. It also may be colored with food coloring or painted after drying. For air-drying, place on a screen for even drying (approximately 3 days). For baking, place in 325° to 350° oven for approximately one-half hour.

Bending and Stretching

Bending and stretching are more specialized, usually being part of another activity. Most of the bending and stretching works better and is more accurate if a guide or fixture is used. For example, if a wire needs to be bent at a 90° angle, drill a hole in a piece of wood about the size of the wire. The wire is

A fixture for bending wire is clamped in a vise. Notice the hole is drilled in the end grain; this prevents the block from breaking when it is hit with the hammer.

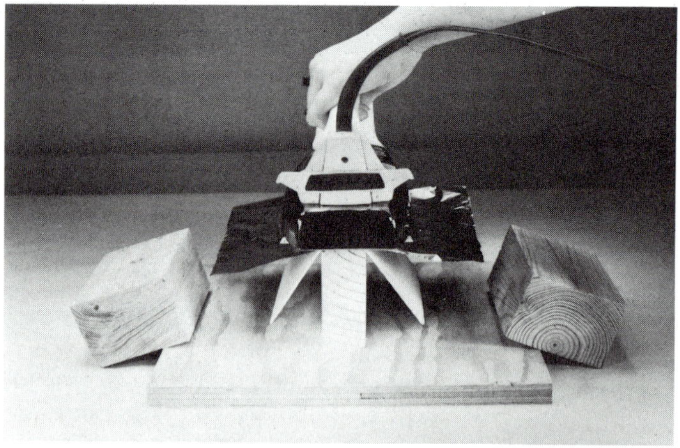

This polystyrene sheet is being bent by heating with an iron (notice the aluminum foil over the plastic). When heated, the plastic is bent down to the shape of the block using the two side blocks. When cooled it is turned over and mounted on a wood base to become a letter holder. This is a two-student operation, under direct teacher supervision.

inserted into the hole to the correct depth and bent flush to the surface of the wood with a small hammer. Fixtures provide control where the students may lack the coordination skills necessary to perform the task.

Rolling

A kitchen rolling pin used to roll out a mixture of salt dough demonstrates the process of rolling.

Weaving

Weaving can be done on commercially made, teacher-made, or student made looms. Looms may be very fancy or very simple.

A student has rolled salt dough into a flat sheet and has cut out a bird with a large cookie cutter. When the salt dough dries, the student will paint the bird and hang it from a branch to make a mobile. In industry, steel and other materials are rolled and stamped into various shapes.

In just a few minutes a spool-weaving tool can be made and students can be weaving. The result is a long rope that can be sewn together to make pot-holders or other items.

One of the simplest is spool weaving. This produces a woven cord as a result of weaving over and under the nails. Illustrated is the step-by-step procedure. This activity may be done in classes as early as the first and second grades. Slot looms and box looms can also be made by the students. Such learning experiences provide an opportunity to use tools and fasteners, gain a better understanding of materials and provide a feeling

Spool weaving is a very simple process and can be taught to students as early as second grade.

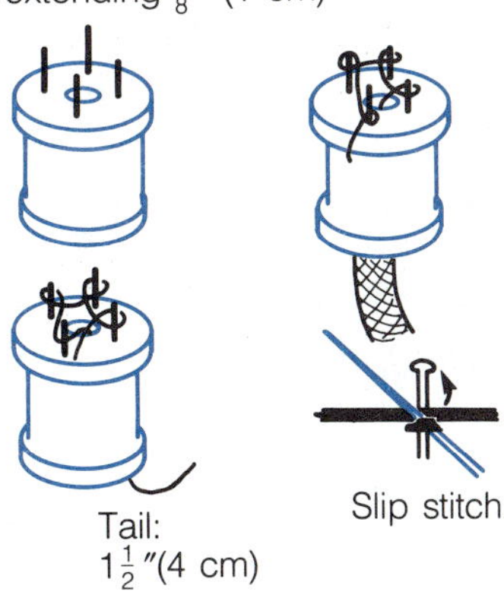

Spool Weaving

1. Run the yarn down through the hole extending out the bottom.
2. Loop the yarn around each brad as shown.
3. As the second loop is made, insert the toothpick in the bottom loop and bring it up and over the top of the brad. (slip stitch)
4. Continue around to each brad.
5. As the loops are made the woven material comes out the bottom.
6. Different colors may be used at any time by tying on a new piece of yarn.
7. The cord may be sewn together to make such items as belts, hot pads, and pillow covers.

This loom has been made in the classroom. Notice the small dowel rods on each side which are held in place by screw eyes. This helps keep the edges straight when inexperienced students are using it. When finished, the rods are simply slipped out of the weaving.

of accomplishment. The students not only have a woven product when they are finished, they also have the tool to create the product.

SAFETY CONSIDERATIONS

As with any classroom activity, safety should always be first and foremost in the mind of the teacher. It is especially important when the work requires students to be involved in several different activities at the same time. The following identifies some basic safety rules that should be observed.

- While planning for the activity, consider safety and anticipate potentially dangerous situations.
- Make sure every student has had a firsthand demonstration by the teacher or another adult with teaching authority. Specific points of safety should be stressed at this time.
- There should be a teacher or a person with teaching authority with the student the first time an activity which requires some skill is attempted. This assures that correct methods and safety procedures are followed.
- The more dangerous tools, such as sharp large blade knives or razor blades, should not be used by the students. Teacher discretion is needed with the use of any tools or equipment because of the wide variation of student abilities between the lower and upper elementary grades.
- Eye protection in the form of safety-approved glasses or shields should be worn by all students where eye safety is a factor, as well as by those in the general vicinity. It is best to check with an administrator on specific eye-protection regulations. Laws vary from state to state.
- Volatile or toxic chemicals should not be used or stored in the classroom. Paints and finishes should be water soluble, and flammable materials (such as alcohol for Bunsen burners) should be handled and stored according to accepted safety procedures. Again, discuss this with an administrator for clarification on rules and regulations.
- All tools, equipment, and supplies should be stored out of sight. This prevents unauthorized use.
- All tools and equipment should be kept in working condition.

- Teachers should always supervise directly the use of a power tool or any hand tool (such as a paper cutter) which might be potentially dangerous.
- Separate the work areas from each other as much as possible.
- Provide proper work surfaces.
- If some students are not involved in the activity at a particular time, make sure there is sufficient seatwork or other activity that will keep them away from the work areas.
- Keep the work areas as clean as possible during the activity. Scraps of material and spilled liquids should be cleaned up continually.
- Make sure there is ample time to clean up at the end of the work session. Open cans of paint or scraps of paper on the floor may be a potential hazard.
- After the activity is completed, make notes about any unanticipated problems. This will help in planning future activities.

SUMMARY QUESTIONS

1. Describe the methods you would use to separate the following materials:
   (1) two liquids of different weight; (2) particles suspended in a liquid; (3) particles dissolved in a liquid; and (4) three sizes of solid granules.
2. Identify the physical differences between a bit and a twist drill.
3. In what direction should wood be sanded? Why do you think this is necessary?
4. Why are different grades of abrasive papers needed to obtain a smooth surface?
5. Identify the difference between casting and molding.
6. Choose three traditional subjects and show how each might be incorporated into a processing unit.

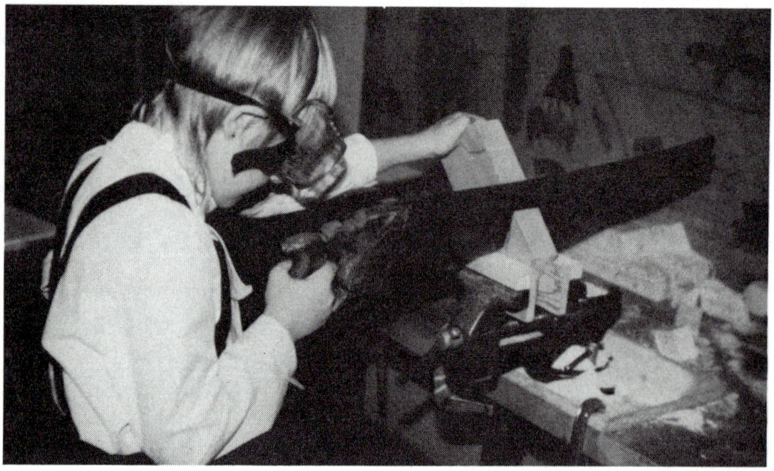

437
270
Subway
BUS

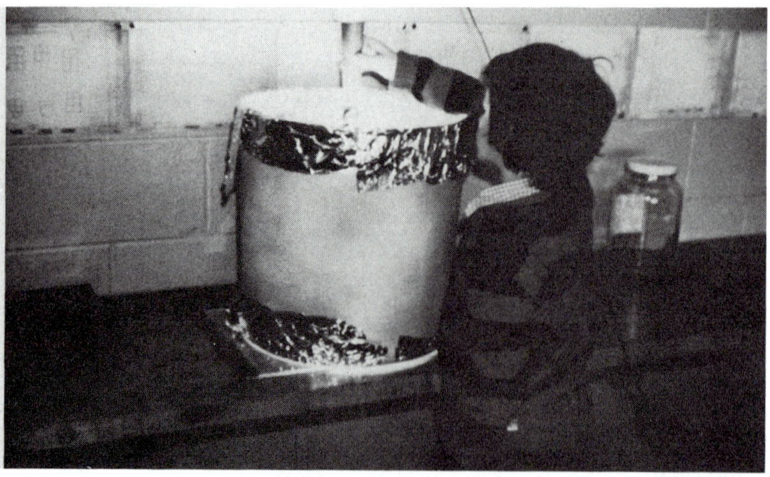

CHAPTER 13

Making and Assembling Products in the Classroom

In this chapter you will learn:

1. The names and uses of tools and fasteners that could be used in the elementary classroom.
2. How planning, drawing, and writing may be used in the study of technology.
3. How concepts may be taught in the classroom concerning product assembly, communication, energy, and transportation.
4. What activities may help develop concepts dealing with technology.

ASSEMBLY OF PRODUCTS

Assembly activities are an integral part of the study of technology. This step represents the pulling together of ideas and preparation of materials into a useful object, simulating the activities of our technological society. Several tools and processes may be involved in assembly which require the mechanical holding of materials.

Holding Tools

A C clamp is easy to use and works well with students at all levels. It may be used to hold material while it is cut or to clamp pieces together which have been glued. It may be used on any material. Soft materials, however, may need a small

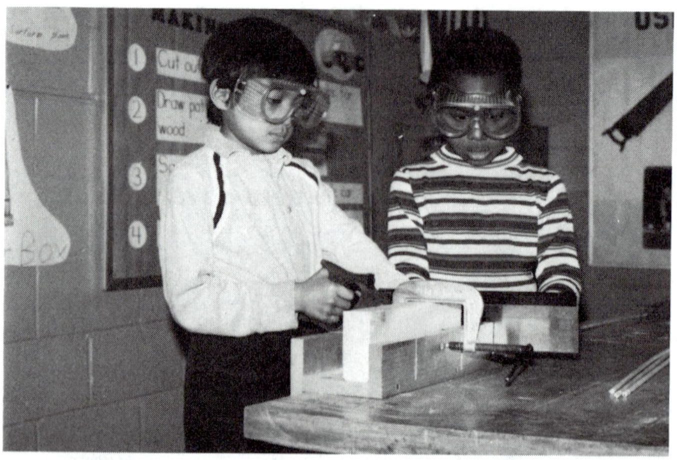

The C clamp may be used in many different ways. These students are using it to clamp a stop block in the miter box. In this way several pieces the same length can be cut without measuring each one.

Hand screw clamps will not damage the surface of materials, even with tight clamping.

scrap placed in the clamp to prevent damage from the pressure.

Hand screws are clamps designed for only soft materials, such as wood. They have the advantage of opening very wide and can hold large pieces of material.

A vise may be used to hold any material. A wood liner should be used in the jaws if wood or another soft material is being held.

A V board is a holding device used to saw thin material with a coping saw. The saw teeth should point down so the pressure in cutting pulls the material tight against the board. These can easily be made of scrap wood.

A bench hook, which can also be made easily in the classroom, provides a raised edge to stop the material from slipping as it is being worked on. This device works well for such operations as sanding and planing.

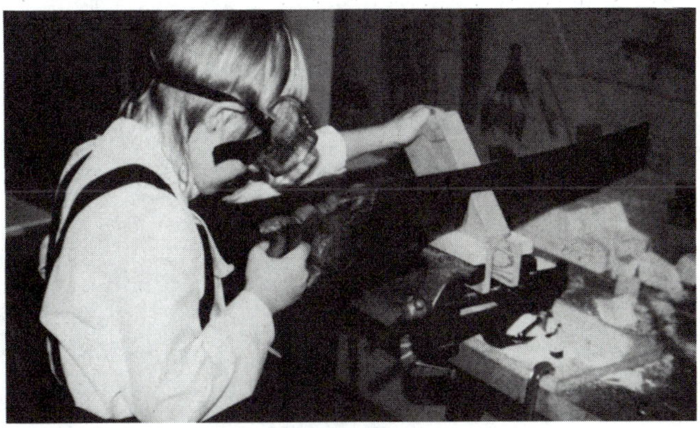

This portable vise can be removed and put away when not in use. The small wood base to which it is attached is clamped to the table or desk top.

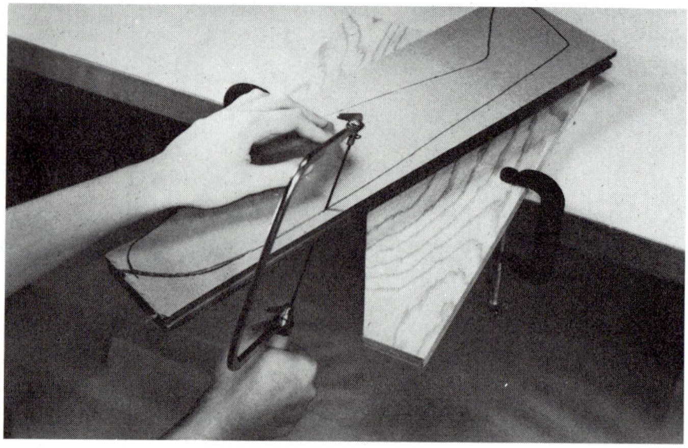

The V board is used as a support when sawing with a coping saw. The blade is placed in the saw with the teeth pointing toward the handle. The handle is under the material and the cutting is accomplished on the downward stroke.

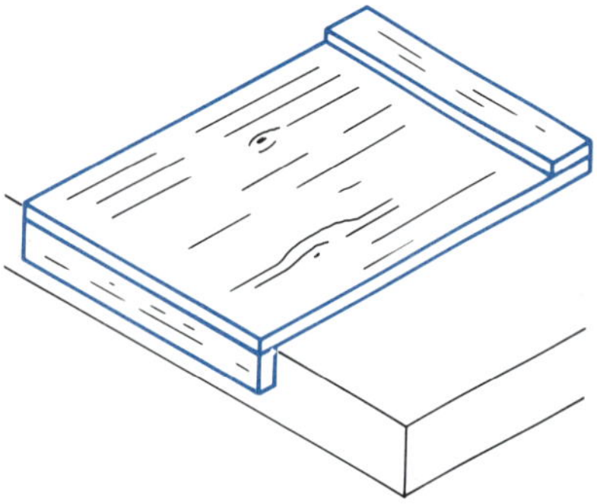

The bench hook rests on the edge of the table. The piece being worked lies on top of it against the raised edge, eliminating the need for a vise or other holding device. All the work (such as sanding), however, has to be in one direction to make use of the stop.

A miter box is a holding tool that can be purchased commercially or made in the classroom. A back saw, which moves in precut slots which are 90 degrees and 45 degrees to the material being cut, is part of the tool. The box serves two functions: it helps hold the material, and it acts as a guide for the saw.

Pliers have a variety of applications, including cutting small diameter wire and bending material, as well as holding small objects.

Open-end and box-end wrenches are used to hold and tighten bolts and nuts. Box-end wrenches require less manual dexterity, thus they are appropriate for early elementary use.

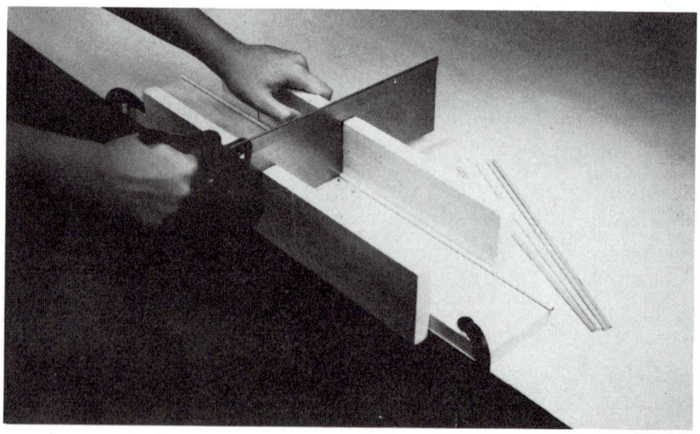

A miter box is helpful in making both 90° and 45° cuts. It uses a back saw and is simple enough that it can be made in the classroom. Notice that it is clamped to the table top to prevent movement.

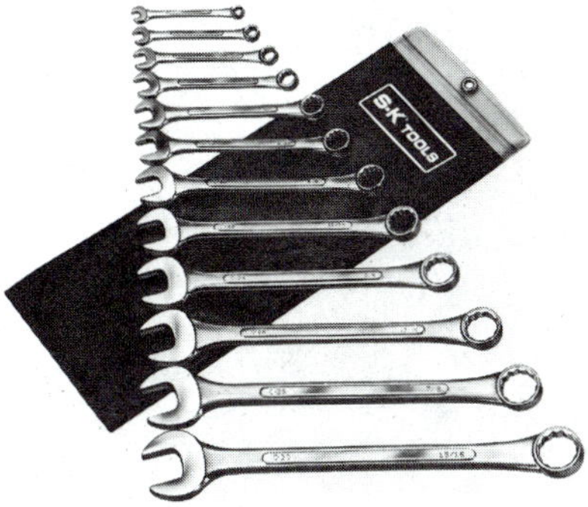

These wrenches have both an open end and a box end.

The adjustable wrench is easy for elementary students to grasp; they come in sizes measured by the length of the wrench.

An adjustable wrench can be made to fit any size of bolts and nuts.

Joining

Fasteners. There are two main types of fasteners: nails and screws. Nails, available in many types and sizes, are normally used only on wood. A good rule to follow is to nail the thinner piece to the thicker piece, using a nail approximately twice the length of the thickness of the thinner piece. In cases where heavier nails are used, pilot holes slightly smaller than the nail diameter should be drilled in the first piece. This prevents the wood from splitting and holds the nail vertically while it is being pounded into place. This procedure helps younger students pound nails successfully. When permanent assembly is desired, nails should be used with an adhesive.

The types of nails most appropriate for the elementary classroom are box nails, brads, and finishing nails. Box nails are thin nails with large heads. They are identified in length by the term "penny" (designated by the symbol d). Brads are wire nails with very small heads, made to be pounded flush with the wood surface. They are sized by their actual length. Finishing nails come in sizes like box nails, but have very small heads like brads. They are to be pounded flush with the surface, or set below the surface with a nail set (small punch). For elementary-level work, however, this is usually not necessary.

The standard hammer for the elementary classroom is a medium-weight claw hammer. It may be used on most materials and has a curved slot for pulling nails. It should be held near the end of the handle (not held close to the head), and the blows should be slow enough to ensure control.

Four major factors affect the holding power of nails: the length, the diameter, the smoothness of the nail, and the density of the wood. There are other conditions, but these four should constitute a general guide for selecting the nail with the correct holding power.

Wood screws usually have greater holding power than nails. They are sized by length and gauge (diameter). In addition, they come in different types of heads. The illustration shows the basic types, with the appropriate screwdriver for each. Most elementary students find the Phillips head screw

Nail sizes should be chosen according to the thickness of the wood. When nailing two pieces of wood together, at least half of the nail should protrude into the second piece. The nail should always be pounded through the thinner piece into the thicker one.

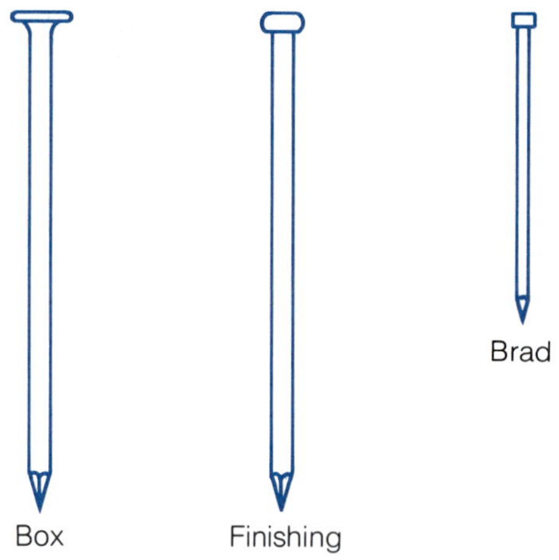

- Box and finishing nails are sized as follows:

2d = 1″	8d = $2\frac{1}{2}$″
4d = $1\frac{1}{2}$″	10d = 3″
6d = 2″	12d = $3\frac{1}{2}$″

- Brads range from $\frac{1}{4}$″ = #20 gauge to 3″ = #10 gauge.

easiest to use, because there is no slot for the screwdriver to slip out of.

Wood screws usually require a hole to be drilled before they can be screwed into place. The exception is where small screws are used in soft wood. Even then, it is advisable to start a hole with a small nail to provide a guide for the screw.

(A) The choice of screw head depends upon whether the surface has to be smooth. The flathead screw is used where the hole has been angled at the top with a countersink; this allows the head to be set below the surface. (B) The Phillips head screw may also be either flat, round, or oval.

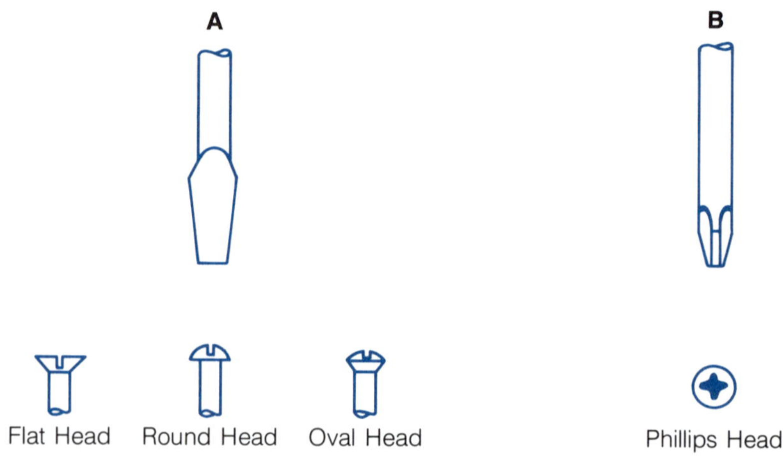

Four steps are usually required in attaching two pieces of wood with screws. If the correct sizes of holes are drilled, the screw will easily thread itself into the second piece.

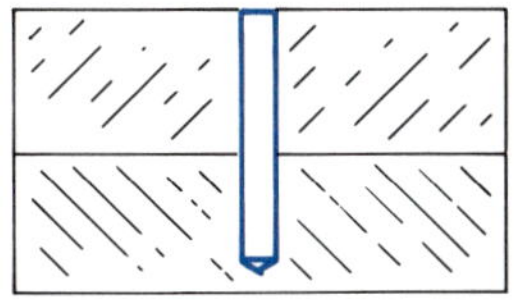

(A)
Pilot hole is drilled through both pieces.

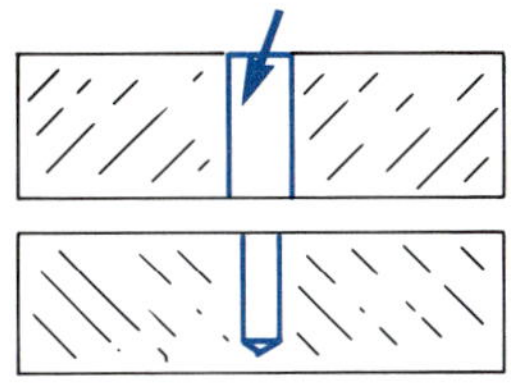

(B)
The shank hole is drilled through the pilot hole in the top piece.

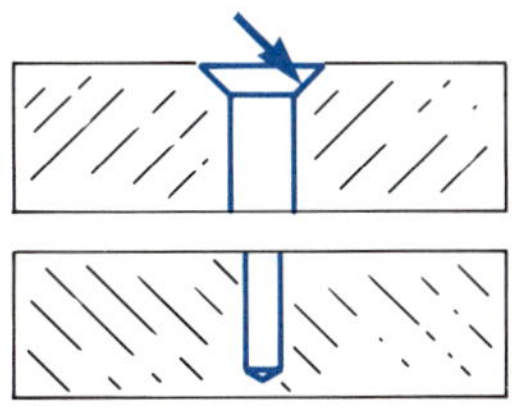

(C)
The countersink is drilled for the flathead screw.

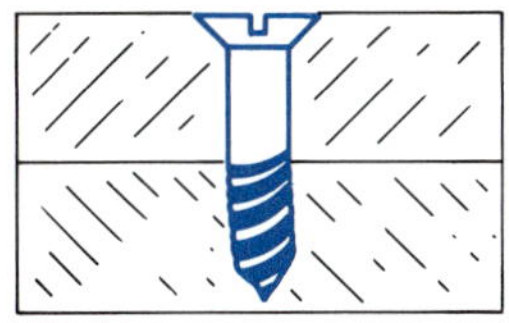

(D)
Both pieces attached with the screw sliding through the shank hole and making its own threads as it is turned into the pilot hole.

Two sizes of holes are required for each screw. The first hole, called the shank hole, is drilled through the first piece and into the second piece at least as deep as the length of the screw. The diameter of the hole should be the same as the inside diameter of the screw. This is the measurement obtained when the screw threads are not included. The second hole, called the pilot hole, is the same diameter of the screw threads. It is drilled through the first piece where the shank hole was drilled. This allows the screw to slip through the first hole and thread itself into the second hole. This prevents any space between the two pieces.

Flathead screws may be made flush with the surface by using a countersink. This cutting tool enlarges the top of the hole and places an angle in it to accept the flathead. Roundhead screws and other fasteners may be made flush with the surface by counterboring. This involves drilling a hole on top of the pilot hole, using a twist drill diameter which is larger than the screw head.

A counterbore is usually made with an auger bit because it makes a relatively flat surface for the bolt head, while the countersink is used for flathead screws.

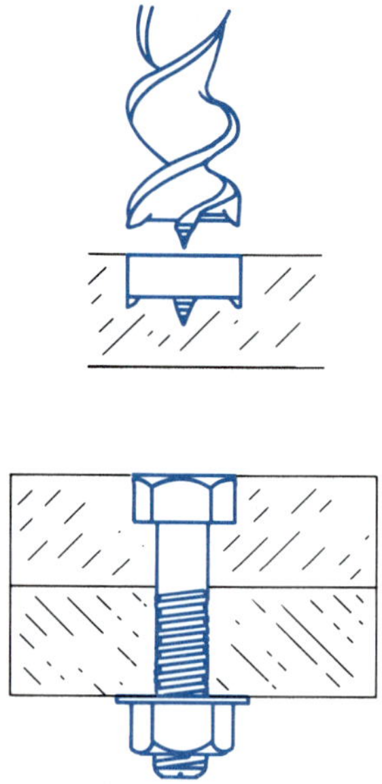

(A)
Counterbore. An auger bit is usually used to make a counterbored hole because it makes a relatively flat bottom for the bolt head. The counterbore is made before the shank hole is drilled.

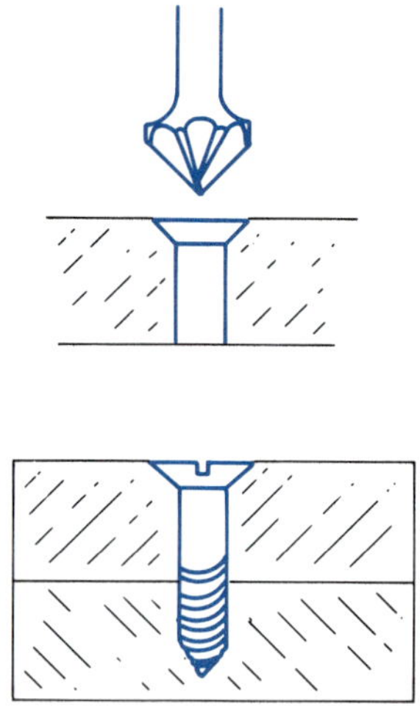

(B)
Countersink. A special tool called a *countersink* is used to make the angled sides of the shank hole. This is done after the hole is drilled.

Adhesives. One all-purpose adhesive works well in the elementary classroom. It is nontoxic, needs no special solvents for cleanup, and dries fairly rapidly. It is called polyvinyl acetate. This glue comes under several trade names, but is commonly known as "white glue." It may be used on any porous material, including wood, paper, fabric, and leather. It is easily cleaned up with a damp cloth before it sets, and it dries clear. For best results, the glued joint should be clamped until the adhesive sets. Setting usually requires about twenty to thirty minutes, while complete hardening usually requires about twelve hours. The adhesive may be used by itself or in conjunction with fasteners such as nails and screws.

Temporary Fasteners. Temporary fasteners are common supplies found in every classroom, such as tape, string, rubber bands, paper clips, and yarn. These may be used with the white glue to hold pieces together until the glue sets. They can then be removed, eliminating the need for nails and screws. Vises and clamps may also be used as temporary fasteners until setting occurs.

Welding. Welding obviously cannot be done in the classroom, but the concept may be developed by comparing melting by heat and melting by dissolving. For example, two sugar cubes could be moistened on one side until the sugar is ready to dissolve. They are then clamped lightly together and allowed to dry. The attachment of molecules during the drying process parallels the attachment and mixing of metal molecules during welding and cooling. It also would be helpful to bring a piece of welded metal into class so the students could see where the metal had been melted and joined.

Organization of Assembly

The organization of assembly processes depends upon whether the students are working individually, in small groups, or as one large group. However, in some cases, the assembly process itself may be the single important aspect of the activity. For example, students will gain a much better understanding of line production and how our products are made if they can experience such a situation.

An example of such an activity might be the assembling of ballpoint pens. The following steps outline the activity.

- Obtain forty inexpensive ballpoint pens which can be taken completely apart, including the spring, plunger, cap, and ring. Many times these can be obtained free from advertising companies. It is helpful if they are different colors, because this process requires color selection in assembly.
- Take twenty pens apart, placing all like parts together on separate paper plates. Take the other twenty apart and place those like parts on paper plates, keeping them segregated from the first twenty.
- Two rows of tables are set up, with assembly stations (plates) organized in sequential steps of assembly, with twenty pens in each row. In addition, the last two stations include testing, where a student has to test each pen and mark an X on a small slip of paper to show it has been inspected. The pen and slip of paper are then passed on where

Concepts concerning assembly and subassembly may be developed by assembling a product such as ballpoint pens. The signs labeling the work stations add reinforcement to the concepts.

they are placed in a small plastic sandwich bag (packaging) and stapled.

- The idea is to have both groups competing in an assembly relay race. This generates interest and gives the students an idea of how important each job is on an assembly line. It also demonstrates the value of having subassembly lines as part of the organization.

The relay may be run as many times as desired. The only consumable materials are the plastic sandwich bags and

Students study the concept of line assembly by assembling small pizzas. The concepts are the same as those involved in ballpoint pen assembly. The big difference for the students, however, is the party after the pizzas are cooked.

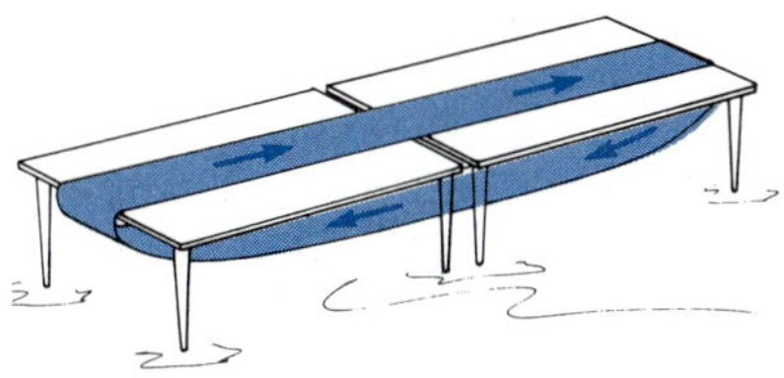

A paper conveyor belt may be made using rolled shelf paper or wrapping paper. It is moved by gently pulling on it.

the small inspection tags, so it is a relatively inexpensive activity. The number of assembly stations will vary according to the number of parts to be assembled. If extra students need to be included, space the tables apart and use some students to transport semi-assembled units from one work station to another. Paper conveyors may even be made (see illustration). Concepts to be developed might include the importance of each job, why "bottlenecks" occur in assembly, why it is important to have a supervisor, why it is important to have quality control, and why the method of line assembly is better than one person doing all the assembly steps.

DESIGNING

The process of designing is the evolution of ideas into reality. The following discussion indicates how students can develop these skills.

Planning

Students need to develop the habit of planning their activities and communicating their plans to others. Because technology involves many materials and processes, planning the step-by-step procedures is very important.

Drawing

Many times a single view of an object does not communicate well. The students should be aware of some basic techniques used in industry to draw objects in three dimensions. There are three types of drawings, called pictorals: isometric, oblique, and perspective.

Isometric drawings are built around an axis which has the same distance between each leg. It is drawn first, with the rest of the lines being placed in the drawing parallel to the legs of the axis. This type of drawing is used in assembly drawings in industry to show people how parts should go together. A good beginning lesson for elementary students would be a work sheet with the axis already drawn in. Using straight edges and pencils, the missing lines are then added by the students, as the teacher follows a step-by-step procedure.

An oblique drawing is probably the easiest for elementary students. It is started with a full front view, with parallel

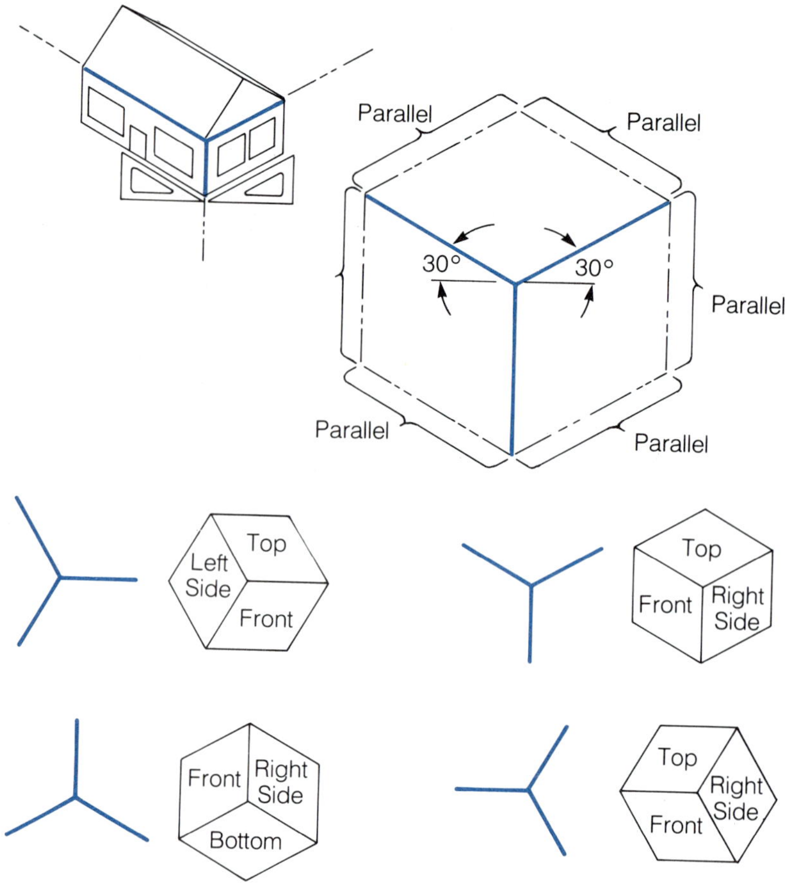

Different sides of an object may be seen in an isometric view by drawing the axis at different angles. It should be noted that all lines are parallel with one of the three legs of the axis.

lines going back from each corner. The cross lines are then placed in parallel with the front view lines. This is a good drawing to use if there is a hole or round object, because the circular part may be drawn in the front as a true circle. This one is best used in sketching and communicating ideas.

A perspective drawing starts like the oblique, with a full front view, but the lines all go back to a common vanishing point. The cross lines are again drawn as in the oblique. This drawing is used in illustrating and advertising, because it appears more like what we actually see.

The development of visualization capabilities can be enhanced by drawing activities such as these, where students must transfer ideas to paper.

Because true length lines running back from the front view give a distorted view (a more rectangular shape) they should be cut approximately in half. This makes the object appear as a cube.

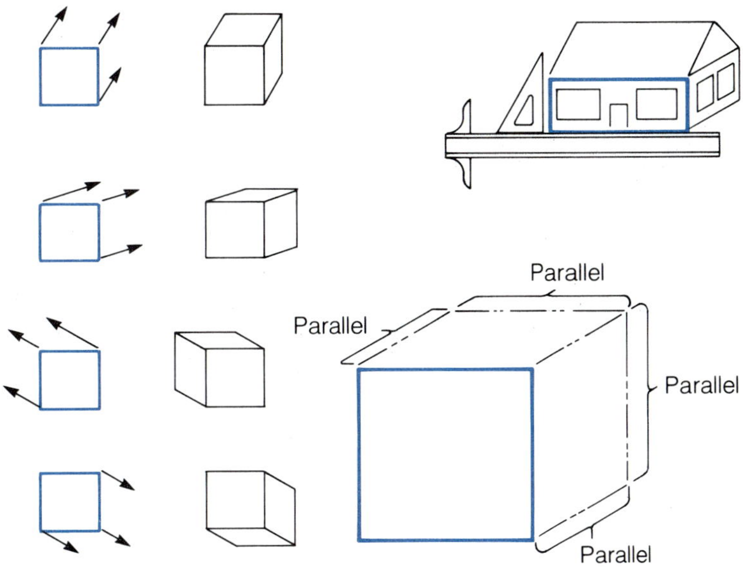

(A) A one-point perspective is used when the view is perpendicular to a surface. (B) A two-point perspective is used when the view is perpendicular to an edge.

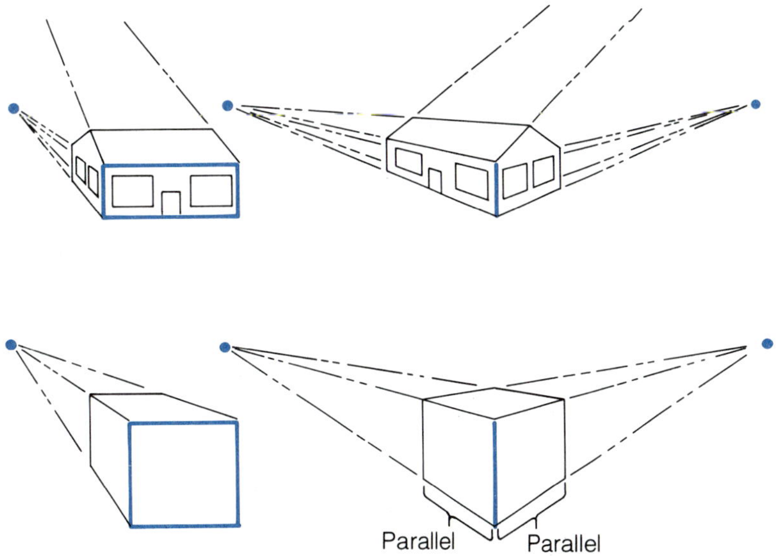

A. One Point Perspective **B.** Two Point Perspective

Writing

An important element in technology systems is the written word. Students should be encouraged to plan in writing. Writing down ideas in a step-by-step sequence forces them to explore more possibilities. These ideas, however, do not have to be a long narrative. On the contrary, industry requires technical writing to be as short and to the point as possible. This means listings are to be encouraged, but the ideas have to be complete, with grammatically correct sentence structure.

PRODUCTS OF VISUAL COMMUNICATION

The classroom activities in communication may fall under the same general categories of communication within the technology system.

Reproduction

Composition. Composition involves the organization and methods of presenting ideas. It deals primarily with the layout and decisions concerning coding, such as size, color, pictures, and wording.

One of the problems many elementary students have in reproducing a drawing of a picture is getting the elements correctly proportioned. Grids may help in this. A grid with squares approximately 3/8ths of an inch (9.5 mm) can be made on overhead transparency material which is taped over the original picture. (If the picture can also be made into a transparency, it may be used on the overhead projector for the whole class to see.) The drawing is done on regular grid paper or paper which has a grid under it. It is made by reproducing the lines one square at a time. This is also an excellent method of reducing and expanding the size, which can be done by using an expanded or reduced grid size. Large drawings, such as those of bulletin-board size, may be done by making the picture into a transparency and projecting it on the wall with an overhead projector. This allows the much enlarged picture to be traced on a prehung sheet of paper.

Newspaper headlines and magazine advertisements may also be cut up to provide letters and pictures for composition. An excellent activity is to have the students identify a theoretical product and design an advertisement for it. Some things to consider would be whom the advertisement is di-

The image is drawn one square at a time. This activity may be used with students as young as first grade. The complexity of the image determines the appropriateness for a particular grade level.

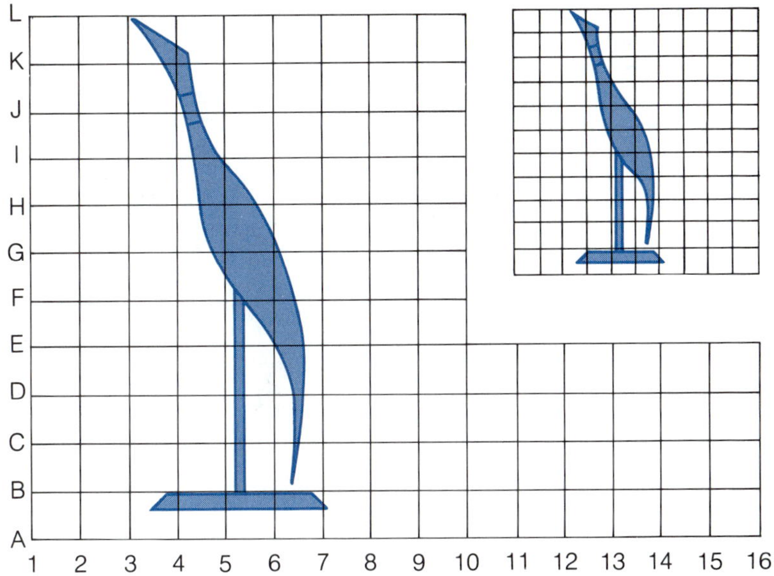

rected toward, what would get the readers' attention, and what information is necessary.

Image Generation. Image generation involves the methods of creating the image or code. Although most students have seen typewriters, many have not had an opportunity to use one. An old typewriter is an excellent addition to the equipment of the classroom. Students can use a mechanical writing device that is very common in our culture. It also provides an opportunity to use a process of transferring ideas by means of a keyboard, which is nearly the same as those found on computer terminals. Many classrooms now have computers with simple word processing software packages, making this activity one that can be integrated with many units of study.

Students should also have an opportunity to experience image transfer. This may be done by simply providing carbon paper. Two concepts may be developed. (1) When an image is transferred from one surface to another, it is always reversed. This can be seen by examining the reversed image on the carbon paper. (2) The image can only be reproduced a few times with each sheet of carbon paper. This is why newspapers and other materials are not printed by this method.

Students used kitchen utensils and a potato in making the print on this fabric, which, when finished, will become a brilliantly colored wall hanging.

Direct printing allows many copies to be made. This can be done by a number of methods, including printing with wood blocks, vegetables, various utensils, and even the fingers. Drawn pictures may also be printed by gluing aluminum or copper craft foil to blocks of wood. The picture is then drawn with a large nail, by pressing hard enough to recess the lines.

The ink is spread on the printing surfaces by dipping into a shallow dish with the bottom barely covered, using a wide paint brush or a brayer (a small rubber printing roller). Tempera paint may be used for the printing ink if a few drops of liquid soap are added. This reduces the surface tension of the water and makes it spread more evenly.

The students should note, however, that the image is always reversed, so some letters or numbers would not be readable. In many methods used by industry, this is taken care of by offset printing—a two-step process which reverses the image twice, making it look like the original. The students may cut shapes or images from pieces of vinyl floor covering and glue them to wood blocks, making the raised surface. They also may write their name with lightweight cord, gluing it down with white glue. The surface is inked with a brayer and pressed against a sheet of spirit duplicating paper. A second sheet is then pressed against the first sheet, transferring the image again. Spirit duplicating paper is used for the first trans-

A screen made of scraps of wood with organdy fabric is stretched tightly and stapled about every half inch (13 mm). The squeegee is a car windshield ice scraper.

fer, because it has a coating which retards absorption. The three images may then be compared.

Screen printing involves stretching organdy fabric tightly on a wood frame. The stencil is made by cutting the images out of adhesive back plastic used for shelf covering. The paper back is peeled off and the plastic is laid down with the adhesive side up. The screen is then placed on top and rubbed hard to adhere the fabric to the plastic. Scraps of paper are then taped around the inside of the frame to prevent the ink from passing through small openings. The printing is accomplished by drawing the ink (which may be tempera and liquid soap) across the screen with a squeegee. This tool may be a piece of heavy cardboard, a thin piece of wood with a rounded edge, or a small rubber squeegee such as found on car windshield ice scrapers. Small screens may also be made by stretching the fabric in a set of knitting hoops.

The advantages of screen printing are that many prints may be made, it can be printed on nearly anything, and it can be made permanent by using enamel or textile ink.

Transmission

Activities which allow students the opportunity to design and carry a communication through to transmission usually cover a wide range. Examples might be a school bulletin board, a classroom newsletter, or certain visual aids for the class that deal with certain topics. Such activities require planning, designing, and producing a communication message in a group endeavor, rather than each student following a single, specific direction of the teacher. Benefits are gained not only from the technical knowledge, but from the social interaction of the group effort.

Because books are an important means of transmitting visual messages, bookmaking can be an excellent classroom activity. It may be very simple or very involved, depending upon the capabilities of the students. It lends itself well to correlation with other subjects, because the topic of the book depends upon the objectives of the teacher.

The students can construct the booklet entirely by themselves, or the teacher, especially with younger students, can do the necessary cutting. Two pieces of heavy cardboard approximately 6 inches by 9 inches (152 mm $\times$ 229 mm) are taped together edge-to-edge with masking tape. Wrap the tape all the way around, leaving approximately 1/8 of an inch (3.2 mm) between the boards. This becomes the cover.

Fold six sheets of typing or duplicating paper in half, making a definite crease. These are then sewn together using needle and thread. The stitches should run down the seam, with each one being about 1/4 inch (6.4 mm) long. The cardboard is then covered with adhesive-back plastic shelf covering. There are many colors and designs from which to choose. An extra inch (25.4 mm) is allowed to fold over the edges and adhere to the inside surface. The outside of the cover sheets on the bound pages are coated with a light film of white glue

In making a book cover, a student lays corrugated board on a scrap pattern to check the size before cutting the plastic. Patterns have the advantages of ensuring accuracy and cutting down on waste.

(including the stitches). They are then pressed to the inside of the hard-bound cover. While drying, the book should be closed and kept flat by pressing with the weight of other books. Wax paper liners in the front and back prevent the book from being glued shut as a result of excess glue.

Storage and Retrieval

Storage and retrieval activities might involve photography in one form or another. Upper elementary students, with the right equipment, can take pictures and process them in the classroom. If the equipment is not available, or if the students are too young to have developed the coordination skills, other activities may be used. One such activity that works well in conjunction with language arts studies is the picture story. Students cut twenty to thirty pictures from old magazines depicting a subject such as transportation vehicles, factories, people, food, sports, energy, or communication. Students working in small groups are to arrange the pictures of their choice into a sequence which tells a story. For example, a typical story might show how we get our food. Pictures would probably include farm equipment and fields, trucks for transportation, factories for processing, workers, packaging, and a grocery store. The pictures rearranged in a different sequence might tell an entirely different story. This activity obviously helps in developing communication skills while dealing with the study of technology. Storage involves the complete set of pictures, while retrieval involves pulling the pictures from the set.

Even lower elementary students should be encouraged to learn how to use the classroom projectors. The study of projection methods provides an excellent learning experience combining technology and science. The study of light bulbs, transparent materials, optics, and reflection all are a part of projection systems. By allowing students to see a slide projector, including the reflector and lenses, a teacher can develop concepts concerning the size of pictures on the screen. Exposed film may also be obtained from a local photography firm for making slides. The students may draw small pictures with permanent, extra-fine-point felt tip markers. The film is then mounted in cardboard slide mounts and projected on the screen. Filmstrips may also be made in the same way, eliminating the need for cutting and mounting.

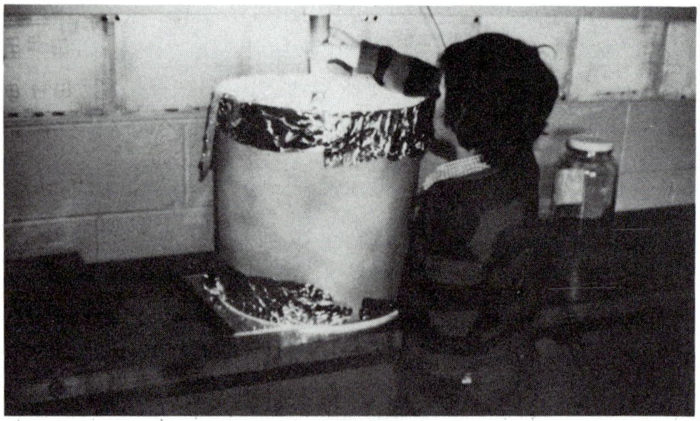

A student exposes a blueprint by means of an inverted table lamp lined with aluminum foil. This has the advantage over the sun in that it can be controlled. Test exposures can be made to determine the best exposure time.

Blueprints may also be made in the classroom with a couple of simple pieces of equipment. Pictures, maps, or floor plans may be drawn on very thin tracing paper. This is exposed to light with a special light-sensitive reproduction paper, which can be purchased from a drafting supply company. The exposure is accomplished by placing them together between two pieces of clear acrylic plastic to hold them in place. Exposure may be done with a spotlight or a table lamp with a shade, which is turned upside down over the paper. Test strips will have to be made to determine the exposure time.

After exposing the paper, it is placed in a wide mouth, gallon (4 l) size glass jar with a sponge soaked in ammonia. The lid should be tight to keep the fumes from escaping. The ammonia fumes serve as the developer, causing a chemical reaction within a few seconds. The development of the image can be seen through the jar. Most reproduction paper of this kind will produce a white background with a blue line. Silhouette images may also be made by placing leaves, paper cutouts, and objects on the light-sensitive paper and exposing it to the sun. Development then occurs in the same manner.

PRODUCTS OF AUDIO COMMUNICATION

Recorders

Although many students have used tape recorders, few can explain how they function. The use of tape recorders in the classroom should be accompanied by the study of the technology. The same can be said of the video tape recorders. Although most classrooms do not have a video tape recorder for regular use, most school buildings, or school systems, have one that could be used on a limited basis. They are simple enough to operate that students could record role playing sequences, or plays, which could be played back and discussed. They might also record a school function, such as an athletic event. By using recorders in the classroom, the students not only learn about the technology but they begin to view the technology as a tool of communication rather than as only a source of entertainment. Recorders also have an educational advantage in that they may be used in many different ways, with no consumable materials being required.

Mass Communication

The study of technology which transmits sound on a mass basis would include radio, television, telephones, and public-address systems. An activity might involve writing and producing a radio program (complete with sound effects and commercials), which is recorded and played back like a prerecorded program. Most telephone companies have telephones which they are willing to loan to elementary classes. If possible, one should be obtained which could be opened for visual inspection. Like tape recorders, most students use the telephone daily, but they have no idea what it looks like inside or how it functions.

Many school systems have a small, single-unit public-address system which could be obtained for a short time. Some tape recorders also have them built into the system. Students could hear their own voices as carried on this equipment.

Students record a radio program that will be played back to the class. Students learn many concepts in the course of writing, planning (including sound effects), and producing a short program.

Students find it easy to make this telegraph. When current passes through the wire, the two electromagnets draw the flexible sheet metal down; when the circuit is broken, it springs back up striking the head of the nail above it making a "click." Thirty to forty turns of wire are needed on each nail for the electromagnets.

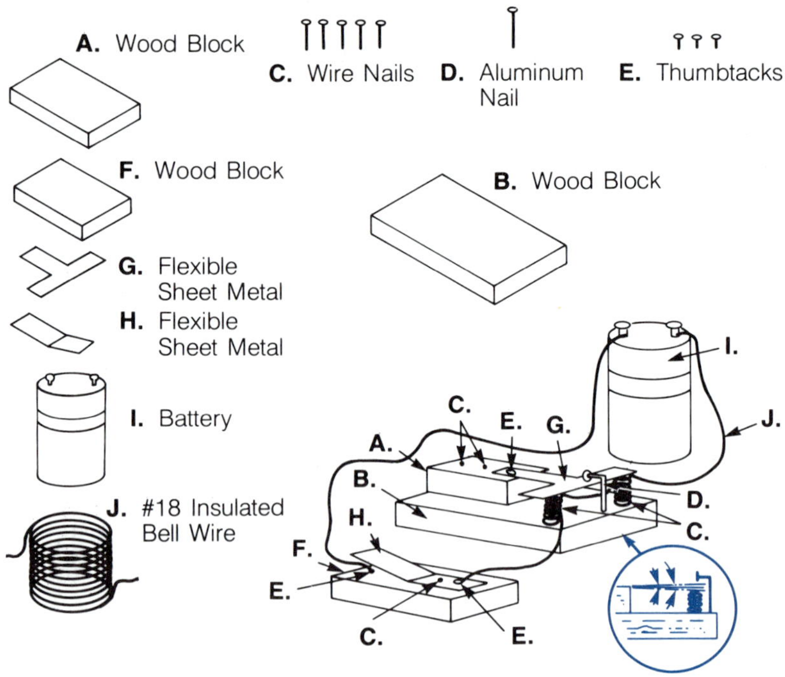

The building of a telegraph system is an activity that involves both the use of tools and a study of history. It also provides an opportunity for a science lesson dealing with electricity. The thin metal can be cut from a metal food can. For younger students, the teacher should do this. Care needs to be taken because of the sharp edges. After the pieces are cut, the sharp edges may be dulled by rubbing with a fine grade of abrasive paper.

PRODUCTS OF ENERGY DEVELOPMENT

Electrical

Many electrical experiments and projects can be used to develop concepts concerning electrical energy. A simple bulb board with four light bulbs, a six-volt battery, wire, and a button switch can be made by the students to demonstrate series and parallel circuits. Alligator clips connect the wires so the students can easily change the circuit. Bells and buzzers may also be used within the system. Information on how electricity

Although demonstration boards can be purchased, constructing one in the classroom provides a variety of experiences, including planning and the use of common tools.

flows, how a bulb produces light, and how a switch functions can be gleaned in a very short time.

The building of a simple electromagnet can also illustrate to students the magnetic characteristics of materials and why some would be appropriate in some products but not in others. The electromagnet can also be used in building other pieces of equipment, such as a model electromagnetic crane with a boom, a small electric motor, and simple magnetic switches.

It is extremely important that concepts be developed concerning the electrical energy found in our houses. Safety in using outlets and electrical appliances, as well as the differences between direct current (batteries) and alternating current (found in outlets) should be stressed.

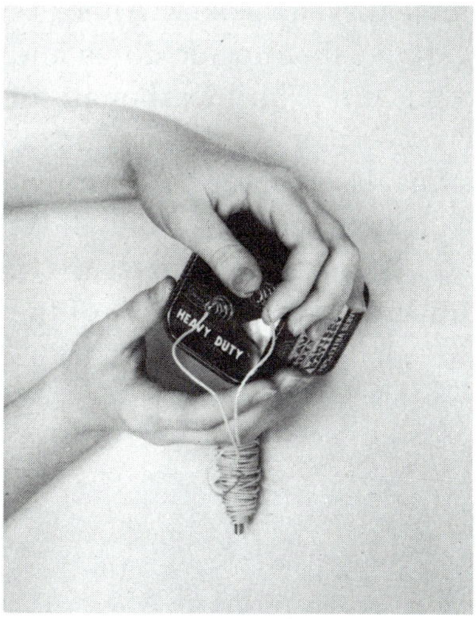

Students can make an electromagnet by using a battery, a large machine screw, and # 18 bell wire, and in the process, develop concepts concerning various materials and their characteristics.

Solar

Small solar cells now on the market can be used in the classroom to demonstrate how light can be transformed into electricity. Small motors can also be purchased to use with them. A class can make a solar water heater by lining a flat box with small pieces of foam insulation panels and black construction paper. Students then set a container of water in the box and a thermometer in the water so it may be easily read. By taping a piece of glass or clear acrylic plastic over the top, the students provide an airtight seal to hold the heat. After placing the box in the sun, students record the temperature every five minutes to see the rate of heat rise. They may then draw a chart to show the data in a graphic form.

Chemical

The release of energy by chemical reaction can be studied through activities which involve the mixing of plaster of Paris. Both the hardening of the material and the heat created during the hardening are the result of chemical reaction.

The creation of a gas by chemical reaction can also be observed by constructing a fire extinguisher. Obtain a metal food can with a plastic lid (approximately one quart (l)), a plastic foam cup, a drinking straw, tape, glue, baking soda, and vinegar. Glue the plastic cup to the inside bottom of the can, and cut a small hole in the plastic lid for the straw. Place 1/4 cup of baking soda in the metal can, while pouring 1/4 cup of vinegar into the plastic cup. Tape down the lid and tape the straw to the lid with about one inch (25.4 mm) inserted into the can.

Close teacher supervision is necessary while the extinguisher is in action because of the small fire. Place a few pieces of shredded newspaper into a small pan or beaker. After the paper has been ignited, turn the extinguisher upside down and hold the straw near the base of the fire. The bicarbonate of soda is an alkaline, while the vinegar is a weak solution of acetic acid. When these two chemicals combine, a chemical reaction occurs which produces a clear gas called carbon dioxide. It is a very heavy gas that can be poured. The pressure in the can forces the gas out the straw and covers the base of the

This solar hot dog cooker can also be made in the classroom. It consists of a V-shaped box lined with mirrors at a 45° angle. The glass or plastic lid is taped to make it as airtight as possible, and the box is tilted to the direct rays of the sun. Depending upon the amount of sun, cooking time is about twenty to thirty minutes.

***For younger students, a teacher demonstrates how a fire extinguisher puts out a small fire. Only older elementary students should be allowed to participate in such a demonstration and only then under* direct teacher supervision.**

fire. Since it is heavier than air, it keeps oxygen from the fire, causing it to go out. This activity could also be correlated with the study of fire safety and with the correct use of commercial extinguishers.

Mechanical

Simple machines, such as belts, pulleys, gears, levers and wheels are involved in mechanical activities. A demonstration board shows how belts can not only transfer energy, but also determine the direction of rotation. A board that allows the wheels to touch, causing a reverse rotation, demonstrates how gears operate when they are placed together.

A model water wheel may also be made to show how rotary action can be generated and how it can be changed to reciprocating action. The study of mechanical toys would enhance students' understanding of wheels and the action they generate. Many toys, which may be broken and ready to be

A demonstration board shows the direction of rotation when energy is transmitted by means of a belt. The poster board wheels are mounted on thread spools which rotate on long machine screw axles. The belt, which may be obtained from a fabric store, is elastic.

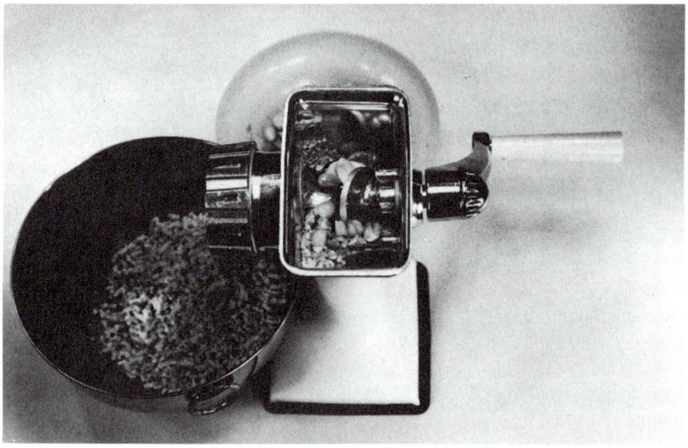

The energy developed by a simple screw can be seen in a food grinder. This is the same principle used in producing plastic products by extrusion; a large screw forces the heated plastic through a die, giving it the same shape as the die opening.

discarded, have usable parts, such as motors, gears, and levers. An important concept to develop in the study of mechanical energy is the mechanical advantage which is possible.

Wind

Wind as an energy source can be studied by building models of sailboats, pinwheels, and mobiles. The concepts, however, should relate to how the movement of air can be used to accomplish work.

This study lends itself very well to learning about aircraft flight. A small wind tunnel can easily be made with a fan and rolled newspapers. A lightweight model aircraft is at-

A wind tunnel can be used to demonstrate how an aircraft flies and can allow students to test a model they have made. The model has an expanded plastic insulation body and heavy paper wings. The rolled newspapers are necessary to obtain straight air movement.

tached by the wings to two posts, allowing the flow of air to pass over the wings causing lift. The wings may also be constructed to allow the flaps to be bent down or up, demonstrating the differences in lift. The purpose of the tunnel is to change the turbulent air caused by the fan into a straight line movement, simulating the flying of an aircraft.

PRODUCTS OF TRANSPORTATION

Models

Much of the activities related to the study of transportation involve the construction of models. Model cars, boats, airplanes, and other vehicles can be made of wood. The complexity of the design will depend upon the grade level of the students. Although precut wheels are commercially available, they may be cut with an electric drill and hole cutter. If several are needed, this should probably be done by the teacher before the project starts. The students should not use the hole cutter unless the electric drill is held in a drill stand and the wood is securely clamped.

A second activity might involve the construction of transportation networks, such as highways, bridges, airport

Displays and dioramas can be made with expanded plastic insulation, which can easily be shaped with abrasive paper. Cereal boxes painted or covered with construction paper can be used for buildings, while weeds and small branches stuck in clay suggest landscaping.

runways, or a combination of several different networks. They may be built on a piece of plywood, a table top, or the floor. (They may even take up an entire corner of the classroom.) Larger networks allow the inclusion of many different kinds of technologies, such as traffic signs, relay towers, and electrical power lines.

The number and kind of materials which may be used in model making are almost endless. For model cars, dowel rods may be used for axles, thumb tacks for lights, and white adhesive-back plastic for windows. The networks may involve plastic foam insulation (which is easily cut and shaped with abrasive paper), plaster of Paris, cement, wood, clay, and even small weeds and branches for landscaping.

Working models, such as cars with moving wheels, may be raced down an inclined board two at a time to see which car is the fastest. A working rocket can be made by running a string through a drinking straw and stretching it tightly across the room. A balloon with an elongated shape is inflated and taped to the straw. When it is released, the rocket will shoot safely across the room. Generally, working models should be made of lightweight materials, because it requires less energy to make them function. Also, lightweight materials are usually less expensive and easier for the students to manipulate.

SAFETY CONSIDERATIONS

The same general rules apply as those identified at the end of the previous chapter. It would be well for the reader to review

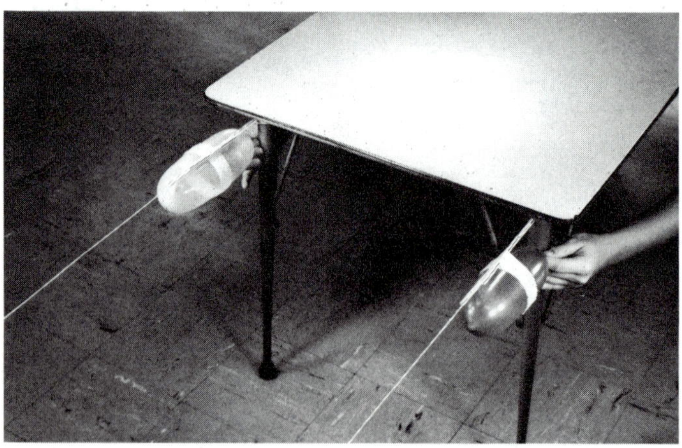

Balloon rocket races generate interest as they provide an opportunity to learn about energy and rocket propulsion.

these safety rules once again, in the context of making and assembling products in the classroom.

SUMMARY QUESTIONS

1. Explain why two sizes of holes are required when screws are used. When would they not be required?
2. Name two positive benefits students may derive as a result of learning how to draw three-dimensional objects.
3. Identify three problems a teacher might anticipate in planning an assembly project.
4. What is the advantage of using an offset printing process (hand or machine)?
5. How would you teach elementary students the difference between a series electrical circuit and a parallel electrical circuit?
6. Name three advantages of building models in the classroom.

An important element of teaching is planning and organizing classroom experiences that foster the development of concepts.

PART IV

Planning For Instruction

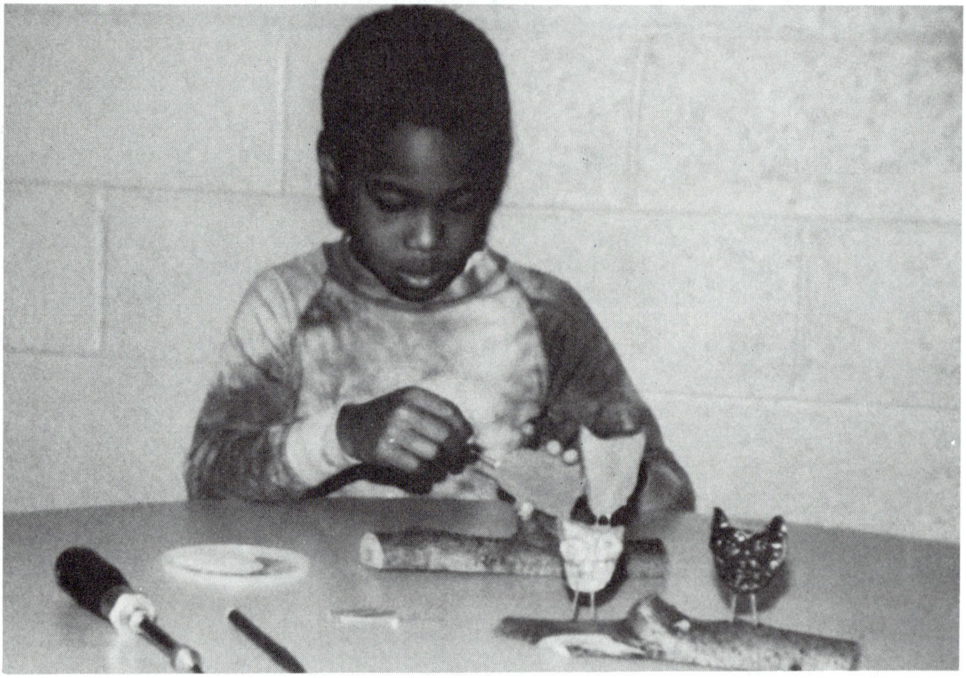

5. hat
6. penny

CHAPTER 14

Developing Concepts

In this chapter you will learn:

1. **The definition of a concept.**
2. **How concepts are formed as information is processed.**
3. **How learning is a growth process occurring in three basic categories called domains.**
4. **Why concept development is important to the students' educational growth.**
5. **How conceptual learning is accomplished in the classroom.**

EXPERIENCE AND THE DEVELOPMENT OF CONCEPTS

Definition

A concept in its simplest term is an idea. The idea, however, must incorporate some form of insight. When a person sees an object he or she has never before seen, it is perceived that it is something, but a concept is not formed until the object can be related to past experience. Closer examination of the object broadens the "idea" as insights are gained concerning the similarities of physical characteristics between the object at hand and others. Previously formed concepts help in the identification of the new experience. This does not guarantee, however, correct insight and concept development; the person may identify the object as something it is not.

All our experiences are recorded in our brains much like a photograph captures a situation and permanently records it. Some of these experiences we remember, and some are recorded in our subconscious mind, which is still very active, even though we are not aware of it. Some concepts can even be formed while we sleep, because the subconscious mind is still processing information.

Information recorded in the brain consists of meaning, feelings about the meaning, and symbols which represent the "ideas" in some form of language. If an elementary student becomes excited about something seen under a microscope in the classroom, and if, at the same time, some information is presented on the electron microscope, the word *electron* becomes a language symbol. Thereafter, when the student sees or

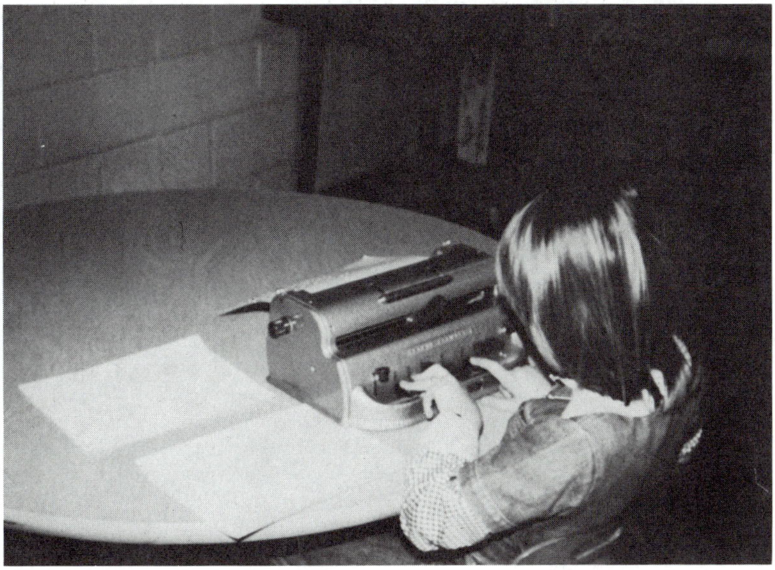

A student taps the keyboard of a Braillewriter. Although the students in the classroom are not blind, they are developing concepts concerning handicapped people and how technology helps them.

hears the word electron (even if it is not in relation to a microscope), the symbol will cause the recall of concepts dealing with the microscope. Thus, we can draw five conclusions concerning concepts:

- Insight must accompany the concept development process.
- Insight is based on past experience.
- Incorrect concepts can be developed if incorrect or incomplete insights are used.
- Recorded experiences may not always be subject to recall, because they may be stored in the subconscious mind.
- Concepts consist of meaning, feelings, and symbols.

How Concepts are Formed

All living organisms interact with the surrounding environment. Humans have the five senses of seeing, hearing, feeling, smelling, and tasting. It is through these senses that humans react to new experiences.

The event, situation, or object with which we react is called a referent. The contact which is made is called the stimulus. The stimulus may be initiated by the referent, such as sound waves coming from a radio, or it may be initiated by the person, such as touching a hot stove. Either way, the stimulus causes an impression to be formed in the brain. This im-

The ability to make correct generalizations depends upon the completeness of concept development. Active involvement causes both concepts and generalizations to evolve at a much more rapid rate.

pression is called a percept, or a perception of receiving the stimulus. These percepts are categorized and stored as they are received, building a backlog of experiences. As percepts are merged into categories of like characteristics, they form a concept. These concepts are then used in responding to new similar situations. For example, when a young child helping prepare dinner in the kitchen, touches a pan on the stove which has a hot handle, a number of simple percepts are identified: the heating element is red, the pan is sitting on the element, and steam is rising from the pan. When the child encounters these conditions a second time, he or she forms a concept that identifies the object as a hot pan. Touch is not necessary.

As a person forms simple concepts and places them into larger categories, he or she is then able to make generalizations. The small child, to continue the example, develops several concepts. The pan may be a dish, the heating element may be different, and there may be no steam. Nevertheless, the situation is recognized, and the child generalizes to include similar situations.

As generalizations are organized, they serve to form principles. These are tapped to assess conditions and situations and provide explanations as to why the conditions are what they are. Again, the child may develop a principle which says that not all pans and stoves are hot, because the position of the stove switch determines the temperature, or that some types of handles may not be as hot as others.

Principles are categorized together to provide the highest form of knowledge application—problem solving. The way problems are solved varies from situation to situation, depending upon the complexity. An older child, in applying principles to solve the problem, may turn off the heat, use two hot pads to grasp the handle, and set the pan on a trivet to cool faster.

It can be seen, then, that learning is not simply built around the accumulation of facts (percepts); rather, it involves the active organization and categorization of knowledge into levels. Just as a computer requires a system of organizing data for recall, so does the human mind.

Several principles concerning this organization and development of knowledge may be indentified. Learning moves in levels from the simple to the complex and from the concrete to the abstract. At the upper levels, symbols may be used to represent pieces of information. For example, the use of mathematical calculations to determine the path of a spacecraft involves symbols.

The linking of these pieces of information (called chaining) may not always be complete. If this is the case, one bit of information may be forgotten or misfiled, preventing it from being used in a higher order of application.

This model illustrates the levels of mental activity involved in the application of problem solving. It should be noted that at least two subelements are required for movement to the next higher step. This simplified model shows a minimum of sixteen percepts; in a real situation the number could easily be in the hundreds.

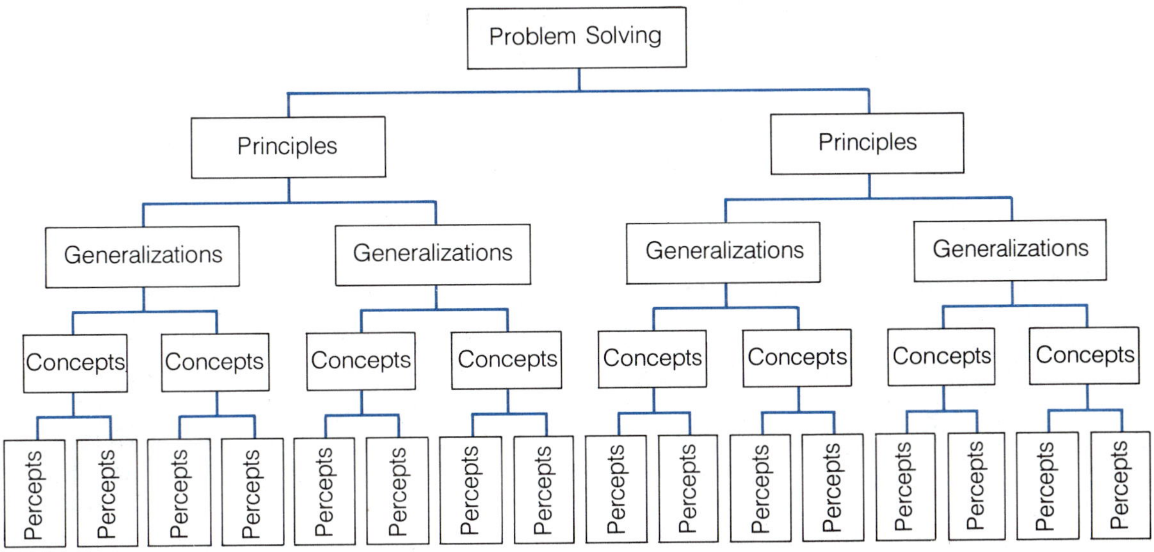

Cognitive, affective, and psychomotor learning were all involved as students studied nutrition and proper eating habits, packaging, sources of foods, and food preparation. The culmination of the unit was a complete breakfast prepared in the classroom by the students.

For complete learning to take place, there can be no major elements missing in the levels below. For example, problem solving cannot be accomplished unless the correct principles are developed, and generalizations cannot be made unless the concepts have first been formed.

The movement to higher levels of knowledge classification depends upon readiness. This is related to both chronological age (because of past experience) and mental age. Readiness, however, can be taught through well-organized sequences, designed to help the student with the chaining process at the various levels. This is best accomplished by providing maximum use of the senses through active involvement.

DOMAINS OF LEARNING

Although learning is a "holistic" process involving the total individual, three basic types of learning can be identified. Such classifications aid the teacher in developing activities to accomplish certain objectives. Specific objectives may be stated for each type, or domain, of learning.

Cognitive Domain

The development of intellectual abilities and skills occurs in the cognitive domain. The classification process of information storage and retrieval is extremely important in the cognitive domain. It is this process which determines to a great extent the level of application possible at a later time. We mentioned earlier that concept development moves from the simple to the complex. The same can be said of the output, or application, of knowledge. There are six major classes to the cognitive domain.

Knowledge involves the recognition, or recall, of basic facts and ideas, such as dates, events, people, and step-by-step procedures.

Comprehension involves the ability to translate meanings into other forms, such as symbols, or vice versa. It also involves both the ability to interpret meanings from several sets of symbols by recombining them and drawing generalizations and the ability to extrapolate, that is, make estimates and predictions. For example, when a student first learns that a traffic light means stop and go, the color symbol is translated into a message. Interpretation involves being able to determine which color means stop and which means go, while extrapolation allows a prediction as to the sequence and timing of the colors as the light changes.

Application is very closely related to comprehension, in that it is a demonstration of what has been comprehended. In other words, the student is sure enough of the situation to initiate some action without being told.

Analysis is a higher step wherein the student not only understands and acts, but breaks down a situation into component parts. This allows a closer look at relationships, principles, and techniques.

Synthesis is characterized by organizing and combining elements into a whole. It results in a structure, or solution, which was not evident before. For example, when students use three or four references on the general topic of energy and from them present an oral report on the various types of power used in the home, they are synthesizing information.

Evaluation represents the highest level. It involves making judgments concerning worth and purpose. It may involve values, economics, or emotion, and the result may be either

The movement of this model is from bottom to top, with each acquired ability contributing to the next, more complex step.

The Cognitive Domain

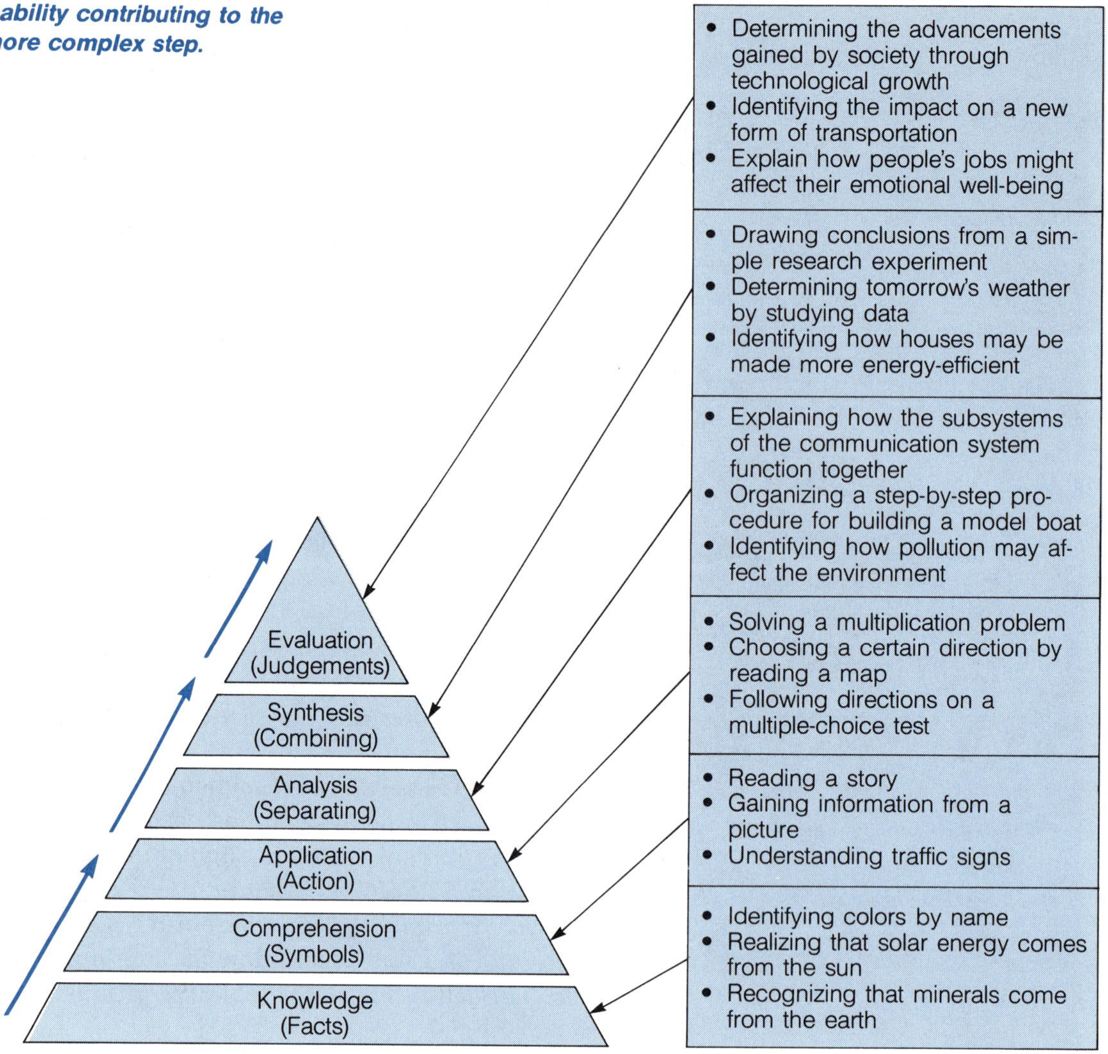

qualitative or quantitative. At each level of the cognitive domain it is necessary to apply the previous levels, and so it is with evaluation. To evaluate effectively, one must have mastered the previous levels.

Affective Domain

The affective domain involves attitudes, values, interests, and appreciations. The components are not as clearly defined as those qualities found in the cognitive domain, and they most certainly are harder to evaluate. They are, however, important, and more should be done in the classroom to develop affective growth.

Thoughts and feelings can not be separated. One cannot act, or be acted upon, without having some kind of feeling or attitude about the situation. Because concepts are developed through thought processes, they also become a part of the affective domain. Four observations may be made concerning the interrelationship of concepts and attitudes:

- One's values determine to an extent the development of certain concepts. Concepts are perceived as positive and strong if one believes they reinforce and enhance personal values.
- As conceptual learning takes place, feelings and attitudes may be altered or completely changed as a result of new knowledge which produces the need for reevaluation.
- Attitudes may develop without the person realizing it. Since there is a direct relationship between concept development and attitude development, attention should be given in the classroom to subject matter and knowledge which can, and should be, learned conceptually.
- Because one can hold attitudes without being aware of their underlying origin or their objective validity, part of education should be the effort to enhance the student's ability to conceptualize—that is, identify, understand, and articulate—his or her true attitudes.
- Because attitudes and values determine behavior, positive growth in the affective domain can be assessed by observing and evaluating behavior.

Psychomotor Domain

The psychomotor domain incorporates the development of motor skills, or the ability to manipulate objects, with the mental activity of coordination, reaction time, dexterity, and steadiness. By motor skills, we mean the actions involving coordinated muscular movements. The term psychomotor indicates a direct relationship between the mind and the body.

Simple motor skills refer to one or two different movements that are performed at one time, while complex motor skills refer to several movements that are required simultaneously. Pounding a nail into a board is an example of a complex skill, because one hand holds the nail steady, while the

***This model illustrates how concept development leads to desirable attitudes. Two key words here are very important for the teacher to remember in planning classroom activities:* positive experiences.**

The Affective Domain

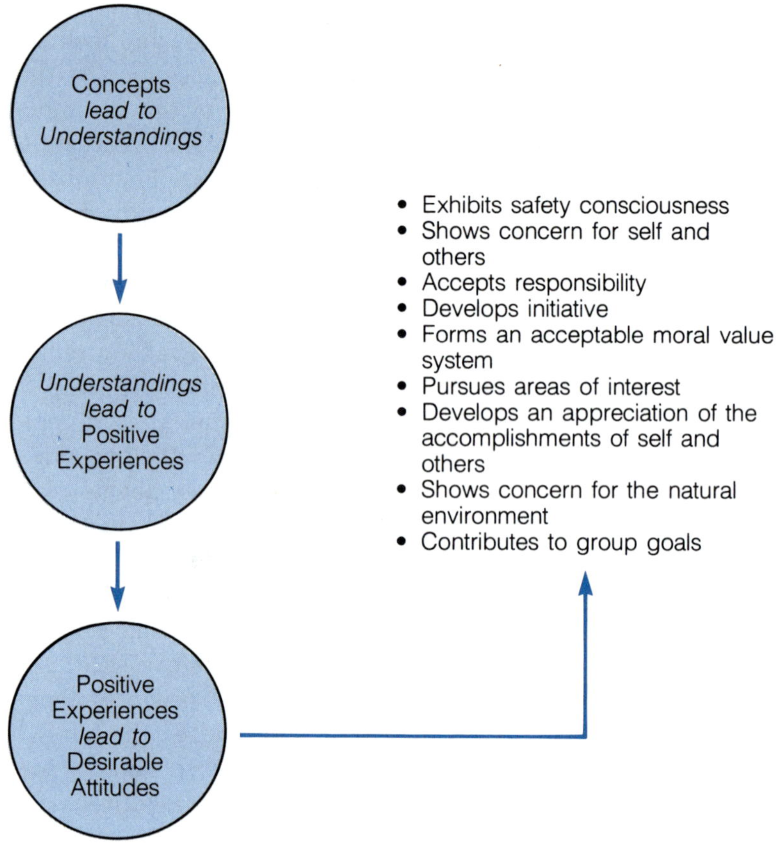

other hand moves downward to strike the nail. The distance has to be judged exactly from the head of the hammer to the hand and, at the same time, from the head to the nail. Large muscle activity involves rapid movements over distance, such as in running or throwing a ball. Small muscle activity involves those fine muscles required in slow, deliberate, very accurate movements, such as in assembling a model car or making a scaled drawing.

Concepts are very much a part of the execution of motor skills. Although simple activities require little thinking, complex activities may require much concentration, the use of concepts up through the problem-solving level of the thought process, and the regulation of movement as conditions change. For example, pouring a liquid from a large-mouth container

Both physical skill and mental processes are necessary if a psychomotor activity is to have meaning.

The Psychomotor Domain

Simple Skills	Mental Application
Throwing a ball	Throwing a ball to a person or location
Basic body movements	Body movements in an active game
Drawing a line	Drawing a picture
Arm movements with a tool	Sawing a piece of wood
Folding paper	Folding paper to make an airplane
Looking through a microscope	Adjusting a microscope
Cutting fabric with scissors	Cutting fabric following a pattern
Turning on a projector	Focusing the image on the screen
Carrying an open container of solids	Carrying an open container of liquid

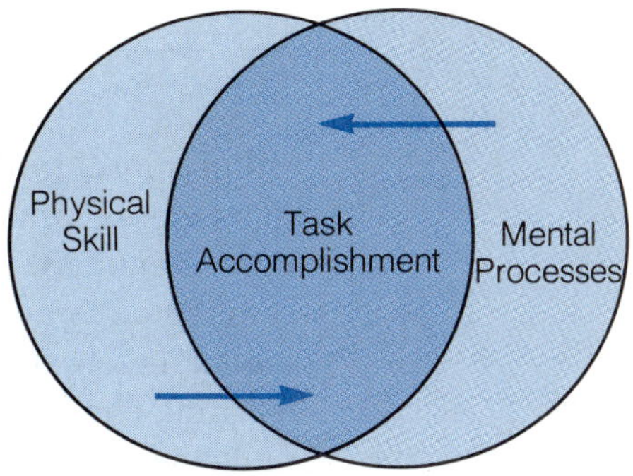

into a small-mouth container involves the concept that as a liquid is poured, the stream gets narrower as it falls further from the container. Also, the angle of the container is important, because a severe angle allows too much liquid at one time, while too small an angle allows it to run down the underside without falling free of the pouring edge. These concepts serve as a mental guide in regulating the movements required. Four observations may be identified which demonstrate the interrelationship of concepts and motor skills.

- The practice of motor skills may help develop concepts, because such practice develops awareness of the individual's abilities and potential and an understanding of the physical environment.
- Concepts may be used to improve motor skills. Skills de-

This progression may be found in nearly any skill improvement. If a new concept or approach is developed during Phase II, the curve could once again make a dramatic rise upward. It is for this reason that concept development is important to skill development.

Learning Curve

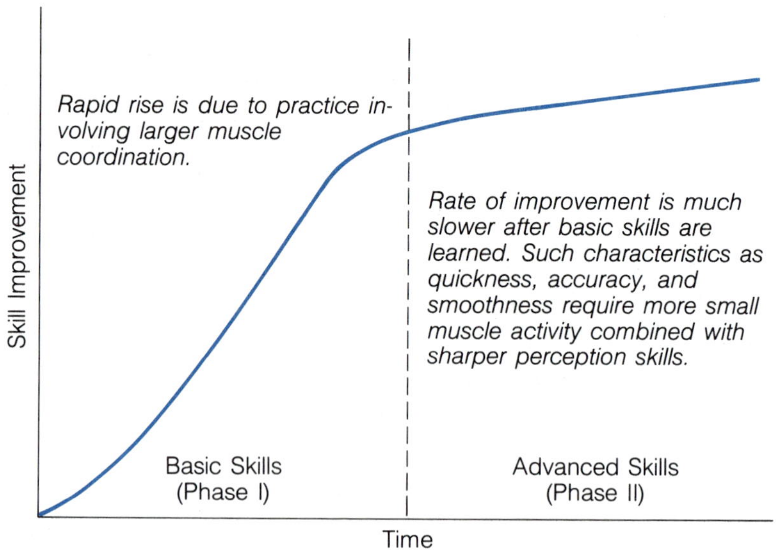

velop more rapidly when someone points out corrective movements and basic concepts which show why those movements are necessary.

- Conceptual understanding becomes more important at the higher levels of motor skills. As a new skill is learned, the learning curve moves up rapidly. After the major elements or movements are attained, however, the curve becomes less pronounced. At that point, conceptual knowledge can help bring about continued improvement by identifying previously overlooked finer points.
- Many types of motor skills can be developed by exercise, until they become a habit, requiring little or no mental activity. Breaking the habit or retraining for a similar skill, however, usually requires new or altered concepts to provide mental guidance in making the change.

TEACHING CONCEPTS ABOUT TECHNOLOGY

The Need for Concept Development

Education and learning is a growth process, which takes place in "building block" fashion. New learning is based on ideas and concepts which were previously learned. Because learning

Because learning builds on past experiences it is important to provide those experiences as early as possible.

evolves from the simple to the complex, it is important that simple concepts (commensurate, of course, with mental and physical levels) be taught, beginning with young students at the kindergarten level or even earlier. During the following years, higher-level mental activity, such as critical thinking and problem solving, will be possible if the proper basic concepts are developed.

Today's highly technological culture is very complex, involving many interrelated elements. When a major change occurs, such as an energy shortage, pollution, or a world food shortage, there are likely to be many cause-and-effect relationships. Concepts that provide an awareness of these relationships help individuals to better adjust to those changes which have a direct effect on them. Helping students develop such concepts will allow them to make more appropriate decisions, which will benefit themselves and society.

The ultimate value of concept development lies in the ability to use the knowledge in planning and controlling the future. Even concepts dealing with the past should serve to provide insight into future events. As a result, the study of the future and the appropriate accompanying concepts should be an extremely important element of the study of technology.

Developing Concepts In the Classroom

Concepts dealing with technology need not be confined to a certain time period each day devoted to the study of technol-

Safety is always a concern when tools and materials are being used. When concepts are learned about how and why they function and how they should be used, safety becomes something to be practiced and ceases to represent something to be feared.

ogy. On the contrary, these concepts should be taught in line with the traditional subjects. What better time to teach concepts concerning the use of lasers than when light is being studied in science? To teach about how computers work than when math is being studied? To talk about pollution problems than in social studies? It is imperative for the elementary teacher to realize that the study of technology is the study of our world today and in the future.

How should concept development be integrated into classroom study? Based on the discussion in this chapter, we can make a number of suggestions. Concepts are best taught through activity-based experiences. The more sensory input the student has, the more complete learning will be. Concepts should be related to expected classroom behavior. Education seeks to modify behavior, as well as to develop new behavior patterns. Concepts which lead to the accomplishment of these goals provide the students with a direction which helps ensure success. Concepts should lead to prediction and analysis levels, which means that planned problem-solving situations, rather than only fact memorization, should be a part of the learning experience. This requires the teacher to be aware of the percepts, concepts, and generalizations which the students need before problem-solving assignments are given. Finally, concepts are more easily developed when reasons for conditions and characteristics are evident.

Presenting cause-and-effect relationships is an important approach in developing concepts. Even the concept that there is such a thing as cause-and-effect is important for students to know. When a fact is stated, it should be accompanied with a condition that relates why the fact is true. For example, in teaching students about vehicles which are used for work, simple facts may state: "Most bulldozers and cranes have wide tracks on which they ride," "the tracks are made like large belts," and "this kind of equipment is used to do very heavy construction work." This would most certainly help the students recognize them upon sight, but such facts may also be stated in such a way that concepts may be more easily developed. For example: "Because bulldozers and cranes do heavy work, they are very heavy themselves," "the tracks are wide like a belt, providing a wider surface on the ground to keep the heavy vehicle from sinking into the soft mud and rocks which are found on construction sites." This approach not only presents the same facts, it presents a number of percepts which allow concepts to be developed.

Concept development requires a conscious effort on the part of the teacher to provide the appropriate experiences. If it is made a part of everyday classroom activities, the effort will provide rewards in more rapid and complete learning, as well as in improved attitudes toward the total learning process.

SUMMARY QUESTIONS

1. Explain the difference between a percept and a concept.
2. Why does problem solving require the development of all the knowledge applications under it in order to be effective?
3. Identify what is meant by the phrase "learning should move from the concrete to the abstract."
4. Identify the three domains of learning and state a short definition of each.
5. List three reasons why concept development is important to the students' learning.
6. Why is an activity-based problem important to concept development?

ALL PLANTS NEED
PLANTS WE EAT
SUMMER
FALL
SPRING
WINTER
Seed & Pollen DISTRIBUTORS
STOP

CHAPTER 15

Determining, Organizing, and Evaluating Content

In this chapter you will learn:

1. **A definition of the word curriculum.**
2. **How the educational system is organized with many levels or subsystems.**
3. **The difference between goals and objectives.**
4. **How units of study may be organized.**
5. **Why evaluation is linked directly to goals and objectives.**

THE SCHOOL CURRICULUM

The central core of our formal education system consists of the curriculum, a term which can be defined either narrowly or broadly. In the past, it meant the total compilation of subjects and the sequence of activities through which each student was required to pass. Its meaning today is much broader in scope.

Curriculum Defined

As a rationale for a complete definition, the following points serve to identify the important elements. Because students learn as a result of the effects of their total environment, the total school environment has to be considered as "the curriculum." This not only includes the subjects and classes, but such elements as the administration, the custodian, the nurse, the playground, the school bus, and even the amount of lighting in the classrooms. Each has an effect on the physical and psychological preparedness of students for learning.

The curriculum not only includes the experiences, it also includes their timing and sequence. This involves two aspects. Learning readiness is necessary for educational growth, so both physical and mental maturity have to be considered in structuring the sequences. Second, because higher levels of learning depend upon the acquisition of necessary percepts and concepts, sequences have to be carefully developed so these are attained as needed.

The primary responsibility of determining the curriculum lies with the teachers and administration. However, the basic concept of a democratic society and a desire for the curriculum to represent real life experiences demand that input be considered from many sources, including the students and the community.

The curriculum affects not only the students. It also has an impact upon the teachers. It is not only a factor in terms of planning time and sequence of everyday duties, it also provides an opportunity for professional growth and development. New insights into more effective teaching evolve as teachers seek to provide new learning experiences.

In summary of these four points, we define curriculum as the combined effort of administrators, teachers, students, staff, and community to provide both a formal and informal school environment that contributes to educational growth.

Curriculum Development

Identifying the content and activities of the classroom involves the past, present, and future. The past includes the thinking and philosophy of scientists, psychologists, learning theorists, social philosophers, and others, whose wisdom provides a solid base of accumulated knowledge for curriculum development. The needs of the present have changed considerably from the past as a result of evolving technology, demanding that curriculum development be an ongoing activity for the

There are many advantages of joint curriculum planning and classroom activities between two or more teachers. It allows sharing of information, teachers' skills, and work areas.

teacher. The same may be said of the future. All learning should be for the future. By the time the students apply their knowledge, it will be the future, whether it is one week later or several years.

Curriculum development cannot be done in a vacuum. It affects, and is affected by, many exterior elements that require direct consideration. Just as technology can be analyzed with a systems approach, so can curriculum development. All elements of the suprasystems (larger) and the subsystems (smaller) have to be considered in order to incorporate these exterior elements.

The System

The purposes of curriculum development and instruction reflect various levels within society. The highest aim of education is to benefit society as a whole—to help students attain a democratic philosophy, develop good citizenship qualities, and acquire the ability and willingness to work hard. These general purposes respond to needs at various subsystem levels, from the national to the state, local, and individual community level. As students mature and become contributors to society at each of these levels, it becomes clear to what extent the curriculum was successful in nurturing the desired characteristics. Thus, teachers should be mindful that the decisions they make every day in the classroom should advance the purposes of education at the school, community, local, state, and national levels of society.

A special commission appointed by the American Association of School Administrators identified nine elements which it called "Imperatives in Education" (AASA, p. i):

1. To make urban life rewarding and satisfying
2. To prepare people for the world of work
3. To discover and nurture creative talent
4. To strengthen the moral fabric of society
5. To deal constructively with psychological tensions
6. To keep democracy working
7. To make intelligent use of natural resources
8. To make the best use of leisure time
9. To work with other peoples of the world for human betterment

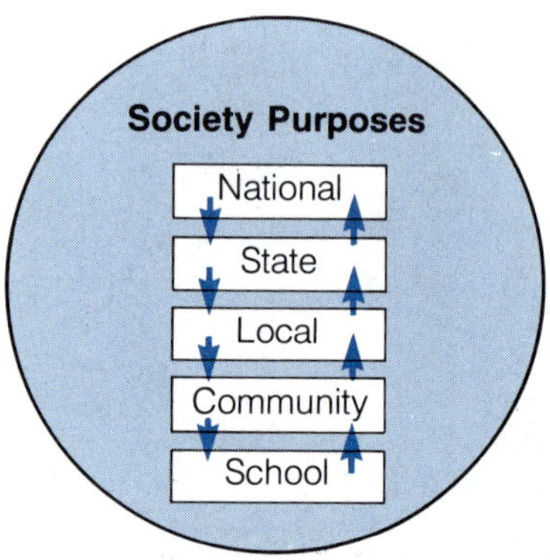

Arrows at each educational level point both upward and downward, indicating that educational responsibility flows in both directions.

These elements were formulated as a guide to help align the educational system with the current needs of society. These imperatives are appropriate at all levels of society.

Terminology

Many times the terminology of curriculum development becomes confusing because terms are applied interchangeably. Below, we identify the terms and place them in context with each other to show similarities and differences.

Purposes. In previous paragraphs, the term purpose connoted very broad statements that would serve as guides or motives. Such guides or motives are based on values and needs and cannot be considered objectives. Objectives, in contrast, describe specifically how broad purposes will be accomplished. For example, a purpose might be to develop an awareness of our environment. An objective to help satisfy this need might be: "After reading the story 'America On Wheels,' the student will be able to list six different transportation vehicles."

Program Goals. Program goals are more specific than purposes in that they indicate exactly what is to be accomplished as a result of the students' experiences in the classes and activities of the program. For example, a goal of the social studies program might be: "As a result of building a simple water-distillation unit, the students will be able to explain the process of distillation and how it is used in producing some of our products." A program goal is more specific than a purpose, which might read: "The students will learn about how we obtain our products."

Program goals serve curriculum development in three ways: (1) because they are identified before the more specific lesson objectives are formulated, they provide guidance for developing the rest of the program; (2) because they identify the knowledge, attitudes, and skills which the students will come to demonstrate, they provide for a complete analysis of the program; and (3) because they provide an organized outline, no important elements are inadvertently left out of the program. Program goals have several characteristics:

- They indentify specific knowledge, attitudes, and skills (for example, "The student will write three examples of how . . ." not "The student will know about . . .").
- They are stated in behavioral terms (for example, "The student will identify . . ." not "The student will work on . . .").
- They are identified as student-involved activities rather than as teacher involved (for example, "The student will build . . . " not "The teacher will provide material for building").
- They identify the desired behavior.
- They require two or more specific performance objectives to evaluate the goal attainment.
- They may be broad enough to be used in formulating performance objectives in more than one course, unit, or lesson.

The use of behavioral verbs—such as identify, explain, demonstrate, relate, and write—is important because these actions can be measured in some way. Nonbehavioral terms, such as "develop an appreciation," "gain an understanding," and "attain feelings, " cannot be measured directly.

Performance Objectives. Performance objectives are specific, written to ensure that the program goals are met. They relate to daily lesson planning, providing a method of planning as well as evaluation. Three characteristics of performance objectives may be identified:

- The performance objective states what the student is expected to do (for example, "The student will define . . ." not "The student will learn . . .").
- The conditions state what parameters the student will be operating within, such as the amount of time needed or the tools required (for example, "Given the microscope, the student will demonstrate the correct method of focusing . . ." not "The student will relate the correct method of focusing . . ."). This also provides equality in student evaluation, through the use of a standardized format.
- The criteria identify the minimum acceptable performance (for example, "Given chalk, the student will list at least six methods of communication on the chalkboard within three minutes." not "Given chalk, the student will list several methods on the chalkboard.").

How do purposes, program goals, and performance objectives relate to the study of technology? To answer this question one has to review some of the characteristics of technology: technology cuts across all levels of society; the technology system involves many subsystems which reach all people; and technology will affect our future.

Because the study of technology can be so broad, an outline or master plan is necessary to incorporate those elements critical to the educational development of the students. The program goals provide this outline. In addition, there is a wide range of ability in students between kindergarten and sixth grade. Concepts need to be taught at all levels, but not in the same way or with expectation of the same results. By stating the objectives in behavioral or performance terms, the teacher analyzes the potential of the students to achieve them and designs a means of obtaining feedback. This, in turn, indicates whether higher levels of learning should be initiated. Because many of the learning situations involving technology are activity-based, tools, equipment, and materials are often needed. Stating objectives in behavioral terms may also serve as a checklist for what will be needed in the way of equipment, supplies, and physical facilities.

In addition, program goals and performance objectives play a role in the professional development of the teacher and in the upgrading of the classroom curriculum. The teacher may find the program goals and performance objectives useful to

Parents should be informed about the curriculum and the objectives of the teacher, particularly when technology is an important element in the curriculum, because many people do not understand either the content or the reasons for teaching it.

show administrators, parents, and other interested persons. An activity-based curriculum can sometimes appear untraditional to those who are not familiar with it. When the program goals and performance objectives are in writing, it makes the explanation much easier.

Developing Units of Study

There has always been disagreement in education as to what should be taught and how it should be made a part of the learning experience. One of the primary differences of opinion affecting the development of modern programs has been whether they should be subject-centered or child-centered. The subject-centered program uses a highly structured format which ensures that each student passes through the same content and masters the same skills. It is characterized by practice, drill, and memorization. The child-centered program is less structured, emphasizing the needs of individual students by independent study, experimentation, and self-motivation. It cannot be denied that both are extremely important. The problem arises when the teacher begins to plan. Which is most important? Which should come first?

This dilemma is at least partially resolved through the use of a "developmental activity program." It is based on sound educational theory and child development, including the intellectual, biological, social, and emotional development of the student. It may be characterized as a program in transition, because it takes into consideration the needs and abilities of the students as they pass through the various stages of growth and development. Using the organizational outline as a guide, it attempts to adjust each learner to the appropriate learning experiences which will lead to concept development.

In general terms, the developmental activity program divides the school day into roughly two parts. Half of the day is subject-centered and half is child-centered. They are not totally separate, however, because, as much as possible, the subject-centered activities should grow out of and be motivated by the developmental activities. This organizational structure has certain advantages:

- Because the study of technology is, by nature, activity-based, the subject matter is easily introduced into the program through the developmental phase.

- It allows the beginning teacher, or one who has not actively taught technology, a chance to gradually move in that direction. For example, one might start with a developmental activity one afternoon a week and, with experience, introduce ongoing activities that would later extend into each school day.
- It ensures the learning of basic skills such as math, spelling, and writing.
- It is especially useful at the beginning of the school year, because the teacher is able to become familiar with each student's abilities in the two segments, and it gives the students a clear idea of what is expected of them at certain times.
- As the year progresses, flexibility can be gradually introduced as the teacher becomes more comfortable with the structure and brings the two closer together.

Units of Study

Units are essential in planning the educational program. They are used as guides to make sure critical elements are taught. The use of daily lesson plans by themselves helps get the teacher through the day, but may not provide a program of purpose and continuity. There are many types of units, and many different names have been given to them. Two which represent the extreme ends of the unit continuum are the subject-matter unit and the experience unit, which represent the subject-centered and child-centered programs just discussed. An alternative to these two, the planned-experience unit, would be more likely found in the developmental activity program.

Subject-Matter Unit. The subject-matter unit is built around a specified subject, with the outline coming from a logical sequence or grouping of topics taken from the subject. An example of a sequence might be math, as it moves from addition to subtraction and on to multiplication. An example of topic grouping might be a social studies unit presented as "The Home And Family," "Our Cities," or "Transportation." They cover the subject thoroughly, but they do not provide adequately for the needs of individual students, because the primary emphasis is on cognitive learning in only that specific subject.

A sample of a planned-experience program unit illustrates the major topics to be considered in planning a unit of study. Program units may be more simple or more complex, depending upon the variables involved.

Planned-Experience Program Unit
Huntington Elementary School

Teacher: C. Wilson *Grade:* Five *Semester:* Winter

Program Unit Title: Our Products and How They Are Assembled

Time Required: One week

Related Subjects: (1) Social studies (2) Language arts (3) Math (4) Science (5) Art

Program Goals:

As a result of classroom discussion and assembly activity the students will be able to identify:

(1) How the use of mass assembly methods changed our society. (Social studies)
(2) Both positive and negative factors of the three assembly methods as they relate to the individual worker. (Social studies)
(3) How the parts of a small assembled product might be attached to each other. (Science)
(4) Why packaging is important. (Social studies)

As a result of the assembly activity the students will be able to:

(5) Experience line assembly work. (Social studies)
(6) Express their feelings toward this kind of work. (Social studies)
(7) Identify common terms related to assembly of products. (Language arts)

As a result of classroom discussion and math assignments the students will be able to:

(8) Use common math procedures in addition, subtraction, and division to solve practical problems. (Math)

As a result of the lettering assignment the students will be able to:

(9) Devise a layout and develop large lettering. (Art)

Types of Learning: Cognitive (Goals #1, 2, 3, 4, 7, 8)
Affective (Goals #6)
Psychomotor (Goals #5, 9)

Number of Teaching Units Required: Six

References and/or Learning Materials: Social studies text
Math text and duplicated work sheets
Spelling words on duplicated sheets
Film: "Where We Get Our Products"

Special Supplies/Equipment:

16 mm projector
Paper for signs
Felt markers
Ballpoint pens with six parts (50)
Paper plates (12)
Staplers (2)
Pencils (2)
Small squares of paper (100)
Plastic sandwich bags (100)
Small electric table lamp for disassembly

Special Room Arrangements: Desks aligned in two rows of six. (Friday)

Consultants, Speakers, Visitors: None

Schedule:

	MORNING		AFTERNOON	
Monday	Social studies	Discussion: Student involved planning	Math	
Tuesday		Spelling		
Wednesday				
Thursday	Film		Art (Lettering signs)	
Friday	Science (Disassembled lamp)	Spelling Test	Assembly of pens and discussion—student-involved evaluation	

The planned experience unit provides for creativity and input from the students, but it also involves guidelines and direction from the teacher.

Experience Units. The experience unit is much broader and flexible than the subject-matter unit. The primary emphasis is on the development of the student in relation to the environment and society. It is largely unplanned by the teacher ahead of time. Rather, it is characterized by pupil-teacher planning, based on the individual needs of the students. The unit hangs together as a result of the psychological relationships formed by the students as the activities take place. That is, the resulting successes psychologically motivate both the teacher and student to link the activities into a meaningful whole. Experience units take into consideration individual differences, so a number of units may be carried on at one time to serve several different students' needs.

Planned-Experience Unit. Although both the subject-matter unit and the experience unit have educationally sound principles, they are largely impractical in their pure form. The subject-matter unit fails to provide flexibility in dealing with student needs and individual differences, while the experience unit is too time consuming for the teacher's busy schedule. A solution to the problem is the planned-experience unit. This takes the best elements of both and combines them into one. It is teacher planned and guided, with student input, has flexibility, emphasizes both subject matter and learning experiences, and plans for cognitive, affective, and psychomotor learning.

An example of a planned-experience unit might be one dealing with telecommunications, involving reading, writing, math, and the building of a model communication satellite. This might be a part of the study of communications within the social studies program. The unit contributes to the broader program goal, but zeros in specifically on telecommunications. There also may be additional units, which emphasize other types of communication, so that all the program goals are met. Some units may be very short, two or three days, while some may run several weeks or even months, depending upon the program goals.

Teaching Units

Teaching units are not primarily concerned with planning and organization, rather they deal with identifying specific techniques and procedures required to obtain the objectives set down in the program goals and the units of study. They also, however, have their own performance objectives.

Unit teaching is developmental in nature. That is, there may be two or more units, each one building on what was learned in the previous one. This provides vertical movement of the students upward through percepts and concepts to problem solving. It also is able to provide varied experiences to develop cognitive, affective, and psychomotor learning. Teaching units may be one hour in length or several days. However, all contain very specific information. The chart that follows illustrates an outline suggesting those elements necessary for a complete teaching unit.

The activity described in the chart involves a relay with two teams competing to see who can assemble, inspect, and package all twenty-five pens first. If the pens are different colors, the inspector makes sure the correct colors are used and the pens are working. An X is placed on the inspection tag to indicate acceptance. When the inspection tag and pen are placed in the bag, it is then ready for stapling (packaging). More students may be involved by spacing the tables apart, which requires transportation of the subassembled parts to the next assembly station. The teaching unit helps the teacher think through both the procedures and supplies needed, thus serving as a checklist and reminder.

This teaching unit represents the sixth and last activity in the program unit.

Teaching Unit No. 6

Teaching Unit
Huntington Elementary School

Program Unit Title: Our Products and How They Are Assembled

Teaching Unit Title: A Line Assembly Relay Race and Unit Evaluation

Time Required: One and one-half hours.

Performance Objectives:

(1) The students will be able to correctly assemble all fifty ballpoint pens within five minutes. (Psychomotor learning)

(2) In a class discussion the students will be able to verbally identify a minimum of five potential problems, and possible solutions, which might arise in an actual line assembly situation in industry. (Cognitive learning)

(3) The students will write a short statement indicating whether they would like to work on an assembly line and identify a minimum of two reasons why their answer is positive or negative. (Affective learning)

References and/or Learning Materials: None

Special Supplies/Equipment:

Student lettered signs	Pencils (2)
Ballpoint pens (50)	Small squares of paper (100)
Paper plates (12)	Plastic sandwich bags (100)
Staplers (2)	

Preparation:

(1) Line desks in two rows of six.

(2) Line six paper plates on each row of desks.

(3) Disassemble pens. Place the parts for twenty five in the plates on each row. (Each plate contains all the parts which are alike; these become the assembly stations.)

(4) Place a pencil, small squares of paper, bags and stapler at the end of each row.

(5) Place signs for subassembly and assembly areas on the desks.

Procedure:

(1) Review the concepts learned concerning line production.

(2) Choose four relay teams.

(3) Explain the rules of the relay.

(4) Identify the specific assembly jobs.

(5) Run the relay with only one pen to demonstrate the process.

(6) Run the actual relay (two teams).

(7) Discuss some of the problems encountered.

(8) Disassemble the pens and set up for the second two teams.

(9) Run the second relay (other two teams).

(10) Discuss any problems and identify whether any previous problems were solved the second time.

(11) Have a student recorder list on the chalkboard problems and possible solutions which might occur in a real line assembly operation. These are identified as the class discussion takes place in summarizing and evaluating the total unit.

(12) Have the students write a short statement indicating their feelings about working on an assembly line.

(13) Discuss sampling of answers and reasons given.

Future Changes in the Teaching Unit: (To be filled out at the conclusion of the unit.)

EVALUATION

Feedback is very important in teaching. It is the indicator of whether adequate learning has really taken place. As a result of feedback, decisions are made to reteach, modify experiences, or move on to the next unit.

Evaluation provides this feedback. Not all evaluation, however, is solely for the teacher's use. Students should be actively involved in the evaluation process so they are able to assess their own performance and abilities. In addition, many times both the teacher and student learn in critiquing a project together.

Program goals provide a guide in developing performance objectives suitable for a particular unit of study. This, in turn, guides the even more specific performance objectives of the teaching units. Thus, the objectives at both the unit of study level and the teaching unit level are stated in performance terms, facilitating evaluation. Just as the performance objectives move downward through the levels of the instructional system, the evaluation moves upward: if objectives are met at the teaching unit level, they should automatically satisfy the objectives at the unit of study level and those at the program goal level.

An activity-based program requires activity-based evaluation. The performance objectives are written with evaluation in mind. The criteria used in making evaluations

Evaluation with the students provides a learning experience in itself. Among other benefits is the development of students' self-evaluation abilities.

will vary from grade to grade and sometimes even within a single classroom, depending upon the students' abilities and what is being measured. Cognitive and psychomotor learning can be more easily accomplished by objective means. Either the students are able to accomplish the required task or they are not. In the case of affective learning, evaluation becomes more subjective. The following examples show how the wording might be changed in performance objectives to provide evaluation in each of the three domains of learning.

- Cognitive domain (knowledge). "The student will be able to accurately write, with an accuracy of 100 percent, the step-by-step sequence of how the project functions."
- Affective domain (attitudes and values). "The student will verbally identify three things which he or she most enjoyed and three things he or she least enjoyed in building the project."
- Psychomotor (skills). "Given a metric scale, the student will accurately measure, within two millimeters, the wood to be cut for the project base."

Evaluation should be a continual process, taking place before, during, and after the learning sequence. This gives the teacher and the students feedback, making them both accountable to each other as well as to themselves. It also helps the teacher with accountability to the administration and to parents.

SUMMARY QUESTIONS

1. Why is the curriculum considered to be more than just subject matter?
2. Explain the difference between program goals and performance objectives.
3. How do units of study determine the daily organization of the classroom?
4. State one advantage and one disadvantage of each of the subject-matter unit and of the experience unit.
5. State the primary difference between a planned-experience unit and a teaching unit.
6. Why do the methods of evaluation have to be closely related to the performance objectives?

Reading
Music
Spelling
P.E.
Science
Career Education
English
A Season

CHAPTER 16

Determining the Learning Environment

In this chapter you will learn:

1. The role of the elementary teacher.
2. How students may be organized for classroom activities.
3. Traditional types of activities which may be planned for teaching technology.
4. How the classroom may be physically organized for conducting activities.
5. Appropriate tools and equipment for the elementary classroom.
6. Important elements to consider when planning activities.

TEACHING STRATEGIES

The elementary teacher wears many hats during the course of one day. To involve the students in an activity-based curriculum the teacher must be a specialist and a generalist, a manager, and a resource person. As a specialist, the teacher has to be familiar with specific subject matter being taught. He or she may not be an expert, but must have a solid background in the traditional subjects such as math, science, and language arts. At the other end of the continuum, a generalist has to have a broad background, when the content includes technology. The study of technology incorporates all other subjects, thus, it requires a generalist who can pull together the various aspects of several fields of study. As a manager in an elementary classroom, the teacher plans many activities, some of which may occur simultaneously, assigns various duties to the students, and supervises ongoing work. In the past, the teacher was an authoritarian figure who made all the decisions. Today, the teacher still retains the authority, but the role is directed toward classroom management, toward making the learning experiences as efficient as possible. Likewise, today's teacher is no longer a mere expounder of facts. Society is too complex for the teacher to attempt to learn all the facts to which elementary students should be exposed. Instead, the teacher acts as a resource person, guiding students to suitable books and references, helping them plan so that objectives are met, and providing appropriate supplementary classroom materials.

Teachers should view themselves as informed resource people, arranging frequent trips to the library or resource center so that students can plan their projects and find suitable research material.

Discovery and Directed Approaches

The approach taken by the teacher in allowing student freedom in activity-based learning depends upon many variables, such as the facilities available, the time needed for completion, and, probably most important, the physical, mental, and social capabilities of the students.

Allowing students the freedom to plan, execute, and evaluate their own activities, to make mistakes and formulate solutions might appear to be the ideal learning situation, but this method, called the discovery approach, has two major disadvantages. In its truest sense, it may be nothing more than trial and error, which is highly inefficient, and considerable time is required to develop the student attitudes that may make it successful.

In contrast, the directed approach casts the teacher in the role of major decision maker. From a classroom management standpoint its particular advantage is that the teacher can allow as much freedom as the students are able to handle, while still retaining control. From a learning point of view, it also has an advantage because it allows the teacher to design learning experiences appropriate for students at any given time. For example, at the beginning of the school year, the students are not familiar with the classroom and the expectations of the teacher. For this reason, early activities should

Almost all activities have to be teacher-directed, whether they are simple ones or complex, like laying a brick wall. The student makes his or her discoveries within the planned framework.

probably be entirely directed by the teacher. This allows a pattern to be set, which provides responsibilities to be assumed by the students as they become capable of handling more freedom and making their own decisions. The teacher should retain the major decision-making authority, while gradually allowing the students to make increasing numbers of minor decisions. For example, if the class is involved in mass producing an item during a unit in manufacturing, the teacher may identify "what, when, and where," while the students might determine "who" and "what color." The teacher makes the decisions essential to the success of the activity, while the students make the decisions which directly concern them, such as choosing what part they would like to play in the activity and what color to paint the object.

Organization of Students

Activities need not always be a major undertaking for the teacher or the students. One of the primary advantages of an activity-based curriculum is the flexibility it provides in planning and providing experiences for individual needs. Activities may involve a whole class or just one student.

Individual Instruction. The term independent study describes individual instruction because an individual student may pursue a topic concerning, for example, technology, which none of the other students would become involved in. Although special students, either advanced or slow, may embark on independent study, it also has great value for any stu-

An activity may be planned for only one student. This provides an opportunity to learn and gain experiences at a level commensurate with his or her abilities.

dent. It provides an opportunity to expand a special interest topic and allows the student to learn at his or her own pace. It allows the pursuit of activities that would not be possible with an entire class. It encourages student resourcefulness, because the student must learn self-reliance as there are no others involved, and it helps establish independent work habits. The teacher should work with the student in establishing the objectives and step-by-step sequence of activities. The performance objectives should be very specific, so the student realizes exactly what is expected in terms of the mode of operation and the final results of the activity.

Small-group instruction has many advantages, particularly when demonstrating the use of tools and materials with which the students will be actively involved.

Small-Group Instruction. The teacher can spend more time helping and instructing a small group than individual students. Small-group instruction allows the teacher to group students who have similar interests and abilities, allows topics to be pursued which would not be appropriate for the total class, helps establish behavior based on peer-group pressure, develops responsibility because contribution to the group is essential, and provides an opportunity for social growth during the process of group interaction. As with individual instruction, the teacher clarifies specific performance objectives and procedures for the group. Periodic evaluation, by the students as well as by the teacher, is also desirable to assess whatever progress is being made.

Class Instruction. The method most commonly used, although not necessarily the most efficient, is class instruction. Many activities do not lend themselves to total class involvement because of space and tool limitations. A compromise plan involves breaking the activity into phases, such as cutting, fabricating, and assembling a project. One group of students may start phase one, while the other students are working at their desks on another assignment, which may or may not be related to the activity. When phase one is completed by the first group, a second group starts phase two. The first group may then return to seatwork or move on to phase

The entire class is studying the names of tools by outlining and coloring them and labeling them with the correct name.

three. The alternating of assignments by groups allows all students to experience the activity, but they are not all working on the same thing at the same time. However, the related information and specific topics of discussion dealing with technology may be covered as a total class.

Among the advantages of class instruction are the following: (1) All students receive the same instruction; (2) the teacher has complete control of the learning situation at all times; (3) informal leadership has the opportunity to develop among the students as the activity progresses, (4) the teacher may more easily relate the teaching of technology to the traditional subjects; and (5) the teacher may more easily evaluate the students' performance because all are accomplishing the same task.

Classroom Activities

It is not our intention to identify and discuss each of the many types of classroom activities which elementary teachers use every day. A few selected activities are offered, however, as examples of how the study of technology may be easily brought into the daily program.

Demonstrations. Demonstrations provide an active, visual element which supplements the learning of abstract ideas and

After a demonstration by the teacher, the students should be allowed to try the procedure under the teacher's direction, if it is appropriate. These students are making paper pulp.

are used with nearly every subject taught in the elementary grades. They offer an excellent means of interfacing technology with the traditional subjects. For example, if the growth patterns and cell structure of trees are being studied as a part of the science program, an assortment of wood products could be brought in to demonstrate how technology produces items made from solid wood. These products are useful because of the nature of the material: wood is light in weight, it is strong, pieces can be nailed together, it is a good temperature and electrical insulator, it will not rust, and it can be bent. In addition, there are many processed wood products, such as charcoal, turpentine, and alcohol. All of these products are possible because of the physical characteristics of trees. These characteristics, processes, and products of wood become important in concept development in the study of science.

Other products, such as paper, may also be brought into the demonstration. Add small pieces of bathroom tissue to a bowl of water as the mixture is whipped with a manual or electric mixer. Keep adding paper until the solution is very thick. This is called pulp, a stage all paper goes through during the steps of processing. Drain the water and lay the pulp out to dry. When it is dry, the wood and other plant fibers can be seen with a magnifying glass or low-power microscope.

Items may be brought into the classroom and demonstrated as a method of generating ideas for creative writing, they may be used to show practical applications of math concepts, and they may be used as an incentive to encourage outside reading. Some demonstrations are better suited to teacher

Role playing provides learning experiences not possible to obtain in any other way.

presentation, while some materials and processes may be easily demonstrated by the students. Active participation by the students automatically involves them in the study of a specific subject, many times before they even realize it, and should be encouraged when possible.

Role Playing. Role playing places the students in a position where they experience conditions representing a real-life situation. For example, a line production project, whether it is producing wood products or double-decker lunch meat sandwiches, gives the student an idea of what it is like to work on an assembly line. The production of a class newsletter to be sent home to parents also gives them a feeling for what is involved in producing a daily newspaper. These kinds of activities add reality to the learning experience.

Simulation Games. Many simulation games can be obtained commercially. They place the student in a role-playing situation, but they provide the added dimension of a game, making it more interesting. These work well for individual and small-group instruction, because it is not necessary for the teacher to be with the students at all times, allowing the teacher the freedom to work with other students.

Experimentation. Experimentation should be organized so students have a specific direction and specific procedures to

A small, portable oven is a valuable tool for the teacher. It may be used for a variety of demonstrations and student experiments. This one belongs to the school and can be wheeled from room to room.

follow. It should not degenerate into trial and error. Tasks may be as simple as comparing the weight of various materials on a balance scale, or testing the effectiveness of three brands of liquid soap on soiled fabric. Tasks may be more complex, such as assembling low voltage electrical circuits of lights, bells, and buzzers, or testing certain household chemicals for acid or alkaline content.

Field Trips and Classroom Visitors. Long an integral part of the elementary education, field trips and classroom visitors can help illustrate aspects of modern technology. In addition to learning about what people do in their jobs, students can see the equipment and materials used in various occupations. A frequent visitor to the classroom is the fire fighter, whose usual lecture concerns traditional safety rules. However, the visit may mean more to students if additional aspects are covered, such as why a fire extinguisher is able to put out a fire, how internal combustion occurs, why some materials burn rapidly, why some materials produce toxic vapors which should not be inhaled, and why a person should stay close to the floor when a room is full of smoke. In addition, the technology of firefighting equipment may be discussed. Where does the water come from in the fire hydrant? What produces the pressure for the water hoses? Why is it necessary to have a driver at the rear of a long ladder truck? Such are the technological questions that

There are many resource people from outside the school who can offer expertise and first-hand knowledge about technology and its relationship to the subject being studied. This also contributes to good public relations between the school and community.

may be addressed by a fire fighter in the classroom, or in the course of a trip to the fire station. Whether the classroom visitor or person talking to the students on a field trip is a police officer, airline flight attendant, farmer, custodian, or manufacturing representative, the technology involved in the particular job should be discussed. If the learning experience contains facts about how things function, there is a much greater understanding of why certain requirements and procedures are used. A few suggestions by the teacher ahead of time will help the resource person plan the discussion to include several elements of technology.

THE LEARNING ENVIRONMENT

The classroom in which technology is a part of the everyday subject content and associated activities need not look any different from other well-organized classrooms. The traditional bulletin boards, mobiles, and table-top displays may be seen, as may the usual desks and chairs. There also may be individual or small-group learning centers. There are, however, two major differences. The physical organization of the classroom is flexible, many times changing daily, depending upon the requirements of the activity, and the subject matter integrates concepts about technology and its effects on society with those of traditional subjects.

Learning Centers

Learning centers may be very small, accommodating only one or two students. They may be marked by a single desk and chair separated from the others, or there might be a small room divider to provide more privacy. Traditionally, learning centers were used for individualized reading and study, but they also may be organized for simple laboratory experiments or the construction of three-dimensional objects. The proper equipment, tools, and supplies, as well as specific recorded or written instructions, are necessary at the center. Learning centers are useful for advanced students, slower students, or those with extra time. For reasons of both safety and proper use of time, it is essential that the students know exactly what is expected of them, since they are working without direct teacher supervision.

A learning center may be no more than an isolated spot in the room for individual study. This divider is made of a lightweight wood frame covered with corrugated board. It can be easily made in the classroom and can serve as both a divider and bulletin board.

Work Areas

Classroom. Work areas are specific locations set aside for manipulative activity, such as experimenting, fabricating, assembling, or finishing. They are equipped with the proper tools, materials, and work surfaces for the activity. Tables and work surfaces should be flexible enough to allow most of the classroom to be used for major activities, or only one small corner for individual and small-group activities. The small work area is best placed in a well-lighted corner as far away as

A work area may be set up in a corner of the classroom. This allows other students who may not be involved in the activity at the time to continue their work at their own desk or table.

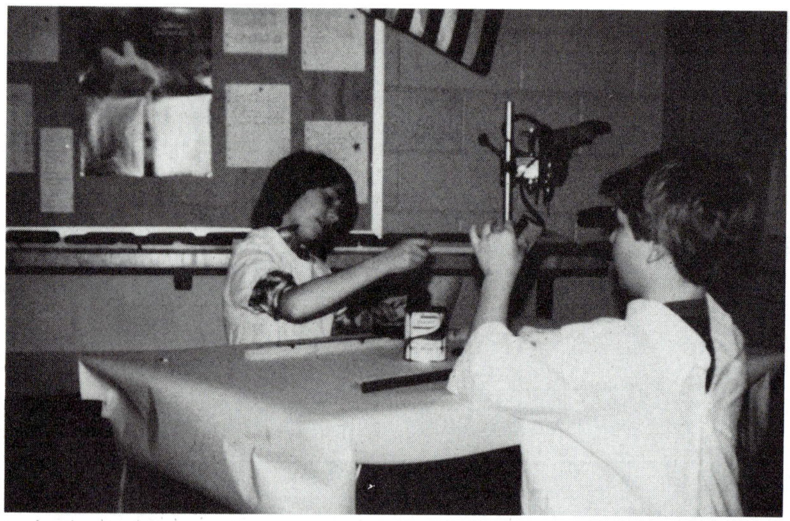

possible from the rest of the students and their activities. It may be a permanent part of the classroom, being used as a learning center when specific activities are not underway, or it may be set up only when needed, with the tools and materials stored elsewhere.

One of the prime considerations in determining the size of area and the general organization of the classroom is student safety. Activities that require eye protection, such as nailing and sawing, should be as far away as possible from unprotected areas. If the room is small, pieces of corrugated board may be used as a shield to separate the areas. Of course, any students participating in activities, where there may be airborne particles, and students in close proximity to the activity, should wear eye-protection.

Multipurpose Rooms. Some schools have a separate room for active games, art projects, and other similar activities, which may suit large-scale projects better than the classroom. Multipurpose rooms provide more space, more appropriate furniture, and, usually, clean-up facilities. Because they are large, more than one class can become involved in a common activity.

All tools and supplies have to be removed or stored securely when not in use, so other students using the room will not have access to them. It is also mandatory for the teacher to be in the same room with the students at all times while they are working. If only part of a class has been taken to the multipurpose room, a responsible person with the proper authority has to be left in charge of those students remaining in the classroom.

Work Surfaces

The work surfaces of desks and table tops work well for most activities, assuming good judgment is used and proper clean-up procedures are followed. The exception is where heavy cutting and fabricating is being done, such as sawing wood and light metal, or assembling items with glue and nails. A low woodworking table works best, but there are other surfaces which may work better for the individual classroom with limited space. A relatively heavy and sturdy classroom table may be quickly converted to a work table by clamping a piece of hardboard or plywood to the top with small C clamps. This prevents chips, dents, and scratches, and at the same time provides a work surface at the correct height for the students. Holding tools such as vises and miter boxes may also be clamped to the top with small C clamps.

A temporary work surface for more than one student may be made by laying pieces of lumber or heavy plywood across two sawhorses. Nailing and sawing should be done directly over the sawhorses to provide the most support. A specially designed sawhorse, which has an extra-wide cross piece, makes an excellent work surface for one or two students. A number of them may be used as work stations, with each one being designated for a specific operation, such as sanding, nailing, or sawing.

It is important that work surfaces be securely fastened together and attached to the legs or support members. Also, any holding tools should be securely attached by screws or C clamps. Although a finish is not required, a work surface should be sanded smooth so there are no splinters or rough edges.

Tools

The tools used in the elementary classroom from the kindergarten level to the upper elementary levels should always be good quality, authentic tools. Toy tools and inexpensive tools should not be used because they are more dangerous, being subject to breakage, and they require more physical strength, often making it difficult for students to achieve satisfactory results. They are also more easily damaged, and most cannot be sharpened or repaired.

Most reputable companies produce a line of tools in several sizes. The tools should be small enough for the students to manipulate, but large enough for the job to be easily accomplished. As a general rule, striking tools, such as hammers, should be of medium weight, while saws should have relatively fine teeth. Smoothing and shaping tools should be of medium size, because smaller tools require finer muscle coordination, which is undeveloped in most elementary students.

The kinds and number of tools will vary according to how the teacher plans to use them. Generally, more multipurpose tools, such as hammers and saws, are needed than are single-purpose tools, such as screwdrivers and wrenches. For the study of technology through an activity-based curriculum, the following chart would serve as a good starting point.

Name	*Number Required*	*Description*
Panel saw	2	20 inch (508 mm) long, with 10 teeth points per inch (25.4 mm)
Back saw	2	12 inch (305 mm) long, with 12 teeth points per inch (25.4 mm)
Coping saw	2	standard size
Hack saw	1	standard size, with 18 teeth points per inch (25.4 mm)
Tin snips	1	10 inch (254 mm)
Claw hammer	3	12 ounce (.34 kg)
Surform plane	3	two 10 inch (254 mm), one pocket plane
Brace	1	standard size
Bits (set)	1	1/4 inch (6.4 mm) to 1 inch (25.4 mm)
Push drill	1	standard size, with drills
Hand drill	1	standard size
Twist drills (set)	1	1/16 inch (1.6 mm) to 1/4 inch (6.4 mm)
File	3	two half round and one flat, 10 inch (254 mm) long, with handles (not included in length)
Pliers	2	combination, 8 inch (203 mm)
C clamp	8	4 inch (102 mm) opening
Try square	2	8 inch (203 mm)
Steel rule	2	12 inch (305 mm)
Miter box, with back saw	1	standard size
Vise	1	4 inch (102 mm) jaw

In addition, a number of pieces of science equipment, such as a thermometer, balance scale, microscope, magnifying glass, alcohol burner, and several beakers and test tubes are desirable. It is not necessary, however, to have all the above tools and equipment before implementing activities. Some may be borrowed for the length of time required to complete the ac-

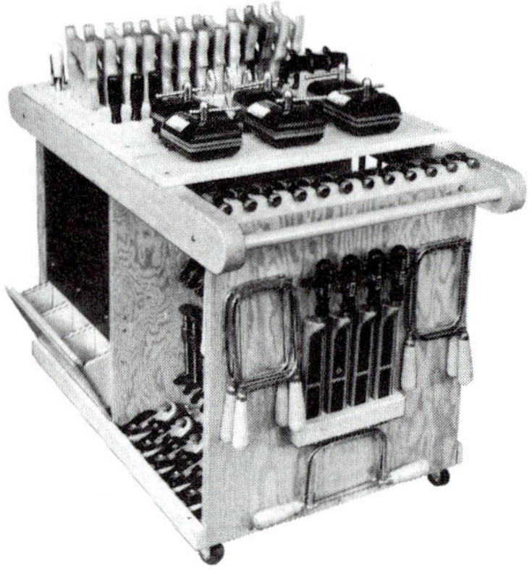

A storage cart has several advantages: it has a place for each tool, so they are easy to keep track of; it can be wheeled from room to room, requiring only one set of tools for several teachers; and it can be locked in a storeroom or closet when not in use.

tivity and individual tools can be added to the classroom one at a time as the financial situation allows. And of course, additional tools may also be added as the need dictates.

In general, tools are stored in a cabinet, which has drawers or special holders for each tool, or a portable cart, which can be rolled into a closet or room that can be locked. All tools must be locked up when not in use to ensure safety as well as security. Several types of commercial storage units are available, or they may be built from standard materials to meet the specific needs of a classroom. They need not be elaborate. On the contrary, if there are only a few hand tools, a couple of standard-size tool boxes work very well. They may be stored in a closet or cabinet and brought out when needed.

Supplies

It is nearly impossible to stock all the supplies which might be needed. Those used on a regular basis, such as adhesive tape, glue, and finishes, can be kept on hand. Others, such as wood, corrugated board, fabric, and expanded plastic, may have to be obtained as needed because of limited storage.

A good grade of abrasive paper, aluminum oxide or silicone carbide, can be used over many times and is usually less expensive in the long run than some of the cheaper materials. Fine, medium, and coarse grades will be needed. Each sheet should be divided into four pieces, making it last much longer.

Vinyl glue (white glue) works well in classroom situations. It is fast drying, requires little or no clamping, dries clear, and may be cleaned up with water. Care should be taken when other glues are used. Teachers should always read the instructions to determine a glue's suitability for classroom use with students.

Tools and supplies may also be kept in plastic containers. These containers are stored in a portable cabinet that doubles as a work surface. Inexpensive plastic dishpans and wastebaskets may also be used for this purpose.

The finishes available on the market today make this part of any activity much easier. The latex finishes have a water base, which means they may be cleaned up with water as long as they have not dried. Latex enamels can be purchased in many colors, and clear latex finishes allow wood grain and other materials to show after sealing.

It should be noted that even though the finishes are water soluble, all paint rags should be either discarded or stored in a closed metal container to prevent internal combustion, as should all paints and finishes.

If a commercially available storage cabinet cannot be obtained, a simple solution is the use of plastic kitchen containers, such as small dishpans. These work well for such materials as glue bottles and abrasive paper. The compartmentalized plastic silverware holders and drawer dividers work well for nails, screws, and other small items. The containers can be neatly stored on a shelf and brought out when needed.

Other materials, such as wood, plastic, and corrugated board, must to be stored according to each classroom or school situation. In general, sheets of material and long narrow pieces are more easily stored in a vertical position, while smaller pieces are more efficiently stored by stacking on a shelf, in a cabinet, or in a closet. Efficient storage may be built, but the specific design would be dictated by the individual classroom.

PREPARING FOR A CLASSROOM ACTIVITY

The preparation for an activity is extremely important. In many cases, it could serve as the determining factor in how well the objectives are met. The three following categories of teacher preparation, student preparation, and classroom preparation illustrates some of the elements to be considered. They also may be used as a checklist in planning an activity.

Prior to small-group instruction or demonstration, a teacher explains an activity to the entire class to prepare them for what they will be expected to do.

Teacher Preparation

Teacher preparation is probably the single, most important element in ensuring success. The following suggestions should be considered:

- The objectives should be clearly identified and a detailed outline of the activity should be written down.
- A detailed list of tools, equipment, and supplies should be made and used as a checklist when they are obtained.
- The complete activity should be practiced ahead of time to identify problems and situations where the students might have difficulty.

Student Preparation

Student preparation may be started several days ahead of time and might even involve some of the students in the final preparations. Some of the topics which should be covered with the students are as follows:

- All safety rules and classroom behavior should be discussed and thoroughly understood.
- The students should know the purposes of the activity, including the learning objectives and how they will be evaluated.
- They should be aware of exactly what responsibilities will be expected of them and approximately how much time will be involved.
- Necessary instruction dealing with required skills should be done prior to the activity, unless it is a part of the activity and identified in the objectives.
- The students should be instructed to have appropriate work clothes, or adequate protection, such as aprons and work tops. An excellent work top is an adult long-sleeve shirt, which is worn backwards and buttoned down the back.

Classroom Preparation

The students may also help in preparing the classroom, providing a feeling of involvement and responsibility toward the success of the activity. The following items should be considered.

- The classroom work area should have only the amount of tables, chairs, and work surfaces needed for the activity. All others should be placed together near an unused wall or corner.

- All necessary tools and equipment should be laid out, ready for use, in an organized manner to provide easy access as needed.
- All materials should be distributed to the work areas where they will be used.
- Appropriate cleanup materials, such as newspapers, rags, paper towels, cleaning solvents, and soap should be placed in an easily accessible location. Where it can be done, newspapers laid on the work surfaces and the floor aid in cleaning up at the conclusion of the activity.
- Safety glasses should be provided for all students who will be working with materials which could become airborne, such as nails, wood chips, and dust. Generally, if students are using tools or liquid other than water, they should be wearing eye-protection.

At the conclusion of the activity, the teacher should evaluate the reaction of the students, how smoothly the sequence went, and how well the objectives were met. It is helpful if comments are written down soon after the activity while they are still in mind. These should include the identification of problem areas, how much material was actually used, and how much time was required for completion. If this is done, the next time the activity is run, problems can be anticipated

Students build a structure with carpet tube, rope, and plastic. Outdoor activities need to be well planned and well supervised, and students need to be prepared in advance for what they will be doing.

and changes made. This makes the activity run as smoothly as possible, making it more enjoyable for both the teacher and the students.

SUMMARY QUESTIONS

1. Name one advantage and one disadvantage of individual instruction, small-group instruction, and class instruction.
2. Traditional activities discussed in this chapter have been used for many years. Why do they many times fall short of achieving their maximum educational goals?
3. Identify three considerations when determining the size of a work area for a specific activity in the classroom.
4. Why are good quality tools and equipment needed for elementary student use?
5. Name three considerations in teacher preparation for a classroom activity.

Bibliography

American Association of School Administrators. *Imperatives in Education*. Washington, D.C.: American Association of School Administrators, 1966.

Bame, E. Allen and Cummings, Paul. *Exploring Technology.* Worcester, MA: Davis Publications, Inc., 1980.

Banathy, Bela H., ed. *Systems Education: Perspectives, Programs, and Methods.* Seaside, CA: Intersystems Publications, 1983.

Bloom, Benjamin S., et al. *Taxonomy of Educational Objectives: Handbook 1: Cognitive Domain.* White Plains, NY: Longman, Inc., 1977.

Boren, Henry C. *The Ancient World: An Historical Perspective.* Englewood Cliffs, NJ: Prentice-Hall, 1986.

Broekhuizen, Richard J. *Graphic Communications.* Bloomington, IL: McKnight, 1979.

Bruner, Jerome S. *Toward A Theory of Instruction*. Cambridge, MA: Harvard University Press, 1966.

Carin, Arthur A. *Activities for Elementary Science.* Columbus, OH: Charles E. Merrill Publishing Company, 1980.

Carin, Arthur A. and Sund, Robert B. *Teaching Modern Science.* Columbus, OH: Charles E. Merrill Publishing Company, 1985.

Clements, B. S. and Evertson, C. M. *Orchestrating Small Group Instruction In The Elementary Classroom* (Report #6053). Austin, TX: The Research and Development Center for Teacher Education, 1982.

Cook, Earl. *Man, Energy, Society.* San Francisco: W. H. Freeman and Company, 1976.

Dammann, George H. *Illustrated History of Ford, 1903–1970.* Glen Ellyn, IL: Crestline, 1974.

DeVore, Paul W. *Technology: An Introduction.* Worcester, MA: Davis Publications, Inc., 1980.

Mager, Robert F. *Measuring Instructional Results.* Belmont, CA: Pitman Learning, Inc., 1984.

Mager, Robert F. *Preparing Instructional Objectives.* Belmont, CA: Pitman Learning, Inc., 1984.

McLuhan, Marshall and Fiore, Quentin. *The Medium is the Message.* New York, NY: Random House, 1967.

Michaels, John U. *Social Studies for Children: A Guide to Basic Instruction.* Englewood Cliffs, NJ: Prentice-Hall, Inc., 1985.

Michaels, John U., et al. *New Designs for Elementary Curriculum and Instruction.* New York, NY: McGraw-Hill, 1975.

Mumford, Lewis. *Technics and Civilization.* New York, NY: Harcourt, Brace, & World, 1963.

Oliver, John W. *History of American Technology.* Ann Arbor, MI: Books on Demand—Xerox Information Resources Co., 1956.

Orlich, Donald C., et al. *Teaching Strategies: A Guide to Better Instruction.* Lexington, MA: D.C. Heath, 1980.

Palardy, J. Michael. *Elementary Education: An Anthology of Trends & Issues.* Lanham, MD: University Printers of America, 1983.

Posner, George J. and Rudnitsky, Alan N. *Course Design—A Guide to Curriculum Development for Teachers.* New York, NY: Longman, 1981.

Pytlik, Edward, et al. *Technology, Change, and Society (2nd ed.).* Worcester, MA: Davis Publications, Inc., 1985.

Reidesel, Alan C. and Clements, Douglas H. *Coping with Computers in the Elementary and Middle Schools.* Englewood Cliffs, NJ: Prentice-Hall, Inc., 1985.

Rochlin, Gene I., ed. *Scientific Technology and Social Change.* San Francisco, CA: W. H. Freeman and Company, 1974.

Saylor, Galen J., et al. *Curriculum Planning for Better Teaching and Learning.* New York, NY: Holt, Rinehart, and Winston, 1981.

Schwaller, Anthony E. *Energy Technology: Sources of Power.* Worcester, MA: Davis Publications, Inc., 1980.

Shaftel, George, and Shaftel, Fannie. *Role Playing in the Classroom.* Englewood Cliffs, NJ: Prentice-Hall, 1982.

Susskind, Charles. *Understanding Technology.* Ann Arbor, MI: Books on Demand—Xerox Information Resources, Inc., 1973.

Teich, Albert H., ed. *Technology and Man's Future.* New York, NY: St. Martin's Press, 1981.

Ticotsky, Alan. *Who Says You Can't Teach Science?* Glenview, IL: Scott Foresman and Company, 1985.

Toffler, Alvin. "Future Shock". New York, NY: Random House, Inc., 1971, ed. *Learning for Tomorrow—The Role of the Future in Education.* New York, NY: Vintage Books, 1974.

Walker, John R. *Exploring Drafting—Basic Fundamentals.* So. Holland, IL: The Goodheart Willcox Co., 1982.

Doll, Ronald C. *Curriculum Improvement—Decision Making and Process.* Boston, MA: Allyn and Bacon, 1981.

Duke, Daniel L., ed. *Helping Teachers Manage Classrooms.* Alexandria, VA: Association for Supervision and Curriculum Development, 1982.

Eisner, Elliot W., ed. *Learning and Teaching the Ways of Knowing: 84th Yearbook.* Chicago, IL: University of Chicago Press, 1985.

Ellul, J. "The Technological Order." In Mitcham, C. and Mackey, R., eds. *Philosophy and Technology.* New York, NY: The Free Press, 1983.

Eurich, Nell. "The Humanities Face Tomorrow." In Alvin Toffler, ed. *Learning for Tomorrow—The Role of the Future in Education.* New York, NY: Vintage Books, 1974.

Evertson, Carolyn M., et al. *Classroom Management for Elementary Teachers.* Englewood Cliffs, NJ: Prentice-Hall, Inc., 1984.

Fuller, R. Buckminster. "Utopia or Oblivion." In Watkins, B. and Meador, R., eds. *Technology and Human Values—Collision and Solution.* Ann Arbor, MI: Ann Arbor Science Publishers, 1977.

Gagne, Robert M. *Essentials of Learning for Instruction.* New York, NY: Holt, Rinehart & Winston, Inc., 1975.

Gronlund, Norman E. *Measurement and Evaluation in Teaching.* New York, NY: Macmillan Publishing Company, 1985.

Gross, Richard E. and Dynneson, Thomas L. *What Should We Be Teaching in the Social Studies?* (#199). Bloomington, IN: Phi Delta Kappa Educational Foundation, 1983.

Havighurst, Robert J. *Developmental Tasks and Education.* White Plains, NY: Longman, Inc., 1979.

Heiner, Carl W. and Hendrix, Wayne R. *People Create Technology.* Worcester, MA: Davis Publications, Inc., 1980.

Hellman, Hal. *Transportation in the World of the Future.* New York, NY: M. Evans and Company, 1974.

Jacobson, David, et al. *Methods for Teaching a Skills Approach.* Columbus, OH: Charles E. Merrill Publishing Company, 1981.

Jarolimek, John. *Social Studies in Elementary Education.* New York, NY: Macmillan Publishing Company, 1986.

Larson, Charles U. *Communication: Everyday Encounters.* Prospect Heights, IL: Waveland Pr., Inc., 1981.

Levenson, Elaine. *Teaching Children About Science.* Englewood Cliffs, NJ: Prentice-Hall, Inc., 1985.

Lederman, Linda Dostigan. *New Dimensions—An Introduction to Human Communication.* Dubuque, IA: Wm. C. Brown Company Publishers, 1977.

Mager, Robert F. *Developing Attitudes Toward Learning.* Belmont, CA: Pitman Learning, Inc., 1984.

Acknowledgments

A key element of a publication such as this is the inclusion of descriptive photographs. Graphic communication of facts and ideas becomes doubly important when dealing with the subject of technology because the concepts center around human activity.

The authors wish to extend a special thanks to those industries, businesses, and organizations that assisted in providing this important material. The courtesy and helpfulness they extended demonstrates a genuine interest in helping to further the educational process:

Field Museum of Natural History, Chicago; Department of Anthropology, Smithsonian Institution; Atlanta Chamber of Commerce; Department of Anthropology, Smithsonian Institution; National Anthropological Archives, Smithsonian Institution; Rockwell International; General Motors Corporation; Library of Congress; Greenfield Village and the Henry Ford Museum, Dearborn, Michigan; NASA; National Air and Space Museum, Smithsonian Institution; U.S. Department of the Interior, Bureau of Reclamation; Liberty Mutual Insurance Company; Zemper Studio and the Bentley Historical Library, University of Michigan; Uniroyal, Inc.; Department of Housing and Urban Development; National Archives; Colonial Williamsburg Photograph; West Advertising and Public Relations; Fruehauf Division, Fruehauf Corporation; American Iron and Steel Institute; Citrus World, Inc.; Industrial Truck Division, Clark Equipment Company; Texas Instruments; USDA—Soil Conservation Service; Michigan State Police; American Telephone & Telegraph Company; Atari; Schwinn Bicycles; National Coal Association; Shell Oil Company; Boeing Marine Systems; Allis-Chalmers; Amtrak; U.S. Army photo, TACOM; Marathon Oil Company; Industrial Truck Division, Clark Equipment Company; Chrysler Corporation; Caterpillar Tractor Company; General Electric; The General Tire & Rubber Company; American Iron and Steel Institute; Holland America Cruises, Inc.; ALCOA; U.S. Department of the Navy; MerCruiser Stern Drives; Bell Aerospace Canada Textron; Boeing Marine Systems; Sikorsky Aircraft, Division of United Technologies; Bell Helicopter Textron; Rockwell International; Lockheed California Company; Group W (Westinghouse Broadcasting Company); Polaroid Corporation; Compugraphic Corporation; Heidelberg Eastern, Inc.; Xerox Corporation; Itek Graphic Products; Teledyne Rotolite; U.S. Postal Service; Bell & Howell; United Technologies Corporation; Keuffel & Esser Company; Western Union; RCA Corporation; Communications Satellite Corporation; Mattel Toys and Daniel J. Edelman, Inc.; HRB—Singer, Inc.; Exxon Corporation; American Plywood Association; Portland Cement Association; Libbey-Owens-Ford Company, Glass Division; Georgia-Pacific; Pfizer Inc.; Standard Oil Company (Ohio); International Salt Company, A Part of Akzona Inc.; Ocean Drilling & Exploration Company; Citrus World, Inc.; American Iron and Steel Institute; Michigan Milk Producers Association; U.S. Borax & Chemical Corporation; Hershey Foods Corporation; Institute of Scrap Iron and Steel, Inc.; Dow Chemical Company; Hewlett-Packard Company; General Telephone & Electronics Corporation; United Airlines; Cadillac Plastics and Chemical Company; Fostoria Glass Company; Grinding Machine Division, Warner & Swasey; American Plywood Association; Snap-on Tools Corporation; Bridgeport Machines, Division of Textron Inc.; Tektronix, Inc.; Brown & Sharpe Manufacturing Company; ARCO Polymers, Inc.; Unimation Inc., Danbury, Conn.; Gilman Engineering & Manufacturing Company; Lockheed Missiles & Space Co., Inc.; Stanley Tools; Brodhead-Garrett Co.; Chevron U.S.A. Inc.; M.S. Plastics Company; Monitor Sugar Company; State of Vermont Travel Division; Black & Decker (U.S.) Inc.; Eastman Kodak Company; Letraset USA; Harris Corporation; Spectra-Physics; Hobart Brothers Company; Screen Printing Association International.

Index